D0918877

Advanced Work Measurement

Advanced Work Measurement

Delmar W. Karger
Dean Emeritus
Ford Foundation Professor of Management Emeritus
Rensselaer Polytechnic Institute

Professor of Management
University of West Florida

Walton M. Hancock
Professor of Industrial & Operations Engineering
Professor of Hospital Administration
The University of Michigan

INDUSTRIAL PRESS INC.
200 Madison Avenue
New York, New York 10157

Library of Congress Cataloging in Publication Data

Karger, Delmar W.
Advanced work measurement.

Includes bibliographical references and index.
1. Work measurement. I. Hancock, Walton M.,
1929– . II. Title.
T60.2.K358 658.5'42 82-935
ISBN 0-8311-1140-2 AACR2

Note: These authors are deeply indebted to Mr. Karl Eady and to the MTM Association for Standards and Research for a detailed review of the MTM portions of the text. This was done in order to assure an accurate and correct presentation of the MTM material.

ADVANCED WORK MEASUREMENT
FIRST PRINTING

Preface

Advanced Work Measurement (AWM) was written to extend the discussion of work measurement beyond the fundamentals found in conventional work measurement texts. More specifically, the authors of this volume assume that the reader has already learned through reading, study, and practice the essentials described in *Engineered Work Measurement* (EWM).[1] Therefore, EWM is somewhat of a companion text; in fact, its material on MTM-1 (or its equivalent) is necessary for the proper understanding of the AWM chapters dealing with advanced MTM systems. Material on the relationship between the authors of the above two texts will be found in Chapter 1 of AWM.

The primary objectives of these authors were to cover the more advanced areas of work measurement knowledge (Chapters 1, 2, 3, 4, 5, 13, and 14) and to cover or present properly the more advanced Methods-Time Measurement (MTM) systems—MTM-1 is covered completely in EWM, as well as conventional time study fundamentals. Chapters 6, 7, 8, 9, 10, and 12 cover material on the advanced and/or higher-level MTM systems. Chapters 11 and 13 cover computer systems developed for work measurement, especially those based upon MTM-1. Chapter 15 covers additional advanced work measurement material.

These authors are deeply indebted to Mr. Karl Eady and to the MTM Association for Standards and Research for a detailed review of the MTM portions of the text. A reasonable effort was made to assure an accurate and correct presentation of the MTM material. Furthermore, the association was gracious in giving us permission to reproduce many of their illustrations and tables.

We are further indebted to Mr. Karl Eady, Manager of the MTM

[1]D. W. Karger and F. H. Bayha, *Engineered Work Measurement,* Industrial Press, Inc., New York, N.Y., 1957, 1966, 1977.

Association's Midwest Office, for authoring Chapter 13 and acting as a coauthor of Chapter 11.

The authors also wish to express appreciation to Ruth Karger and Charlene Hancock for their encouragement and understanding during the writing and editing of this text.

Contents

Advanced Work Measurement

CHAPTER 1

Introducing Advanced Work Measurement

Work measurement, like other facets of knowledge of related complexity, has been greatly expanding in both scope and technology. It has grown from simple operation charting into time and motion study procedures, on into predetermined time techniques and systems, and now begins a move toward the broad interdisciplinary prediction of almost all major aspects of human performance—aided by expanding computer technology.

It has long ago burst out of the confines of the factory production floor and is now widely applied in the clerical, business, financial, distribution, and social fields of human effort. It has even begun to be applied to highly complex mental work that can be somewhat goal defined—like the work of managers, engineers, accountants, etc. In fact, it seems to have no theoretical limit. However, its so-called standards of measurement become increasingly less precise as its application moves toward the creative and/or professional work.

Advanced Work Measurement (AWM) presents an integrated and comprehensive set of material beyond the fundamentals. The fundamentals, as defined for this writing, are the material contained in a predecessor book *Engineered Work Measurement*[1] (EWM) by D. W. Karger and F. H. Bayha. It perhaps should also be stated that this first chapter will attempt generally to provide background and underlying material, the purpose and thrust of the book, and, in general, set the tone and scene for this text.

EWM covered the origins and fundamentals of traditional techniques, predetermined time systems in general, and Methods-Time Measurement (MTM) specifically (including MTM-1 in complete detail), and the application of this set of knowledge to the more traditional areas of work measurement application.

EWM should be studied as a foundation to the full understanding of this text if the reader is not already fully conversant with MTM-1 and the traditional techniques. Optimum reader satisfaction depends on the reader's background knowledge, hence the above recommendation.

Readers already qualified by the MTM Association for Standards and Research or another national group[2] within the International MTM Directorate

[1]Delmar W. Karger and Franklin H. Bayha, *Engineered Work Measurement,* 3rd ed., Industrial Press, Inc., New York, 1977. Subsequent references in this text will be annotated by "See EWM (footnote 1, Chapter 1)."

[2]The MTM Associations of many countries are associated with other professional groups.

(IMD) as MTM-1 practitioners should be able to proceed without reference to EWM or any other reference text, since the requisite background knowledge usually is given in MTM-1 (the basic MTM system) approved courses.[3,4] Those using this text in conjuction with an MTM Association approved training course covering a higher-level MTM system[5] will probably be able to understand it readily because of the content of such courses or because the presenters make knowledge of MTM-1 a prerequisite. Those with no previous work measurement background are advised to first study EWM.

All of the chapters dealing with the MTM systems utilize a special time unit. Without some knowledge of its absolute value and of its derivation, much reader understanding will be lost. Therefore, for those who do not already have this knowledge and want to proceed in the reading of AWM before referring to EWM, the material which follows should be helpful.

The originators of MTM utilized in their research work high-speed motion picture cameras, which were of the same speed and represented the technology of their day. By applying performance rating in percentage terms, it was possible to find the average time consumed by an average operator in the frame in question.

> However, it was of practical necessity to assign time values in units which had the dual advantages of easy usage and yielded numerical results that could readily be used to supply the input to cost systems. Most industries rely on decimal minutes or decimal hours for their measurement and subsequent costing of labor. Time units such as these, however, would have been difficult to use because of the fact that many of the basic motions were very short in elapsed time of performance and many zeros would have been needed between the decimal point and the first significant digit.
>
> This is clearly illustrated by the film speed times expressed in commonly used time units. Since the film speed used in the original research was 16 frames per second, each frame covered an unrated elapsed time of 0.0625 seconds, 0.0010417 minutes, or 0.00001737 hours.
>
> The obvious way to avoid such unwieldy time units was to recognize that units are arbitrary by nature—the necessity for conversion to other desired units being the

[3]MTM-1 as well as other MTM courses are available from the membership or from the executive offices of the MTM Association for Standards and Research, 16-01 Broadway, Fair Lawn, New Jersey 07410. This is the national governing and research association for the United States and Canada within the International MTM Directorate (IMD), which has general cognizance over MTM around the world and grew from the USA/Canada group to include 11 other national associations. The work and relationships among the various MTM bodies are thoroughly explained in the 1977 edition of EWM. Approved course completion and passing of the national association examination are recognized by an official qualification card issued to work analysts, industrial engineers, and others participating in any of the wide variety of such MTM training courses available.

[4]Approved courses are courses taught as directed by national MTM associations.

[5]These systems are covered directly in this text. The first level MTM-1 system is greatly detailed in EWM. Generic systems MTM-2 and MTM-3 are detailed in this text. Other MTM systems are also described herein in less detail to keep within reasonable space availability and application safeguards. These systems presently include MTM-V, MTM-M, and MTM-C. Also covered to a reasonable degree are such computerized systems as 4M DATA, ADAM, RADAC, and others. Furthermore, along with other recent MTM topics of practical value, MTM standard data procedures are presented. In addition, the text contains work measurement material that is not directly identified with MTM.

only real limitation. Maynard, Stegemerten, and Schwab invented a new time unit known as the Time Measurement Unit and assigned 0.00001 hours as the value of one Time Measurement Unit. Since most wages are in dollars per hour, the TMU can be multiplied by the hourly rate and the decimal point then shifted five places to the left to find the cost of labor directly. Also the hours required to produce 100 motions (or pieces) can be found by shifting the decimal two places to the left.

As a result of the unit chosen, the following time conversions will be valid:

1 TMU	equals 0.00001	hours	1 Hour equals	100,000	TMU
	equals 0.0006	minutes	1 Minute equals	1,667	TMU
	equals 0.036	seconds	1 Second equals	27.8	TMU

Under these conditions, one unrated film frame at the research speed was equal to 1.737 TMU. No motion time less than this value could be set from the film. This proved to be the principal limit on the precision of the original data. When other film speeds, such as the faster speeds used in later MTM research sponsored by the MTM Association, are used to get a finer time determination, it is merely necessary to find the new per frame time in TMU to easily fit the new data into the present MTM system.[6]

Advanced Work Measurement

The initial thoughts about the content of this text were formulated by Professor Karger and Mr. Bayha. Because of an unfortunate accident, Mr. Bayha could not continue. Since both of the original authors had worked with Dr. Walton Hancock of the University of Michigan on work measurement and MTM research it was natural for Professor Karger to ask him to replace Mr. Bayha as co-author. He accepted, the book was naturally restructured to reflect the new combination of views, and the authors hope that the result will be useful to industrial engineers and work measurement practitioners, researchers, students, teachers, and theoreticians.

The title *Advanced Work Measurement* is justified on the basis that it represents, at this writing, the only text that bridges the gap between "the fundamentals" and the later and much more advanced techniques and knowledge. The material presented will build upon the base of EWM as previously indicated, and like EWM it will cover both MTM oriented and non-MTM oriented material. That a wide spectrum of other work measurement items beyond MTM will be covered can be seen by a review of the table of contents.

The Work Study Scene

Predetermined time systems appeared in the 1930s and 1940s and demonstrated the motion-time and methods-time connections. Since then, the stopwatch has more and more been relegated to timing process-controlled and machine-controlled work elements. The manually controlled time can often be

[6]See EWM (footnote 1, Chapter 1).

more easily and accurately found by a clerical process—or even by a computer, as is now routinely accomplished (see Chapters 11 and 13).

Besides standardizing the manual time components to an accepted world-wide norm, this development of predetermined-time systems permitted a shift of study concentration to the proper emphasis—the essential motions/methods needed to minimize the time and cost of the work. Also, analysis costs have been greatly reduced by the efficiency and reliability of standard time data, which are more rapidly developed with this new approach even though it requires better qualified and better trained analysts.

MTM sponsored research can be used to illustrate progress in work measurement; in fact, this research represents the bulk of all such research that is available to the public. After MTM's inception and through the 1950s, research efforts were centered on validating, refining, and enlarging the basic data. Then, application areas and procedures became the emphasis, and special projects sought to evaluate additional human performance parameters for about the second decade. Included were the learning process (Chapter 3), personnel selection and selection tests (Chapter 2), physiological factors (Chapter 4), mental data (Chapter 5), the indirect costs of work measurement error rates (Chapter 12), and computerization (Chapters 11 and 13). The referenced chapters not only cover MTM sponsored research, but also other documented and related research. The treatment of these subjects is in such detail and of such practicality that work analysts will more and more begin to utilize all this knowledge. More recently, the research activity has concentrated upon development of higher-level systems and functional data packages, most of which are described in this text.

It should be noted that the emphasis of research has almost fully shifted from the basic data level to the application level. Concurrently, the MTM research effort became centered less in academic settings and more in industrial and business consortiums. In addition, the efforts were more widely spread among the international associations rather than being confined essentially to the original national group. Application coverage also was extended from mainly manufacturing activities to business and financial areas, building and construction, and, now, farming.

All of these phases were responsive to the various associations' membership needs and desires. However, they represented true work measurement progress, and the MTM effort can be credited with a major share of the advancement of such knowledge.

Looking Ahead for MTM and Work Measurement

The prospect for further advances under the current research emphasis is good, but not limitless. This is because specialized data packages to promote more economical work study tend to emphasize the time dimension at some sacrifice of other dimensions of work study and work measurement. Also, such efforts essentially result in a restructuring of the basic data and, therefore, do not advance basic knowledge of either the underlying MTM-1 or of the other important facets of knowledge related to work study and measurement. A point can be reached where the practical systems development needs have been saturated, especially as far as current capabilities go with the underlying data. Ultimately, if the current trend continues, there will soon exist a satisfying superstructure on the underlying basic data, while the data base has stagnated and will permit no further development until new basic research efforts have been accomplished. A further danger from neglecting basic research is that new or revised basic data which might result from a new effort at such research would pyramid the possible changes needed to keep the superstructure sound.

Future progress seems to require a renewed effort at the basic level. This effort, in addition to striking off into new promising areas, should review the technical points neglected or bypassed. It would also deal with those subjects mentioned earlier that were not fully reduced to practice, often because of missing bits of knowledge. It is easy to think of possible research areas, for example, more body motion data, error rate and error control, further work in the mental area, fatigue measurement, and more sophisticated methods of determining indirect labor.

From the discussion in the previous sections it should be somewhat obvious that this text is trying to provide enough knowledge for readers to begin to utilize the newer research results. Each of the advanced areas could be viewed as a subsystem. The authors believe that enough is now known so that someone or some group could begin to tie all of these facets of knowledge into a major computerized work measurement system. This is another possible research project that will become increasingly obvious as the study of this text proceeds.

It should also be obvious that the dynamism so characteristic of past developments can easily be continued far into the future to achieve a logical extension of existing knowledge and thereby pave the way for continued application progress.

Forces for Change

Work measurement practices and development must always respond to changing demands inside the field itself as well as those from any other agency which shapes the world of work. Forces outside of the field often are ignored until they have created a major new problem. Hence, in this section we will primarily note a representative sample of changes outside of the field that cannot help but have a significant impact on work design.

A force of great impact on the labor supply results from demographic and social changes, which fortunately are well documented and can therefore be predicted rather easily. For example, there has been a significant decline in the growth rate of the U.S. male labor force, caused by a declining birth rate and a consequent "aging" in the United States. This naturally changes the age distribution of workers, causing some additional concentration in the prime group. Another somewhat similar change is the enlarged role for women and other traditionally excluded groups in the active labor force. Both of the above affect work measurement directly and indirectly.

Both the prime-age (25–54) male group and the total male labor force currently are growing about 2 percent annually. In the 1965–1970 period, the total male force grew about 10 percent faster, while the prime-age group grew at only 0.6 percent per year. Now reflect on this implication of change: by 1985–1990 the total work force growth is predicted to decline to near 1.1 percent per year while the prime-age group will continue to grow about at the current rate. The young no longer will be entering the labor force in droves, and the competition for top jobs will intensify within a shrinking total force.

The relationships seem simple, but the effects are often surprising. For example, males 16–24 years old and men over 55 years old both will decline about 2 percent in their proportion of the total toward the end of the next decade. More importantly, the prime-age male group will meanwhile become about 6 percent larger in the total. This all should indicate to the work analysts and industrial engineers that a larger number of more mature, better trained workers are available in most cases. The current problem of absorbing a large number of young men entering the job market will lessen. This might encourage capital investment and help reverse the sometimes perceived present decline of productivity.

The previous discussion about the male work force is further complicated when the continuing movement of women into the labor force is considered. Well before the end of the next decade, women will comprise about one-half of the total worker population. While the growth rate for younger women into the ranks will slow, there will be an increase in the rate of entry for older women. These factors will influence work design and the amount of job training required—factors of significance to work measurement people.

Not only will quantity and mix of the work force change, but it will also evolve qualitatively. There is and will continue to be much change in the collective attitudes, beliefs, and aspirations of the work force within and about the work. The values of workers have changed, and if employers do not meet the challenges, there will be increased labor trouble. However, it is still too early to say with any certainty that the "work ethic" is already eroded. It is a worldwide phenomenon as is illustrated by what happened in Sweden in the past decade. The Volvo Company finally gave up trying to cope with a 15 percent absentee rate and built a new plant designed to make work more pleasing—the major technique employed was based on the concept of "job enlargement."

The changing educational level of workers must also be taken into consideration. In 1976, for example, one of every four members of the 25–29 age group possessed a collegiate baccalaureate degree as compared to only one out of 20 before 1940. The average educational attainment of every age group has been and still is rising significantly. This factor and others certainly affect the values of the workers, which in turn interact with management style. For example, a 1977 poll by the Daniel Yankelovich company indicated that 54 percent of Americans claimed a right to share in decisions affecting their jobs.

Workers in America and in other advanced nations are no longer required to accept just *any* job in order to meet Dr. Abraham Maslow's "lower level" needs of survival, security, and safety.

As new technology continues to invade every area of human endeavor, a basic fact emerges that work becomes increasingly less oriented to "thing" activity and more to "think" activity. Some people estimate that only about 20 percent of the labor force will accomplish the "thing" needs of the modern nation by the turn of the century, while the remaining 80 percent will engage in such activities as distribution, marketing, finance, insurance, transportation, communications, health, welfare, education, entertainment, and government. Presently, approximately 55 percent of the employed work force are in service organizations. Of the 45 percent in manufacturing, only about 2 percent work on assembly lines.

Linking together many of the forces of change is the present and projected future conditions of local and worldwide communications. Knowledge among all social groups, regardless of educational level, is much greater than before and is, moreover, increasing continuously. The effects are both positive and negative, but the effect of the free flow of ideas and information is much more positive than negative—if we can learn to understand and cope with the changes it brings about.

All of these forces of change represent additional research opportunities related to work measurement. If the reader wants more general information on change in our world he is referred to the works of Alvin Toffler,[7] including their bibliographies.

A System Perspective

Work measurement systems are prime examples of engineering design consideration of the various factors affecting system performance. There always are "trade-offs" that represent compromises between the various performance attributes. One cannot have everything, regardless of how much "everything" is desired.

Before discussing parameters of systems it might be well to define a few terms so as to put all parties on the same base. Normally, all usage of specialized IE terms will be in accord with American National Standards Institute *Standard Z94, Industrial Engineering Terminology.*[8]

> *Work design: the design of work systems. System components include people, machines, materials, sequence, and the appropriate working facilities. The process technology and the human characteristics are considered. Individual areas of study may include: analysis and simplification of manual motion components; design of jigs, fixtures, and tooling; human–machine analysis and design; or the analysis of gang or crew work. Syn: Ergonomics, Job Design, Methods Engineering, Methods Study, Motion Study, Operation Analysis, Work Simplification, Motion Economy.*[9]

[7]Works of Alvin Toffler dealing with change in our society: *Future Shock,* Random House & Bantam, New York, 1970; *The Third Wave,* William Morrow and Company, Inc., New York, 1980.

[8]ANSI Standard Z94, Industrial Engineering Terminology, American National Standards Institute, 1972. The standard is jointly sponsored by the American Institute of Industrial Engineers and the Society of Mechanical Engineers, from whom copies can be purchased. A revision of this standard will be available in 1982.

[9]*Ibid.*

Webster's Unabridged Dictionary contains these "system" definitions:

1. *A set or arrangement of things so related or connected as to form a unity or organic whole; . . .*
4. *A set of facts, principles, rules, etc., classified or arranged in a regular, orderly form so as to show a logical plan linking the various parts.*

In ordinary language, a system is any interconnected set of "things" (usually considered to be subsystems) around which an arbitrary boundary is drawn to deal with them as a unified subject. This accords with the free body concept common to all branches of engineering. The "boundary" mentioned earlier can include a large complete subject or overall system or it can be used to encompass a subsystem. Therefore, a number of "systems" belonging to a larger whole system can be described as a "family of systems" when the larger boundary is meant. This term is especially applicable when the "family of systems" consists of a basic system and all the other systems have been derived from it and depend upon it—like the MTM family of systems. The only really basic data that remain in such derivative systems are those of the original generic system.

Figure 1–1 shows many of the design elements and system selection rela-

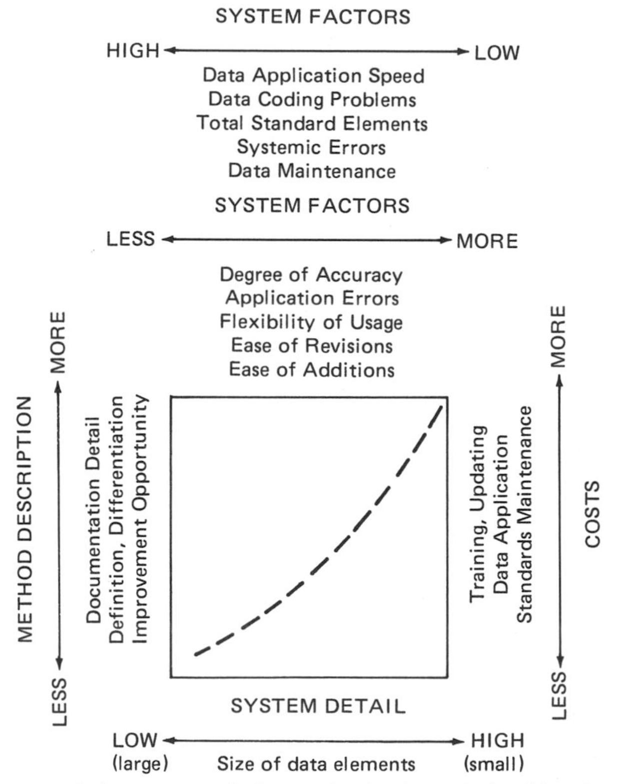

Figure 1–1. General data system design and selection relationships. Reprinted from Engineered Work Measurement, *p. 263.*

tionships associated with predetermined time work measurement systems. This model is a modification of Roger L. Yard, Sr.'s "New Methods of Developing Standard Data."[10] In spite of the figure's work measurement orientation, most of its content applies in some form to many other fields of design and therefore merits considerable attention.

First, note at the bottom one of the more important design determinants—the *system detail.* Low detail means large data elements, whereas high detail represents much smaller data elements. If the total average of a design or a system would be the same (cover the same ground, have the same content), then, obviously, it will take many more basic data elements of high detail to cover the subject than would be the case for larger data elements. The dashed curve shows this relationship in a general way and, obviously, interrelates all ordinates.

Next, examine the top group of system factors. These all have an *inverse behavior* relative to the system detail. For example, low system detail means large data elements are found in "higher"-level MTM systems. MTM-2 has larger data elements than MTM-1, and MTM-3 has larger sized elements than MTM-2. Some higher-level systems, as later explained, are generic, and the remainder are identified as functional—but some of them are also very close to being "specific." These terms are defined in the next few pages. They coincide with less accuracy, less flexibility of usage, and less ease of revision or addition. The large data elements also signify more application speed, more data coding problems, more systemic error. Small data elements coincide with the reverse of the above relations.

The reader should understand that the dashed curve could not possibly show the exact relationship for each of the above relationships or for any of the other indicated relationships in the figure. Given a set of work measurement systems, each such system would vary somewhat from the relationship indicated by the dashed line in the figure. However, the figure *is* useful because it does indicate the generalities—and these important relationships are either unknown, or not properly understood, or ignored by many practicing industrial engineers. The relationships *are important* and should be fully understood before proceeding further in this text.

Elemental Data Combinations

Several important data design terms concern the way of classifying the construction and application of motion and time data—and it is important to understand them since they will later be referenced without explanation.

All standard data consist of elements put together in various ways to accommodate variable application needs. For human performance data, one must focus on the motion description or element content/coverage together with the attendant time value for the element. The size of the elements from a time viewpoint is commonly referred to as being microscopic or macroscopic, usually abbreviated to micro and macro, to denote whether the smallest subdivisions

[10]Roger L. Yard, Sr., "New Methods of Developing Standard Data," *Journal of Methods-Time Measurement,* Volume XI, No. 3 (July/Aug. 1966).

are very short or relatively longer. Elemental content is aimed at describing basic motions, motion sequences, tasks, operations, etc. Time and motion descriptors of elements are collectively termed the *orientation* of the data. So far, the elemental level is meant in this paragraph.

The remainder of this section is taken verbatim from the companion text,[11] as follows:

> When the data elements are assembled for application, the mechanism used is called their *construction* and their resulting scope is classified as to *universality*. The two major construction procedures yield *vertical* or *horizontal* data. Vertical data include elements and element combinations developed for one kind or class of work, or at most a very limited range of work, and therefore fairly specific areas of application. Horizontal data elements are tailored to cover many kinds and classes of work, which means their applicability is quite general at the sacrifice of exact coverage for any given single kind or class of work. The scope of a data set in varying from highly universal to very specialized or restricted, is often termed as being *generic, functional,* or *specific*. Generic data apply almost anywhere due to their generality. Specific data usually apply to rather narrow, or even a single, area of interest or variety of work. Functional data applicability falls between these extremes. These three scopes of data are important enough to justify further elaboration.
>
> A *generic* system is oriented to human behavior, and therefore has maximum universality, with elements recognizable as distinct human actions by the work performer. Such elements can be single basic motions, sequences of serial or averaged motions, or motion aggregates keyed to work actions. All of the motions within the elements are drawn from the basic system, which is itself generic in full detail. Element names are general in that they describe universal human efforts, but do not inherently reveal the nature of the work activity being analyzed. Examples: Reach, Grasp, Walk, Get, Put, Handle, Transport.
>
> A *functional* system is oriented to work actions on and by the parts and tools involved, thus being restricted to similar parts and tools which may or may not be found universally and/or used in the identical way, with elements recognizable as having distinct results produced by the handler of the parts and tools. Such elements can achieve single actions, sequences of serial or averaged actions, or action aggregates keyed to work results. The elements are derived from motions in the basic system and/or elements in one or more of its generic derivatives. Element names are essentially specific in that they use an action verb and an object noun to tie together the operations and tools used, and thus inherently reveal the nature of the work activity being analyzed. Examples: hammer nail, clamp blocks, staple cards, dial telephone.
>
> A *specific* system is oriented to coverage of work goals for particular work tasks, thus lacking universality because all pertinent work conditions must be met for proper application, with elements recognizable as the attainment of the distinct working steps involved. Element derivation may be from basic system motions and/or elements from derived generic and/or functional systems. The size of such elements can vary with the magnitude of the goal, and the element names uniquely identify the narrow or broad sphere of application. Examples: make casting, prepare Form R, machine cam lobe, overhaul engine, balance ledger.

[11]See EWM (footnote 1, Chapter 1).

In the MTM context, the formal systems are either generic or functional, since the users will construct their own specific data from these two kinds of published systems. The existing *generic* systems include MTM-1, MTM-2, and MTM-3 together with certain portions of MTM-GPD.[12] Other MTM-GPD data and all the remaining formal MTM systems are of the *functional* variety. Since the users develop their own MTM-based *specific* systems, and these normally do not enjoy widespread circulation, they may be considered as informal data.

Finally, one can speak of various data *levels*. A level represents either of two ways of classifying a complete data package. One way is with respect to the magnitude of the number associated with the package. For example, in MTM, MTM-3 is a higher level than MTM-2 and MTM-2 is a higher level than MTM-1. The other way refers to the *generation* history of the data. Being the first developed, MTM-1 would be called first generation as to level. Second generation, being next developed and based on MTM-1, would be MTM-2 and much of MTM-GPD. The rest of MTM-GPD and MTM-3 is third generation level as regards its order of development. The number thus affixed to the generation being described therefore indicates a level.

Motion Aggregates

This entire section is taken verbatim from the companion text,[13] as follows:

Regardless of their type and other characteristics, the higher-level systems ultimately rest upon the basic motions in the underlying MTM-1 system. Therefore, to devise larger elements for later generation systems, it is necessary to use a process that can be called "aggregation." With time study, this procedure was severely restricted due to the inability of changing time breakpoints flexibly; but predetermined motions permit complete freedom in selecting breakpoints, so that data developments have mushroomed since their advent. These newer data systems are generated with motion aggregates—new compilations of any desirable portions of the total motions in a work sequence, task, operation, etc., to fit design parameters.

When motion aggregates are developed in various ways, they are thereby defined to a separate existence in terms of method and time information that has been abstracted and amalgamated from the basic source. They can be used for still higher-level data generation in the same way as they were themselves derived from the initial data base of the first level system. In addition, they can be applied directly to determine operational standards in the accuracy and analysis speed range for which they were designed.

The general effect of a higher-level system as compared to the basic system is to gain speed of application at some acceptable sacrifice of accuracy and loss of method detail, because the motion aggregates are larger units of motion and time data having less flexibility insofar as exactly tailoring the operational standard to the actual occurrence of a work event. Depending on the monetary, scheduling, and other control demands on the standards produced with the aggregated system, these drawbacks

[12]MTM-GPD has been, for all practical purposes, largely bypassed by the higher-level MTM systems. Therefore, no attempt has been made to describe it in this text. However, it is a large set of standard data (part generic and part functional) that can be used to cover a wide variety of manual work. It is available from the MTM Association.

[13]See EWM (footnote 1, Chapter 1).

may be minimal; indeed, the final operational standards might still be more accurate and descriptive than actually needed to optimize the production, although generally they are less able to meet such criteria than would the basic system if its cost of analysis could be justified.

Many mechanisms or procedures have been utilized to derive motion aggregates. One of the more successful approaches has been called the "building block concept." It was used extensively in MTM-GPD, which as a result contains motion aggregates that fall on both the generic and functional levels, with a few almost on the specific level. The concept is that, at any given data level or when changing to higher or lower levels, the data will be compatible between systems if the motion aggregates are:

1. Behaviorally based; that is, defined to include physiological wholes which utilize natural motion breakpoints in the basic system.
2. Traceable back to the motions in the basic system and do not violate any of the definitional rules of that system.
3. In accord with the same requirements for elemental independence as is the basic system, for elemental additivity rests upon this foundation.
4. Unique as to content and description, so that classification of operational occurrences will be possible without equivocation or undue analysis time; because analysis decisions using the aggregates must be quick, sure, and free of misinterpretation.
5. Coded without ambiguities so that both adaptation to automatic data processing and efficient auditing are possible.

Other sources may express these ideas differently, but the true essentials are as given above.

When using the "building block concept" or any other viable procedure to develop motion aggregates for higher-level data systems, a number of distinct strategies are available. Not only must the developer understand these, but the user of the aggregated system can also profit and improve its application by knowing them. Basically, four ways can be used to put together elements aggregated from basic motions or lower-level elements:

1. *Sequential*. Using natural breakpoints, the serial motions or elements can be combined with a new name. For example, handling almost any object commonly involves reaching to it, getting hold of it, bringing it to a different location approximately or precisely, and letting go of it. This is the highest frequency of motion occurrence. Whole sequences like this can be easily derived for different ranges of the variables present and sizes of objects involved, then used thereafter as aggregated elements.
2. *Representative*. Motion sequences to do the same job are sometimes highly variable on each occasion with justification. Getting a fair aggregate then means that, after sufficient study of the many possible sequences, a representative sequence can be chosen as an aggregate for all of the occurrences.
3. *Averaged*. When a sequence of motions is desirable for an aggregate and it varies only in key dimensions, an averaging approach can be used. For example, motions can be otherwise identical but require differing time values due to variable distances, weights, etc. If all the points of interest in the range can

be timed with the basic data, then an average for the midpoint, mean, mode, etc., of the range can be found to serve as an aggregate value.

4. *Proportional.* This approach is best when a series of motions to be aggregated has, perhaps, one or more of the motions variable as to case, type, kind, or other characteristic. If observation or analysis can establish a frequency for each of the variations in the motion sequence, then a realistic usage of proportioning or percentage factoring can be devised to find the aggregate condition.

All of these approaches have been used in developing higher-level MTM systems in the MTM family. The important point to note is the really basic contribution the identification of these approaches has made to the standard data practices of industrial engineering, for it is predictable that they will foster other data systems in the future. Especially with electronic computers easily available for simulation and handling of large amounts of data economically, the shackles on analysis and further motion aggregation have melted away!

A Forward Vision

There is a fundamental reason why MTM has so successfully proceeded and continues to afford great promise. It is that *the MTM-1 motions are a biological classification of behavioral motions involved in human work.* The importance of this cannot be stressed too greatly; it is the very heart of the power behind the system. Strangely, many users have satisfactorily used the system without realizing this attribute and why it is so vital and important. Researchers have long been aware of it and appreciate the help it affords in generating new data or investigating basic human behavior and performance.

Consider for a moment what that fact means in terms of information and intelligent data handling: Literally contained in an efficient, quantified code that can be physically reproduced, directly observed, and directly measured is the true behavior as it occurs in actual working environments. One has a microcosm of variables effectively classified and verified in countless applications around the world for over a third of a century. A motion pattern, therefore, is a true picture of the human behavior as work is accomplished.

It should therefore be possible to discover a data abstraction procedure which will yield from a valid motion pattern the broad human characteristics and effects which accomplish work. In fact that exact procedure or concept was used in studying manual learning and is now being routinely applied with or without the use of a computer on a practical level. The motions in a pattern where grouped for time totals within certain major variable categories. These were kind of motion and case of motion; the distance or length and direction of motion need not even be considered directly. These time subtotals are used to apportion validated unit regression equations for learning of the variable categories into a Composite Learning Equation. This equation can be solved for a predicted value of the number of learning cycles needed to attain the time standard associated with the motion pattern. From the procedural algorithm, the learning curve can be plotted or charted to establish a norm against which actual performance can be gauged and controlled or corrected to keep both output and associated cost within planned limits. This obviates the need to guess or estimate start-up costs due to the learning factor, for they are technically determinable as soon as the working method is congealed into a motion pattern. Learning is covered in Chapter 3.

Although a great deal of careful research and validation would be needed, there is no theoretical or possibly even no practial barrier to prevent the extension of this basic idea into further progressive advancements. Since the motion coding contains variables, subvariables, objective dimensions, and real limiting performance qualifiers, other broad human factors might well be quantified for routine assessment from the motion pattern. Factors such as fatigue, stress, strain, heat or caloric expenditure, and metabolic rates are, for example, hidden in the pattern coding and work descriptions (more on this will be found in Chapter 4 on fatigue).

The forward vision presented herein points the way to even more research opportunities. The future of work measurement can be almost unlimited in scope and importance.

CHAPTER 2

Worker Selection and Training

The Methods-Time Measurement (MTM) family of systems is designed to predict human performance under certain conditions. Three of the factors that affect work force capability are the methods of selecting workers, the age of the workers, and the training of the workers. The purpose of this chapter is to present some of the concepts of worker selection and training that are of importance to all work measurement analysts, regardless of the method used to establish the work-time standards.

What Is a Standard with Regard to Worker Capability?

When applying one of the MTM systems, the analyst determines how long it should take the large majority of experienced workers between the ages of 18 and 65 in good health to do the work; this statement also generally applies to all predetermined time systems and to most properly conducted time studies. Figure 2–1 can be used to more fully understand what such a standard repre-

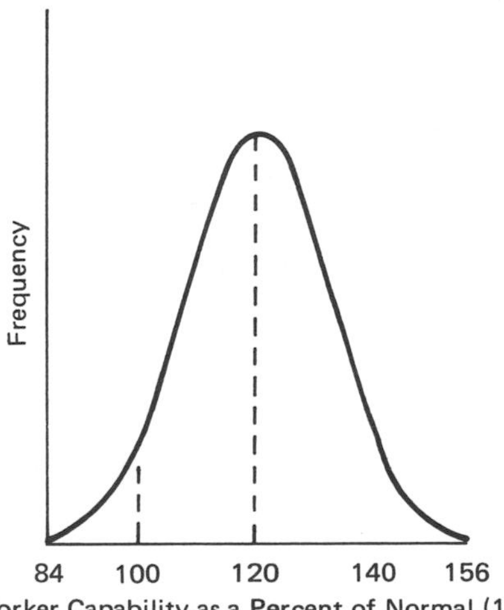

Figure 2–1. A distribution of worker capability where no allowances are considered.

sents in terms of worker capability. Normal performance in a work measurement sense is depicted as 100. The MTM family of data systems is based upon this point; also, this is true for many other predetermined time systems and for most properly conducted time studies. It does not apply to some time standards because another concept of "normal" is used, as explained below. A further examination of the curve reveals that the vast majority of workers are capable of exceeding the described standard. The statement that is frequently made regarding this standard's base is that at least 95 percent of the healthy working population between the ages of 18 and 65 can achieve this type of standard. Such standards are known as Low-Task Standards.

The 120 value is the real average capability of the general working force.[1] Thus, the average capability of the general or total work force enables them to exceed the Low-Task Standards by 20 percent. The figure usually given in the literature is not 20 percent but 15–35 percent.[2,3] More properly, 15–35 percent (the figure most frequently used is 33 percent, but no specific reference could be found) is the value used as the amount workers exceed the normal pace when under incentive payment systems. The difference between 20 and 15–35 is due to the addition of allowances (usually called Personal, Fatigue, and Delay—P, F, and D—allowances) to the basic Low-Task time values. Frequently, the allowances, especially fatigue allowances, are inflated, shifting the rate of performance toward the higher values. Fatigue allowances are discussed in Chapter 4 and in *Engineered Work Measurement*.[4] In addition to the above, there is a self-selection process which is discussed later in this chapter.

The main point made previously is that standards based upon the described "normal" performance concept should be achievable by practically all workers. This is really the basis for calling the MTM system a Low-Task system, because attaining a normal work pace is not difficult for most workers. Most time study analysts subscribe to this concept of "normal."

The American National Standards Institute in their Z94.12 Industrial Engineering Terminology; Work Measurement and Methods[5] standard defines Low Task as follows:

> *Low Task: A term used to indicate that performance rating or production standards are based on daywork levels as contrasted to high task or incentive work performance. Sometimes taken to mean a level of performance below the level expected under measured daywork conditions.*

The same source defines High Task as follows:

[1]Åberg, U. and Hancock, W. M., *Design Criteria of Predetermined Time Systems,* International MTM Directorate, Sweden, 1968.

[2]Mundel, M. E., *Motion & Time Study,* Prentice Hall, Englewood Cliffs, New Jersey, 1950, 4th ed., pp. 304–305.

[3]Barnes, R. M., *Motion and Time Study: Design & Measurement of Work,* John Wiley and Sons, New York, New York, 5th Edition, p. 344.

[4]Karger, D. W. and Bayha, F. H., *Engineered Work Measurement,* Industrial Press, Inc., New York, New York, 1978, 3rd ed.

[5]American National Standard, *Industrial Engineering Terminology, Work Measurement and Methods,* Secretariat: American Institute of Industrial Engineers, Inc. & The American Society of Mechanical Engineers.

High Task: Performance of an average experienced operator working at an efficient pace, over an 8-hour day under incentive conditions without undue or cumulative fatigue. Often stated as a percentage above normal performance (see Normal Effort, Low Task). Synonym: Incentive Pace.

The High-Task Standard is designed to utilize a concept of normal at about the 120 point in Fig. 2–1. When such standards are used, one must take into consideration the same factors as are described in this chapter in the design and operation of such a system. Often High-Task–based incentive systems begin paying incentives at 90–95 percent of normal performance point—in a Low-Task Incentive Program the payment usually begins at the 100 percent point.

If many workers can easily exceed a standard, why not establish the standard on another basis, such as at the High-Task point? The level of performance has to be established in many applications so that no matter which of the workers is placed on a job, the vast majority of them can meet or exceed the "standard" rate of work. Thus, the work output can be predicted.

What happens to the 5 percent that cannot work at the normal rate? In a well-administered employee selection plan, the workers with less than "normal" performance are removed from the work force through the probationary or the selection process.

Why not set the standard so that the lower 5 percent could also achieve the standard? Examination of Fig. 2–1 reveals that the lower tail of the curve goes considerably below 100, with some people's capability as low as 84; therefore, to include the lower 5 percent would result in very low standards. The point where the standard has been set has been found to be a workable point because it excludes a relatively small number of workers.

Note that if the probationary period is used correctly, then workers remaining after it should be able to attain the Low-Task Standards. Figure 2–2 shows

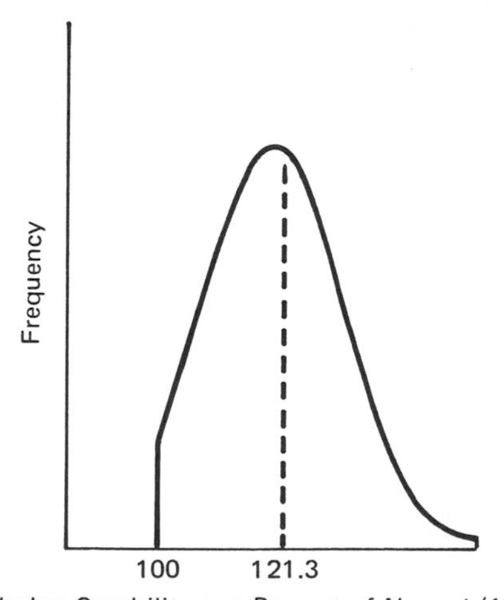

Figure 2–2. Worker capability after the probationary period.

this situation. Thus, 100 percent performance should not be achieved by everyone. The capability of the work force will increase from 120 to 121.3 because the left-hand tail no longer exists. If 100 percent is not achieved, then an immediate investigation should be undertaken to determine the reasons for the low attainment. The most important reasons have been found to be:

1. The workers are not following the methods upon which the standards were developed.
2. The delay allowances have not been properly established or the delays are greater than when the delay standards were established.
3. The workers are inexperienced and need more practice opportunity.
4. Workers who could not meet the 100 percent performance during the probationary period have been allowed to stay in the work force.
5. Materials and supplies used by the workers are not to specification.

The Relationship between Worker Selection and Productivity

How do we get higher than 100 percent productivity? Examination of Figs. 2–1 and 2–2 reveals that the capabilities of the work force are much greater than the level expected by the Low-Task–based standard. In a competitive society, it is the desire of many organizations to have as productive a work force as possible. Why, then, use standards that do not require workers to use their capabilities? The usual response is that equity requires equal pay for equal work. Thus, one has to set the standard that almost everyone can attain. The costs of this approach could be viewed as very high, not only because approximately 22 percent of the capabilities of the work force are not being utilized, but also because many of the more capable workers will not be challenged by the work and will become bored. What then can be done? Two of the possible solutions are:

1. To raise the probationary cutoff point by establishing standards that are more demanding than the 100 percent level. This approach is consistent with the equal pay for equal work concept because we expect all workers to meet the higher performance level, but not necessarily exceed it.

   As an example, suppose we were to reduce the Low-Task Standard concept by 10 percent (make the achievement more difficult), then Fig. 2–2 would be modified as depicted in Fig. 2–3.

   Since everyone would have to be capable of the 110 percent performance, then these people to the left of 110 percent would be excluded from the work force as part of the probationary and selection processes. If we assume that the worker capability is normally distributed, then approximately 20.6 percent would not have the capability of doing the work. The average capability of the work force would increase to approximately 124.4 percent. Interestingly, the capabilities of the work force will be more homogeneous because the variation of capabilities will be less.

   Since approximately 20.6 percent of the people who attempt to do the work will not be able to, high turnover may be experienced during the probationary period

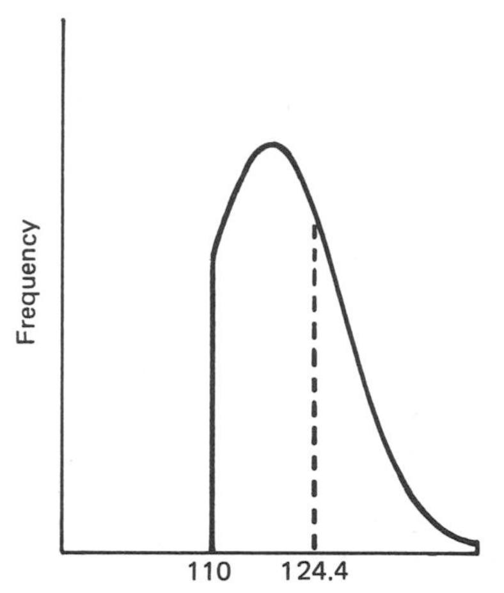

Figure 2–3. The distribution of worker capabilities where the standards are raised to 110 percent.

if the method of selection is to hire the workers, see if he/she can do the job at the standard rate after sufficient practice opportunity, and then hire them permanently only if they can. If this is a problem, then selection tests can be used to substantially reduce the number of probationary employees who will not be deemed acceptable. The antidiscrimination laws in the United States require that a selection test must accurately measure the capability of a worker to do the job that he/she has been hired for. Unfortunately, many selection tests have been found not to correlate highly with the actual job requirements. More will be said about selection tests later in this chapter.

2. Introduce a monetary incentive system where people with more capability will get paid more for higher productivity. These systems are widely used, so little has to be said about them here except to discuss the impact of these systems on worker selection.

   The typical system, when it is initiated, usually has a work force that has been working under a Low-Task system. The capabilities of the work force are usually as depicted in Fig. 2–2 where everyone can achieve 100 percent productivity. Under these conditions, the average increase in productivity will initially be to 121.3 percent, the average capability of the work force. Then, after a period of time, some of the poorer performers (for example, under 110 percent) are likely to self-select themselves out of the work force because they do not have the capability to make the earnings of others in the group. The result will be that after a while the capabilities of the work force will be more like Fig. 2–3 and the average output will increase to 123–127 percent of normal.

Differences between Younger and Older Workers

In the previous sections we discussed worker capability. These distributions are for workers in the age group of 18 to 65 years. Does age affect worker capability, and, if so, how? Age does affect worker capability. Generally, older workers (older being defined as greater than 35 years) have less capability than younger workers. Thus, many, but certainly not all, of the older workers would have capabilities in the 100–120 range of Fig. 2–2, whereas many of the younger workers would have capabilities in the 120–150 range. The main point here is that as a worker ages, worker capability tends to decline. The only worker that this decline seriously affects is the one who first made the probationary cut off when he/she was young and then continued to do the same work throughout his/her working life. Fewer workers will find themselves in this situation under the work scheme depicted by Fig. 2–2 than under the scheme of Fig. 2–3 because there are fewer workers at the probationary boundary. Workers who are on the boundary when young will "select out" of the work force as they age with a much higher probability than workers of more ability originally.

What happens as a person ages that causes capability to decline? Briefly, several of the human systems tend to deteriorate, but they deteriorate at different rates. For example:

1. The decision process used in making choices tends to deteriorate starting at about age 35. Actually, the variation between workers tends to increase, with some people's decision processes deteriorating and some staying the same.[6]
2. Physical capability tends to decrease for some workers starting at about age 45. The decline is gradual and is approximately 10 percent less at age 65. Here again some people deteriorate more than others, thus again increasing the variance between workers with respect to this factor.[7]
3. The main sensors, the eyes, tend to need glasses as they get older. Glasses, when correctly prescribed for the work, minimize many of the effects of aging. Estimates of the percent of the working population that have defective, but correctable, eyesight are as high as 20 percent. Since the eyes are the major sensor used in working environments, poor eyesight can definitely affect the performance of the employees. There exist a number of commercial machines to quickly check a person's eyes. Where an employee complains that he/she cannot meet a standard, the eyes should be checked as part of the investigation.

Studies of the type of work that workers prefer indicate that physical work is preferred to mental work. This preference can partly be explained by the earlier deterioration of the decision processes which causes the person to avoid mental work at the subconscious level.

The deterioration of the decision processes also helps to partially explain why older workers tend to work more slowly than younger workers. To understand why, one must observe an experienced worker using his/her decision/sensory processes in conjunction with his/her motor processes. If you closely observe an experienced worker, you will observe that his/her eyes are practically always ahead of his/her hands. For exmple, if a worker has to reach for an object, his/her eyes will travel to the object long before his/her hands get to the object. If you observe what his/her hands are doing, while his/her eyes are traveling to

---

[6]Welford, A. T., *Aging and Human Skill,* Oxford University Press, Cambridge, England, 1958.
[7]*Ibid.*

the object, you will observe that they are performing another operation such as releasing the previous object. The eye travels and, to a certain extent, the information processing that goes on to direct the hand(s) to the object is limited out (concurrent with manual activity). That part that is not limited out with an experienced worker is included in the basic MTM motion time. For example, the time difference between an R-A and an R-B of equal distance is primarily the time that is not limited out by the parallel use of the decision/sensory processes and the hands.

The act of reaching for an object has been done millions of times by most workers. A preconditioned response has been learned so that the eyes lead the hands without the worker having to consciously think about it. Older workers have more practice at doing motions like reach, so their preconditioned responses may be even more highly developed. Thus, as long as nothing goes wrong, the older worker can equal or even exceed the younger workers' rate of work.

But, suppose that the worker reaches for the object, grasps it, starts to move it, and then drops the object or fumbles. Where are the eyes at the time the fumble occurs? For an experienced worker, the eyes are traveling to the destination of the object in order to collect and process information about the positioning of the object. When the fumble occurs, the eyes must quickly return to the fumbled object in order to regrasp it. Thus, the decision/sensory processes are now the limiting time factor. Older people, in general, take longer to recover from the fumble (or any error) than younger people. Thus, an error to an older person is a much more costly event than it is to a younger person. How does a person avoid these time-consuming events? One way is to slow down so that the chances of making a mistake are less. For example, the older worker may keep his/her eyes at the grasp site slightly longer so that if a problem occurs, a regrasp can occur preventing the more costly fumble. The overall result is that the older worker works at a slower pace. However, the occurrence of fewer errors frequently results in a higher probability that the product will be of acceptable quality.

The Role Timely Worker Training Plays in Achieving High Productivity

Let us define worker training as the purposeful act to condition workers to do the work in the manner determined in advance of the training to be the most suitable method. The sequence for the highest productivity is first to select workers who have the potential to do the work in the prescribed manner and then to train them to use their capabilities to obtain the levels of productivity that are expected.

The vast majority of productivity problems occur because of either poor worker selection or poor training. For high productivity, both aspects (selection and training) have to occur at the proper time in order to achieve productivity at minimum cost. Unfortunately, many organizations do not understand the importance of doing training at the proper time. Perhaps the best way to explain the importance of training at the proper time is to describe the "typical" way of getting a new product line started and the typical negative results when higher productivity is attempted at a later date.

The scenario is that the management is usually interested in getting the line started as quickly as possible, and then after the line is started, getting the productivity up to acceptable levels. Typically, the establishment of production standards is left until the line is running. This procedure is necessary if the standards are set by time study because time study requires that workers be observed while actually doing the work. It is not necessary if MTM or other predetermined time systems are used, because observation of the worker is not necessary. Also, if applicable standard data exist, regardless of its base (time study or MTM), it, too, can be used to set the labor standards before the start of production.

Figure 2–4 is a graph of a typical reduction in the time/piece for a new process as the day's work is increased. This is a form of a learning curve commonly called a production progress function. The output improves (hours/part decreases) because the workers learn, the bugs in the system are corrected, and the organization improves. Studies indicate that during the first couple of days the workers are more concerned with becoming accustomed to the new machinery and to the new environment than they are in productivity. In technical terms, they are becoming used to their environment, thus reducing the uncertainty of the "cognitive" aspects.[8]

After two or three days, the workers start to become more aware of their need to get to know the other employees working in the area. Part of this socialization process is to talk to fellow employees about "what is expected of us?" In the absence of any standards, *the workers and the immediate management have no goals to work toward.* With no goals, the management may not make an all-out effort to get the bugs out of the system. After a week or two, both groups tend to relax, and, unless they hear to the contrary, they assume that the level of productivity is acceptable. The delays and malfunctions now become a "normal" part of the process. Once this happens, the workers assume that the level of output is acceptable. Thus, an implicit standard of productivity is established. The day-to-day variation in output will decrease around the implicit standard point. The level of output, in behavioral science terms, becomes "frozen."[9]

Now, suppose the industrial engineer is asked to set standards in the operations for the purposes of getting productivity to where it should be. This usually means that he/she is to use time study since with a predetermined time system or with applicable standard data the standards would already exist. Frequently, the engineer discovers that productivity could substantially improve by reducing delays, by better machine maintenance, and by instructing employees in the proper methods. All of this must be accomplished before a proper time study of the job can be accomplished. He/she informs the immediate manage-

[8]Likert, R., *New Patterns of Management,* McGraw-Hill, New York, New York, 1961, p. 68.

[9]Munson, F., Griffith, J., and Hancock, W., "Problems of Implementing Change into the Hospital Setting," *AIIE Transactions,* Volume 4, No. 4, December, 1972.

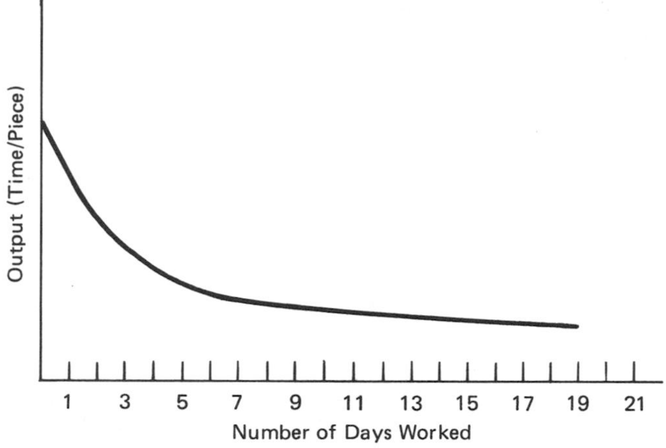

Figure 2–4. A typical production progress function.

ment of the situation concurrently with an estimate of the potential changes in standards. Because the workers and immediate management have become "frozen" at the present productivity, the industrial engineer's statements about higher productivity potential are practically always met with resistance. The recommended changes are viewed by the management as a criticism of what they have been doing. The workers will argue that they are now having to produce more for the same pay, which is unfair. This perception is usually termed a "speed-up." Under these conditions, the industrial engineer will frequently have to "bargain" with the management and/or the employees to get some improvement in productivity. Rarely will he/she be able to convince the parties involved that his/her estimates of potential productivity and/or time study standards can be realized. Even more rarely will he/she be able to get the output consistent with the group's capabilities. There will be considerable pressure on the industrial engineer either to loosen the newly developed standards and/or to accept productivity levels of less than 100 percent.

A typical example of the latter occurs when the industrial engineer knows that 100 percent of standard can be attained, but the group is only attaining 70 percent. Under resistance and time pressure, he/she may "bargain" to get 85 percent productivity. Of course, output is improved, and, since no one except the industrial engineer *knows* what the group is really capable of, the organization accepts 85 percent as being reasonable. Of course, there is a 15 percent loss in productivity that may, over time, erode the competitive position of the organization, since a competitive organization may manage to get 100 percent productivity by using a better approach.

How do we get high productivity (100 percent) without causing all of the bad feelings and resistance? The answer is to establish the production standards before the process is started. If we do this, then we have knowledge of what the production group is capable of. This goal then provides a basis for determining where the group is at any point in time. Likert states that if the goal is established, if capable technical support is supplied to attain the goal, and if the goal is reasonable, the group will strive to attain the goal until it is met.[10] This statement translated into engineering terms means that standards must be established (goals), industrial engineering support must be given (technical support), and the higher management must keep insisting that standards are to be attained so that the workers and immediate management will not freeze at the lower levels of output.

Organizations who have followed the above process do not experience employee and immediate management resistance, the accusations of speed-up, and the lower productivity. Of course, they need an engineering staff with the time and capability to establish standards as an integral part of the production design process. Predetermined time systems such as MTM are almost a necessity for establishing standards and methods prior to production unless *applicable* standard data are available. Standard data, by themselves, have little or no methods analysis capability, so they cannot be used on work that has not previously been standardized as to methods and time.

If time study is used, the production standards cannot be set prior to production, but an attempt can be made to set the standards as soon as the production is started. This method is superior to waiting until the workers "freeze" on output levels. However, there is a major professional problem of which the industrial engineer should be aware. Referring to Fig. 2–4, suppose we make a time study on the second day of production. Generally, if the normal time study procedure is used for rating the elements, we will find that the ratings will be higher than they should be, thus resulting in a standard which is "loose." The reason is that the worker, because he/she is still on the steep portion of the learning curve, will be using his/her decision/sensory functions much more frequently than when he/she is fully experienced. The time study person will perceive that the worker is working harder than normal, and indeed he/she is, but the perception of hard work combined with the fact that the worker is taking longer to do the work will result in an inaccurate statement of the standard time.

DeJong[11] and Hancock and Bayha[12] have proposed a method of adjusting time study standards that were determined using inexperienced workers to predict the standards of experienced workers. Briefly, the method is to take time studies twice, on different production days of the same worker, in order to get an estimate of the slope of the learning curve. Then, by utilizing previous experience obtained from data concerning the number of cycles a worker needs to become experienced, a standard can be predicted for the experienced worker. The method is cumbersome because several studies have to be made early in the production process. It is less accurate than using predetermined time val-

[10]Likert, R., *New Patterns of Management,* McGraw-Hill, New York, New York, 1961.

[11]DeJong, J. R., "The Effects of Increasing Skill and Methods-Time Measurement," *Time and Motion Study,* February, 1981.

[12]Hancock, W. and Bayha, F. H., "The Learning Curve," *Handbook of Industrial Engineering,* John Wiley, New York, 1981.

ues, and it is rarely used. However, it is far better than allowing the worker to "freeze" his/her output, and then try to improve productivity.

Job Evaluation and its Relationship to Productivity

Few people realize the relationship between job evaluation and high productivity. The problems of older workers who can no longer meet the standards can, in many cases, be greatly helped if the job evaluation program is designed to do so. The situation that may exist in an organization where standards are set above Low-Task Standards is that a worker may find the work harder to do as he/she gets older. Decreases in capability, as previously discussed, may exist, but a more frequent case occurs when the worker has a health problem. Under these conditions, it is important to have a number of acceptable methods available to enable the worker to stay in the organization but not to adversely affect the productivity. If a standard exists and if everyone must meet or exceed the standard, then, if a worker cannot make the standard, pressure will be brought on the organization to reduce the standard. Lower productivity will usually result, because once the standard is reduced, the reasons for changing the standard will be forgotten and the output will be at the lower level even if the worker is reassigned.

Typically, the worker is given a medical restriction so that he/she can only be assigned to jobs where the health problem will not be restrictive. This is certainly an acceptable approach so long as there are enough jobs for the workers to do. This is not the case in many situations. The situation is probably going to get worse because of the increase in the age of the work force and because of the federal laws that have extended the retirement age.

In practically all production or service operations, there exists a number of "indirect" jobs that are not nearly as demanding as the "direct" jobs. If these jobs are evaluated, especially in the experience aspects, so that the older, more senior employees have the best chance of getting them, then there will be one more method of providing for the employment of long service employees without reducing direct labor productivity. In fact, the worker's experience often is a distinct and tangible advantage in many service or indirect operations. Yet, if the job evaluation system "glosses over" this aspect, this advantage of the experienced worker may be lost to both the company and the worker.

Some agencies may react to the proposal that the job evaluation system should not be "biased" to allow certain employees to have a better chance of getting assigned to the jobs. They will argue that the organization may have to pay more for the jobs that are designed in this manner. The argument may be correct, but there is also a cost to allowing workers to continue on jobs where they cannot attain the standard—and there is a cost to causing workers to select themselves out of the work force. This cost is intangible, but it can be argued that if this selecting-out process is perceived as being a frequent occurrence for many of the workers, then the workers will perceive the organization to be a less desirable place to work. Also, such workers often opt out by allowing themselves to be dismissed, and this does cost in terms of unemployment "assessments."

Selection Testing

A Brief Discussion on Testing

In the United States selection tests have been used for years in attempts to provide for a better fit between a person's capabilities and the requirements of a job or occupation. The mores of American society and the laws of the United States are not uniform regarding the use of selection tests. For example, many employees are required to take a medical examination, and the physician may recommend rejection of employment even though little or no data are available as to whether the condition of the patient will cause him/her not to be able to do the work—a person who has had tuberculosis may be rejected for a clerical job because the physician may have a poor understanding of specifics of the job.

Another example is the use of tests to predict the performance of students in college. Here, there are substantial data concerning the validity of the tests. The predictions do aid in the selection of college students, but they are by no means perfect. Simply put, those students who do poorly on a college entrance examination may still be able some of the time to successfully complete a college program. The point is that the American society accepts selection in certain cases without concern and with the understanding that errors in selection can and do occur.

In the area of psychomotor testing, which is the testing of a person's abilities to use the members of his/her body in conjunction with the sensory functions, the laws are much more restrictive. Part of the restriction is due to the fact that many of the selection tests that have been used in the past have had low or unknown reliability. The restrictions also have occurred because of the antidiscrimination laws which prohibit selection on the basis of race, color, or creed. The major problem here has been that different races, colors, and creeds have had unequal opportunities for education. Thus, if a test uses the English language or mathematics, it could be considered discriminatory.

Tests have also been used with cutoff scores. This procedure was widely used in World War II to select soldiers for flight training. If the soldier obtained higher than the cutoff score, he was placed in a pool awaiting training. If he did not meet the cutoff score, he was not put in the pool. Because the tests were not perfect predictors, some soldiers were placed in the pool who should not have been and some were rejected who should have been placed in the pool. In spite of these problems, the selection procedures were considered to be very successful because the failure rates in pilot training were much lower than before the tests were used.

The use of cutoff scores became a major court issue in 1963 when the Motorola Corporation gave an applicant a short verbal and numerical ability test.[13] A test score of "6" was used as a passing score and the employee applicant obtained "4," which was deemed unacceptable. As a result of cases such as these, organizations have been very careful to severely limit the use of cutoff

[13]French, R. L., "The Motorola Case," *The Industrial Psychologist,* Volume 2, No. 3, 1965.

scores. Test results, nevertheless, can be very useful in aiding in the determination of the probability that a worker will be able to do the job for which he/she may be hired.

In an effort to provide MTM users with estimates of whether an experienced worker can perform acceptably at the prescribed level, the MTM Association has sponsored laboratory and industrial studies in an attempt to develop reliable and valid test procedures. Space does not permit a detailed description of the work and procedures developed. The readers are referred to Poock[14] for the laboratory research and initial industrial studies and to Anderson and Edstrom[15] and Foulke[16] for additional studies and explanation.

Examples of the Use of the MTM Selection Tests

Each MTM motion requires the use of a number of human capabilities. Attempts have been made to identify these capabilities and to measure them by simple, easy-to-use testing procedures. In order to predict the performance of a worker on a particular job, the results of the tests are correlated with the specific MTM-1 motions used to predict the capabilities of the worker to do the job. The results are in terms of a predicted mean rate relative to the standard level and a statistical confidence interval. For example, if the acceptable work level is the normal pace (100 percent), and the test results are 110 percent with a 95 percent confidence interval of ±8 percent, then the most likely estimate is that he/she can exceed the MTM-1 rate by 10 percent and that 95 percent of the time his/her true output, if he/she uses his/her capabilities, will be between 102 and 118 percent. Thus, this worker has a high probability (greater than 95 percent) that he/she can equal or exceed the MTM-1 rate of 100 percent and would be a good candidate for the job. Providing the worker uses his/her capabilities, the chances are very high that he/she will successfully complete the probationary period.

As another example, suppose that the result of the selection tests is 100 percent with a 95 percent confidence interval of ±8 percent; this means that the best estimate is that the worker can do the job on the acceptable level of 100 percent, but the actual results will be between 92 and 108 percent, 95 percent of the time. If this worker is hired, there is a 50 percent chance that he/she can do the work at the 100 percent level or above. Thus, 50 percent of the workers hired with this test score will not be able to do the work at the acceptable level. What should be done in this case? Two suggestions are:

1. The worker could be informed what his/her chances are. If the worker has to relocate, he/she may decide to withdraw his/her application. The important point is that they are forewarned.

[14]Poock, G. K., *Prediction of Elemental Motion Performance Using Personnel Selections Tests,* MTM Association for Standards and Research Report Number 115, 1968.

[15]Anderson, D. S., and Edstrom, D. P., "MTM Personnel Selection Tests; Validation at Northwestern National Life Insurance Company," *The Journal of Methods-Time Measurement,* Volume XV, No. 3.

[16]Foulke, J. A., "Estimating Industrial Operator Performance," *Proceedings of 17th Annual MTM Conference,* October, 1969, New York, New York.

2. Since the performance predictions are job specific, the worker can be encouraged to apply for another job where his/her chances are higher of completing the probationary period.

Test Descriptions

Presently, the tests have been used to predict the performance on the most frequently used MTM-1 motions, i.e., the motions involved in bench-type assembly operations. A brief description of the tests, equipment used, and methods of scoring follows[17]:

1. **Decision Test (X1)**
   Description: A choice reaction test to determine the decisionmaking ability of a person by the rate at which he/she can process information from external stimuli where his/her capacity is not exceeded.
   Equipment: A transilluminated screen mounted in a black chassis. A response panel with four buttons, and a control unit with timers.
   Scoring: Total time in tenths of a second for 1 Bit Test, where 1 bit is defined as information needed for response by two equally likely alternatives.
2. **Puzzle-First Trial (X2)**
   Description: The learning ability based on the ability to collect, store, rearrange, and retrieve data from memory and carry out the proper sequence of motions in the shortest possible time.
   Equipment: Puzzle board containing 30 blank pieces, the first of which was permanently positioned in the lower left corner of the board.
   Scoring: First trial time in hundredths of a minute was synthetically determined by fitting a least squares line through the second to fifteenth trials.
3. **Maze (X3)**
   Description: The ability of the hand to follow a path having high constraints as rapidly as possible, to determine eye–hand coordination.
   Equipment: A stylus held by subject is moved through a path starting at upper right corner ending in lower left corner. A micro switch in start–stop holes kept time to a tenth of a second.
   Scoring: Average time of two trials through maze recorded in hundredths of a second.
4. **Pencil Flip (X4)**
   Description: Finger dexterity of the more-complex functioning of the rotational ability of thumb, index, and middle fingers about the longitudinal axis of the forearm.
   Equipment: Unsharpened 6-inch pencil.
   Scoring: Average time of two trials in tenth of a second. Each trial required 25 flips of pencil.

[17]Anderson, D. S. and Edstrom, D. P., *op cit.*, pp. 11–12, reproduced by permission of *The Journal of Methods-Time Measurement.*

5. **Tactile (X5)**
Description: The relative ability of a person to use the sense of touch by determining the relative size of paired holes in a steel plate.
Equipment: A metal plate with five sets of four holes each. Each set progressively harder to discriminate as the subject ranked holes—large to small—in each set.
Scoring: Total number of mistakes in ranking holes within sets.
6. **Tapping (X6)**
Description: Finger dexterity of the index and middle fingers alternately and separately tapping up and down to test the vertical motion of those fingers.
Equipment: A telegrapher's key attached to a recorder–timer.
Scoring: Total taps made in a 5-second trial, expressed as the average of two trials.
7. **Puzzle-Final Trial (X7)**
Description: A measure of one's recording ability (see X2) but may be limited by one's finger dexterity.
Equipment: Same as X2.
Scoring: Average time to perform puzzle from the last three trials of a total of 15 trials recorded in hundredths of a minute.
8. **Visual (X8)**
Description: A test of near acuity (fineness of vision in close work) to perform desk work.
Equipment: Bausch & Lomb Ortho-Rater. Although this machine can measure near and far acuity, muscular balance, depth perception, and color blindness, we were concerned only with near acuity, as mentioned above.
Scoring: Total score on slides 4 and 5.
9. **Kinesthetic (X9)**
Description: "Muscle memory" or ability to know position of arms and hands without visual aid.
Equipment: A board, approximating the desk work area, containing three circular areas with concentric rings numbered from zero in the center to eight in the outer ring. A shield was used to block the target area so that the eyes could remain open. Using a rubber finger containing a tack as a point, the subject made repeated trials to target area.
Predicting Worker Performance: In order to predict the performance of the worker, two approaches have been tried. Poock developed a set of linear equations. An example is as follows[18]:

$$\text{R6A–1H} = 3.82 + 0.266 \text{ (one bit).} \tag{1}$$

His results were a correlation of 0.46 between the predicted results and his industrial studies. This is significant at the 0.025 level, and this result is considered quite good as compared to previously achieved results in the field.

Anderson and Edstrom, in their attempts to use the selection tests in their organization, used Poock's equations, but obtained far better results by developing stepwise regression functions where the test results were correlated with the total time the workers used to perform their tasks. The results are given in Fig. 2–5.

[18]Poock, G. K., *op cit.*

Work Activity	Correlation Coefficient	Tests Used
Keypunch–Typing	0.87	X1, X6, X9, X8, X4
Compare–Take Action	0.83	X7, X9, X8
General Paper Handling	0.83	X5, X1, X8

Figure 2–5. The results achieved by Anderson and Edstrom.

As a further indication of the results achieved, Fig. 2–6 is a plotted comparison between the predicted and the actual performance of general paper handling.[19] Although differences do exist, the differences between predicted and actual curves are quite small.

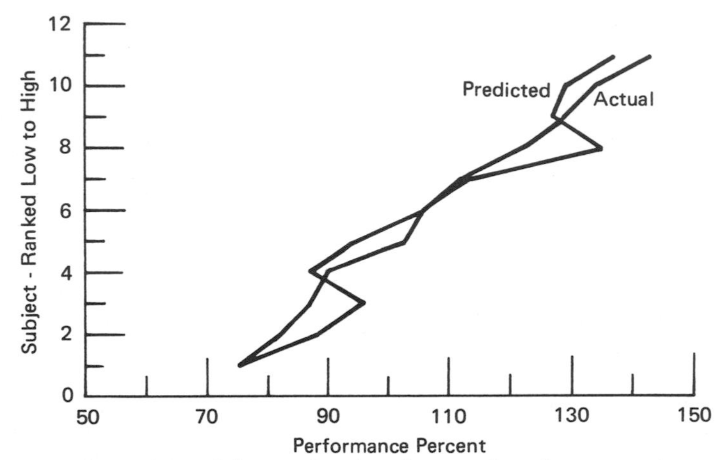

Figure 2–6. A comparison of the predicted vs the actual performance of two workers doing general paper handling. Reference 15, page 16. Reproduced by permission of The Journal of Methods-Time Measurement.

Test Availability

The MTM selection tests described herein have been available for over 10 years; however, their use has been quite limited. The test equipment is available from the MTM Association or the cost of constructing the equipment using more recent solid-state technology would probably be less than $5000. Because of the limited use of the tests to date, validation studies should be done to make sure the prediction equations are correlated significantly with results on the job.

[19]Anderson, D. S., and Edstrom, D. P., *op cit.*, Figure 7, p. 16, reproduced by permission of *The Journal of Methods-Time Measurement*.

CHAPTER 3

The Learning Curve Methodologies

This chapter is divided into two sections: The first is a section reprinted by permission from the *Handbook of Industrial Engineering* titled "The Learning Curve." This section is reprinted because it provides the reader with a general overview of learning curves and how they can be used in the establishment of standards and learning times. The second section entitled "The MTM-1 Learning Curve Methodolgy" is presented to enable the reader to develop more precise determinations of learning curves. Once the particular curve is determined, the methodologies described in the first section can be more precisely used. Because the first section is reprinted, references and equation numbers are not sequential. They are specific to the section where they are used. References are listed at the end of the section.

Before proceeding to the specific discussion of the two kinds of learning discussed, several important statements should be made.

Learning, as applied to individual workers performing tasks coverable by MTM or ordinary time study standards, does not automatically proceed at the optimum rate. The worker needs to be informed of the standard and regarding his/her performance in trying to achieve the standard. Encouragement by supervision and/or the method of payment also helps activate optimum learning. Also, proper preliminary instruction should occur, since allowing the operator to discover "the method" through trial and error impedes the learning rate. Finally, people do differ as to possible learning rates.

The second kind of learning discussed is that of organizational learning, often called "the production progress function." However, do not be deceived by the last descriptive title. All elements of the organization are involved. Progress is usually measured in terms of total cost per unit.

The rate of such learning exhibits large differences between organizations. To change *overall* organization learning one must either change the Chief Executive Officer (CEO) or somehow change the direction and effectiveness of the existing CEO. A slow learning organization in competition with a fast learning organization is doomed to ultimate failure in the product area involved unless the learning rate is changed—often in conjunction with a redesign of the product that presents better "features" and/or is producible at a lower cost.

The Learning Curve*

A learning curve is the phenomenon where, as the number of cycles increases, the time/cycle or the cost/cycle decreases, for a large number of cycles. Learning curves can be separated into two major areas: the learning that occurs while a person is doing a task repetitively, and the learning that occurs by an organization that produces a number of the same product. The former topic will be called human learning, and the latter will be called the production progress function, which is the term generally used in the literature. Fortunately, the mathematics are the same for both topics. Therefore, the appropriate equations will be presented, and then human learning and production progress functions will be discussed. Examples of the use of the equations will be given.

The Mathematics of Learning

Learning curve equations generally take the form of:

$$Y = KX^{-A} \tag{1}$$

where

Y = the time/cycle
K = the time for the first cycle
X = the cycle number
A = a constant for any given situation. The value is determined by the learning rate.

If we take the logarithm of both sides of equation (1), we get a straight line.

$$\text{Log } Y = \text{Log } K - A \text{ Log } X \tag{2}$$

Thus, if we plot the equation on log-log paper, A will be the slope and K will be the intercept. One of the useful properties of this equation is that every time X (the number of cycles) is doubled, Y (the time per cycle) decreases by a fixed percentage. This is the source of the commonly used term "percent learning curve." For example, every time the units double, the value of Y for a 90 percent curve is 90 percent of the previous value. Suppose the first unit had a cycle time of 10 minutes, then we would get the results shown in Fig. 3–1 by successively doubling the number of units.

Figure 3–2 is a plot of a 90 percent curve using linear coordinates, and Fig. 3–3 is the same equation plotted on log-log paper. Examination of Figs. 3–1, 3–2, and 3–3 reveal that rate of learning decreases as the number of cycles increases. Learning, however, continues in many situations for a very long period of time. This fact has a very important impact on the setting of production standards and on the potential for cost reduction of high volume production. In the past, the equations were usually solved by graphical means, but with the widespread use of hand calculators that can compute logarithms, graphs are no longer necessary. The following are examples of the most typical calculations.

*Chapter 4.3, *Handbook of Industrial Engineering,* G. Salvendy, Ed., Copyright © 1982, John Wiley & Sons. Reprinted by permission of John Wiley & Sons, Inc.

Number of Units (*X*)	Cycle Time (*Y*)
1	10.0
2	9.0
4	8.1
8	7.3
16	6.6
32	5.9
64	5.3
128	4.8
256	4.3
512	3.9

Figure 3–1. An example of the cycle times for a 90 percent curve.

1. What is the average cycle time for units N_1 to N_2? Assume cycle times are in minutes. Let AV = the average time, then:

$$\mathrm{AV} = \frac{\int_{N_1-1/2}^{N_2+1/2} KX^{-A}}{N_2 + 1/2 - (N_1 - 1/2)} = \frac{\dfrac{K(N_2 + 1/2)^{1-A}}{1-A} - \dfrac{K(N_1 - 1/2)^{1-A}}{1-A}}{N_2 - N_1 + 1}$$

$$= \frac{\dfrac{K[(N_2 + 1/2)^{1-A} - (N_1 - 1/2)^{1-A}]}{1-A}}{N_2 - N_1 + 1}$$

$$= \frac{K[(N_2 + 1/2)^{1-A} - (N_1 - 1/2)^{1-A}]}{(1-A)(N_2 - N_1 + 1)} \quad (3)$$

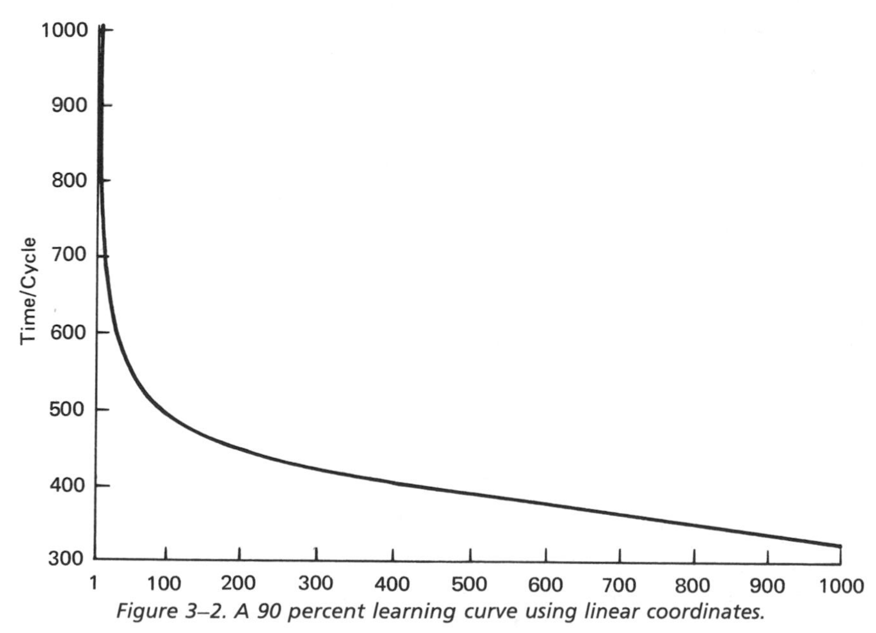

Figure 3–2. A 90 percent learning curve using linear coordinates.

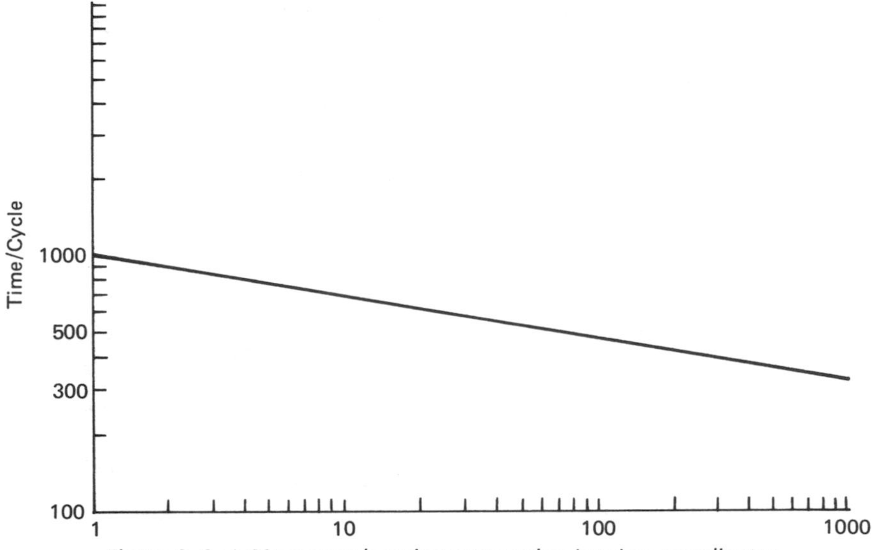

Figure 3–3. A 90 percent learning curve using Log-Log coordinates.

where N_1 is the first unit of the run and N_2 is the last unit.

Using the conditions of Fig. 3-1, where $K = 10$ minutes, and a 90 percent curve, what would be the average cycle time for the first 50 cycles? We would proceed as follows:

A. Determine the value of A using equation (1), and any value of X except 1, where the solution for A is indeterminate.

$$Y = KX^{-A}, \text{ let } X = 2, \text{ then } Y = 9.0$$
$$9 = (10)\,(2)^{-A}$$
$$\text{Log } 9 = \text{Log } 10 - A \text{ Log } 2$$
$$A = \frac{\text{Log } 10 - \text{Log } 9}{\text{Log } 2} = 0.1520$$

B. Substituting in equation (3)

$$\text{AV} = 10\left[\frac{(50\ 1/2)^{0.8480} - (1/2)^{0.8480}}{(0.8480)\ 50}\right]$$
$$= 10\,\frac{(27.82 - 0.556)}{42.40} = 6.43 \text{ minutes}$$

2. What would be the cycle time for the 50th cycle, using equation (1)?

$$Y = (10)\,(50)^{-0.1520} = \frac{10}{50^{0.1520}} = 5.52 \text{ minutes}$$

3. If the standard time is 3.5 minutes, and if K is 2.5 times the standard, how many cycles would it take to attain the standard time assuming a 90 percent curve? Let S = the standard time, then:

$$S = 2.5\,S\,X^{-0.1520},\ X^{0.1520} = \frac{2.5\,S}{S} = 2.5 \text{ or}$$

$$\text{Log } X = \frac{\text{Log } 2.5}{0.1520} \text{ or } X = \text{antilog } \frac{\text{Log } 2.5}{0.1520} = 415 \text{ cycles}$$

4. How long should it take an operator to attain the standard in question 3? We would use the numerator of equation (3) for the answer. Let C = the cumulative time:

$$C = K\left[\frac{(N_2 + 1/2)^{1-A} - (N_1 - 1/2)^{1-A}}{1 - A}\right]$$

$$= 10\left[\frac{(415 + 1/2)^{0.8480} - (1 - 1/2^{0.8480})}{.8480}\right] \tag{4}$$

$$= 10\left[\frac{166.17 - 0.56}{0.848}\right] = 1952.98 \text{ minutes}$$

If there were 7.5 working hours in one day, then

$$\frac{1952.98}{7.5 \times 60} = 4.34 \text{ days.}$$

5. Of the 4.34 days (1952.98 minutes), how much of the time could be considered training costs? Figure 3-4 might be helpful. Since the total time under the learning curve would be 1952.98 minutes, and the time to do the task if the person were performing at standard would be 415 × 3.5 = 1452.50 minutes, the training time would be 1952.98 − 1452.50 = 500.48 minutes.

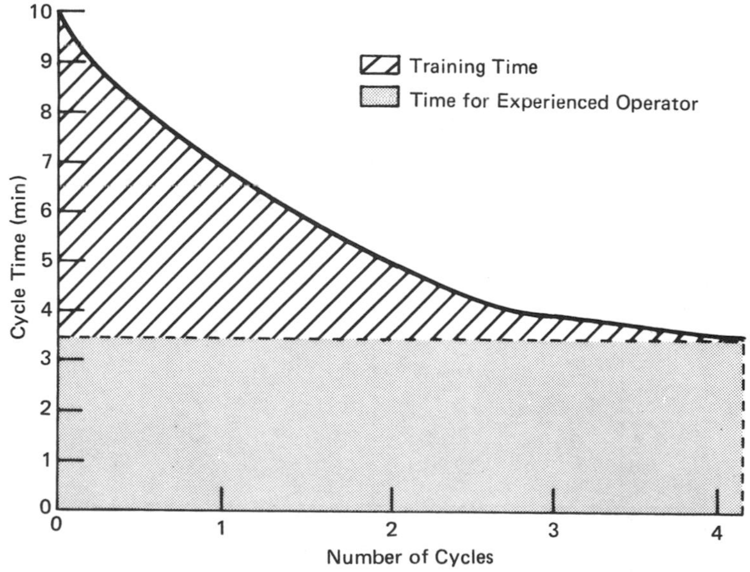

Figure 3–4. The training portion of the learning curve.

6. Suppose an operator would do 50 cycles, and then 2 weeks from now would do another 150 cycles, what would be cumulative and the average times for the 150 cycles? Figure 3–5 depicts the situation. The problem is somewhat complex because the operator will forget a certain amount that he has learned as a result of having a break. This is called remission. The amount of remission has been found to be a function of where the operator is on the learning curve when the break occurs. An approximation to the point of remission where the operator will start the second run can be made by drawing a straight line between the time for the first cycle and the standard time (S). The equation for this line is:

$$R = K - \frac{K - S}{CS} X_i, \tag{5}$$

where R is the time of the first cycle after the break, CS is the number of cycles to standard computed for the first run of 50 cycles, and X_i is the cycle number of the first cycle after the break. Since the second run started at the 51st cycle, then $R = 10 - 0.0157(51) = 9.20$ minutes. For the 150 cycles, 9.20 then becomes the new value for K. The value of A also changes because the rate of learning increases.[8] A is estimated by assuming that S will be attained in 365 cycles. Thus:

$$3.5 = 9.20(365)^{-A}$$
$$A = 0.1638$$

The times for the 150th cycle would be:

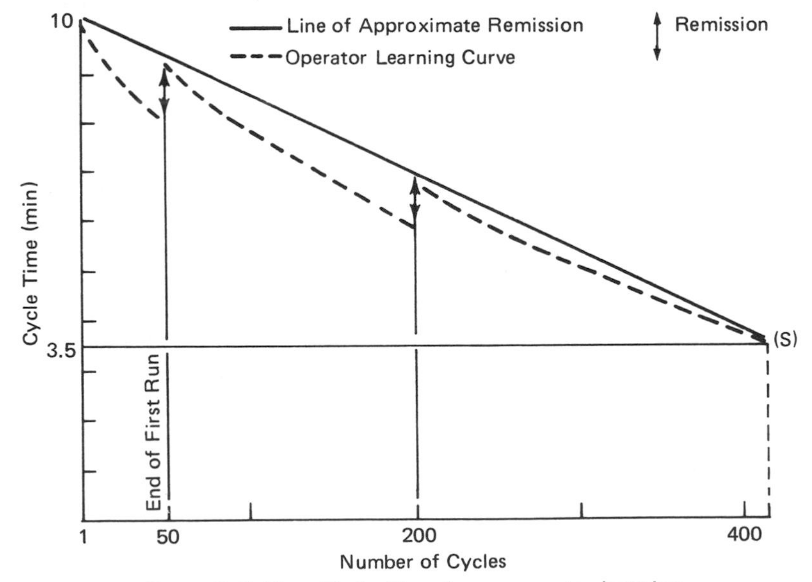

Figure 3–5. The effect of breaks on operator learning.

$$AV = \left[\frac{(150 + 1/2)^{0.8362} - (1 - 1/2)^{0.8362}}{0.8362\,(150)}\right]$$
$$= 4.81 \text{ minutes} \quad (6)$$
$$C = AV \times CY$$

where CY is the number of cycles involved, and C is the cumulative time. $C = 4.81 \times 150 = 721.50$ minutes.

Human Learning

There are a number of factors that affect the rate that people learn to do repetitive jobs. The complexity or chance effect inherent in task performance of the job to do affects the rate of learning. The capabilities of the worker also have an impact. Much has been written about both topics, and the results will only be summarized here. (See Chapter 2 for the basic concepts of psychomotor performance.)

1. The complexities of the job—Job complexity can be examined as a three-dimensional situation, because there are three major variables that affect complexity from a learning viewpoint. These are:
   A. Cycle length—Normally, longer jobs are considered to be more complex, because the worker will forget more between repetitive acts. Cycle length is at least partially accounted for in the learning curve equations, because the cumulative times increase as a function of cycle length.
   B. The amount of uncertainty in the motions involved—Uncertainty is generally measured by the number of higher skilled motions involved such as the more difficult positions, simultaneous motions, and grasps. The more uncertainty in the task, the longer it will take the operator to learn to do the job. This aspect has been investigated using the MTM-1 system.[2]
   C. The amount of prior training—In many operations the operator may have developed high skills at certain subtasks. Once a person has sufficient practice opportunity to become skilled, the remission rate is very low. An example is the ability to hit a ball with a bat. Typical examples in working situations are the ability to use a calculator, a soldering gun, or a micrometer. The MTM-1 learning curve methodology also takes this aspect into account.[8]

2. The capabilities of humans—For the vast majority of working situations, the human capabilities of primary interest are the psychomotor capabilities. These capabilities, which are the ability of people to use their hands and feet in conjunction with their sensory functions, vary between people. The ability to learn to use one's psychomotor capabilities also varies. Some of the factors that affect learning are:
   A. The person's age—Many, although certainly not all, older people learn psychomotor skills slower than younger people. The rate of learning, due to age, seems to be relatively constant from approximately 18 to 35, where it then starts to decline for many people.[3,10]

B. The amount that people have been required to learn in the past—There is some evidence to indicate that if a person stops learning to do new jobs as he/she grows older, then his/her ability to learn decreases. Decrease means that it will take longer to learn to do a new task, but that it can still be learned.[10]

C. A person's nervous system and physical capabilities—The quality of a person's nervous system declines with age. People who have good nervous systems when they are young tend to show less decline with age than people who have relatively poor nervous systems when they are young. A person's physical capabilities decline with age also, but they start to decline at a later age, and the rate of decline is not nearly as critical to job performance. This is why many people tend to choose physical jobs over jobs with high information content as they grow older.[10]

3. How humans learn—With all of the above complexities, how can the learning rate of a person or a group of people be predicted? The answer is that, although the various factors affect learning, the range of learning rates, as measured by the percent learning function seems to lie, generally, between 88 and 92 percent learning curves for those parts of a job that have to be learned. If one has older people in the work force, then it would probably be desirable to choose the higher number.[8]

Computation of the number of days to learn a job may provide estimates that are considerably shorter than are presently used in many situations. The reason for the difference will vary, but one important aspect is the method used to teach the people "threshold" learning. This is the learning that occurs prior to the operator barely knowing how to do the job without external assistance. Threshold learning is not included in the learning curve equations, because the methods used are so variable that they cannot be predicted with any accuracy. (See Chapter 2 for the effect of training methods.) Thus, the threshold learning time has to be added to the time predicted by the equations. The learning curve equations contain the conditioned learning time, which is the time that it takes to learn after a person barely knows how to do the work.

Observation and measurement of threshold times have provided considerable evidence that trial and error learning, where the operator is told to figure out to do the job himself without any formal instruction, results in the longest threshold learning times. Unfortunately, this is the method most commonly used in industry. Considerable cost savings are possible by using formal instruction by people who are thoroughly familiar with the operations to be performed, and have the ability to properly teach the method sequences to the operator.

Although the learning curve equations suggest that people learn at a smooth rate, this is not the usual case. Figure 3–6 is an example of a person learning a task.

The actual curve appears jagged, primarily because the number of eye movements are decreasing, but at an uneven rate during the learning process. When humans begin to perform a motor task, they need to use the eyes very frequently to get information; as they continue to repeat the task, they "chunk" information. Chunking is a process where they get more information per eye fixation, and thus

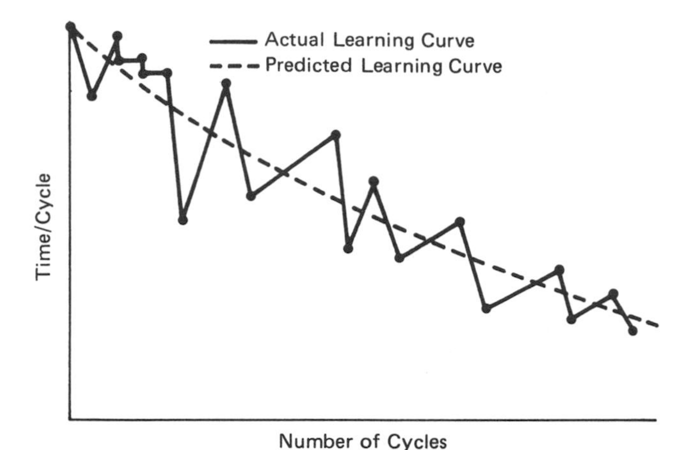

Figure 3–6. A comparison of actual learning with the predicted learning rate.

need to use their eyes less and less with increasing numbers of cycles. The jagged curve occurs because in the process of attempting to reduce eye fixation humans frequently reduce the number too much and then have to slow down, and when this happens they will use more eye fixations the next cycle.

Humans not only attempt to reduce cycle times by chunking information, but they also attempt to get the information they need by using their lower order senses, especially their kinesthetic senses (sense of position) and their tactile senses (sense of touch). The motivation for this effort is that the use of the eyes consumes a large amount of time as compared to the lower order senses. Observations of experienced people performing a task reveal that they use their eyes a minimum amount. Therefore, the industrial engineer can tell the experience level of a person by watching the worker's eyes. The operator who can look all over the place and still work without interruption is a skilled worker. This situation bothers many managers because they feel that the workers are not paying attention to what they are doing, and sometimes discipline them for it. Of course, if the workers respond to the discipline and use their eyes more often, their productivity will probably decrease.

4. The effect of human learning on time standard developed using time studies—An understanding of human learning is very important for the industrial engineer responsible for setting production standards. For the engineer using time study it is especially important because of the consequence of time studying an inexperienced person. Referring to example 1 (pp. 35–36), the average time for the first 50 cycles was 6.43 minutes, whereas, the average time for the next 150 cycles was 4.81 minutes (example 6, pp. 38–39). Thus a time study taken on a person with limited practice opportunity will result in a loose standard, unless the worker or any other worker will not have any more practice opportunity than when the time study was taken.

   The situation is further complicated by the fact that while a person is learning, he or she appears to the industrial engineer, who is not knowledgeable about learning curves, as having a high pace. The results are much higher pace ratings than should be given, which inflates the standard further. Using example 2 (p.36) where the 50th cycle is predicted to be 5.52 minutes, a person at the point on the curve would usually appear to be exerting high effort. If the industrial engineer were to rate the person at 130, then the standard would be 5.52 × 1.30 = 7.18 minutes! The best procedure is for the time study engineer to estimate the pace, which will usually be high, and the skill level, which will usually be low (too many eye fixations, hesitations, and possible fumbles), and average the two estimates.

If we assume, as many time study people do, that the standard applies to an experienced person, then how many cycles must be performed before a person becomes experienced? The cycle time, of course, decreases and does not reach a constant time. Studies of people doing repetitive cycle work for several million cycles find that the cycle time continues to decrease as shown in Fig. 3–7. Thus a time standard presumes certain run lengths. If the run lengths are longer, the standard will be loose and if the standard run length is less, the standard will be tight. If run length stays constant, but turnover increases, the standards will also be harder to achieve.

What then is an appropriate procedure to follow? Perhaps the most practical method for many types of operations where time study is used is the following[4,5]:

A. Determine the acceptable period of time that a person should have to be able to attain the standard. (The examples will show the importance on the resultant standard time of establishing this period carefully.)

B. Subtract from A above the time the person should take to start the conditioned learning phase.

C. Take two or more time studies at different numbers of cycles for several operators who are in the conditioned learning phase. Performance rate the studies, being very careful to compensate for the higher effort and low skill that occurs during learning. Also be careful to record the cycle number where the time studies are taken.

D. Obtain the value of A and K for the learning curve using the following equations where Y_1 and Y_2 are the normal times at X_1 and X_2 cycles, respectively:

$$Y_1 = KX_1^{-A}$$
$$Y_2 = KX_2^{-A}$$

Number of Cycles	Cycle Time
1	10.0
100	5.0
1,000	3.5
10,000	2.5
100,000	1.7
1,000,000	1.2
2,000,000	1.1
3,000,000	1.0

Figure 3–7. The reduction in cycle time as a function of the number of cycles where K = 10 and a 90 percent learning curve.

E. Solve equation (4) for N_2 where C = the conditioned learning time found in B above. Assume $N_1 = 1$:

$$N_2 = \left[\frac{C\,(1 - A)}{K} + 0.5^{(1-A)}\right]^{\left(\frac{1}{1-A}\right)} - 1/2. \tag{7}$$

F. The standard time is the cycle time at N_2. We obtain the standard time using equation (1) where N_2 = X.

As an example of the above methodology, suppose we have the following data:

A. Normal time at 50 cycles = 6.23 minutes
B. Normal time at 250 cycles = 5.01 minutes
C. Conditioned learning time = 2400 minutes (40 hours)

Then: $6.23(50^A) = K,\ 5.01(250^A) = K$

$$\frac{6.23}{5.01} = \frac{250^A}{50^A}$$

$$\text{Log}\,\frac{6.23}{5.01} = A\ \text{Log}\ 250 - A\ \text{Log}\ 50$$

$A = 0.1354$

Substituting, $K = 6.23\ (50^{0.1354})$

$K = 10.5810$

Solve equation (7) for N_2

$$N_2 = \left(2400\,\frac{(1 - 0.1354) + 0.5^{(1-0.1354)}}{10.5810}\right)^{\left(\frac{1}{1-0.1354}\right)} - 1/2$$

$= 449.20$ cycles.

The time standard will then be $Y = 10.5810\ (449.20)^{-0.1354}$

$= 4.63$ minutes.

5. The prediction of learning rates using Predetermined Time Systems—Of the Predetermined Time Systems, the MTM-1 system has a methodology to determine how long conditioned learning should take to attain the standard. The methodology is explained in detail in Refs. 6 and 8, and there are computer programs available from the MTM Association to aid in applications. Linear prediction equations are used for most of the MTM-1 elements. They can easily be added to attain linear approximations to equation (1). The rate of learning is a function of the relative frequency of the various MTM motions in the standard. Research on the learning rates of the various MTM motions has found that the higher order motions, such as grasp and position, take much longer to learn than the lower order motions, such as reach and move. The methodology provides for submotion sequences that are repeated within a cycle, as well as those subsequences where practice opportunity exists from other work situations. A methodology for handling repeated work opportunities while the operator is learning is also included. Once the prediction equations are developed for a particular application, the number of cycles to standard, the cycle time for a given cycle, and the average time for a number of cycles can be computed. The value of K is usually assumed to be 2.5, whereas, in the time study methodology, K is usually computed. The value of K is

assumed to be 2.5 because in the industrial studies of the application, the learning curves have indicated that the first cycle is usually performed at a cycle time that is approximately 2.5 times the MTM-1 standard. The methodology can also be used if another value of K is found to be more appropriate.

6. Determining the standard and the learning times for a work group—One of the situations that may be encountered in the application of learning curves by time study is the fact that the normal standard time, or the MTM-1 standard time, is what is known as "low task" time. That is, they are standard times that can be attained and sustained for long periods of time by a large majority of the healthy, experienced working population. Figure 3–8 is a histogram of the relationship between the standard time and the performance of the working population while working at the maximum sustainable pace.

   Different authors have slightly different histograms of the average output of the work force under high motivation conditions. Figure 3–8 is derived from research data and from industrial experiences, where the allowances are less than or equal to 5 percent.[7] Other authors[1,9] estimate the average motivated performance of the work force to be 125–130 percent. Unfortunately, the allowances that are used are not stated, but they are probably in the 10–15 percent range.

   Since a large percentage of the population can equal or exceed the normal times, then most of the people who have just learned sufficiently to attain the normal times will not be fully learned. Thus, the number of cycles found in the time study example of 449.20 (example F, p. 43) cycles is the number of cycles that the average person should take to attain the low task standard. Since in a statistical sense 50 percent will learn faster, and 50 percent slower, the following questions might be asked:

   A. How long will it take to learn the job so that the large majority (95 percent or more) can attain the standard of 4.63 minutes?
   B. What should be the normal time standard if 4.63 minutes is the average accomplishment after 4, and everyone should be able to accomplish the standard at this point?

   The answer to A is as follows:

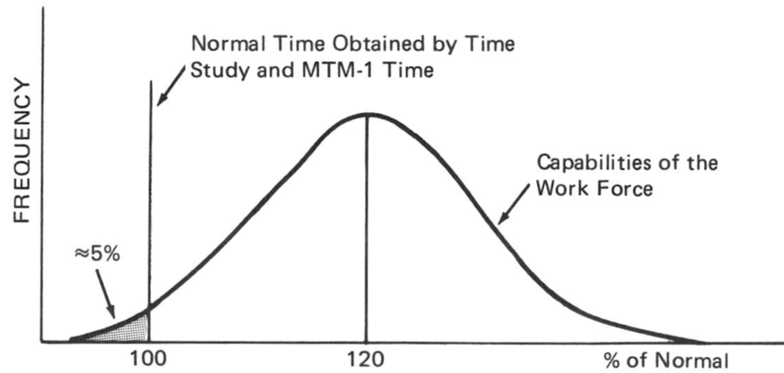

Figure 3–8. The relationship between the motivated performance of the work force and "low task" standards.

1. $\frac{4.63}{1.20}$ = average population capability = 3.86 minutes;
2. $3.86 = 10.5810\ X^{-0.1354}$, $X = 1727$ cycles

Thus, 1727 cycles is the estimate of the number of cycles that it would take for approximately 95 percent of the work force to attain at least 4.63 minutes.

3. Using equation 4 to get the cumulative time:

$$C = 10.5810\left[\frac{(1727 + 1/2)^{1-0.1354} - (1 - 1/2)^{1-0.1354}}{1 - 0.1354}\right]$$

$$= 7698 \text{ minutes}$$

Thus, 7698 minutes or 128.30 hours will be required for 95 percent of the people to do the task at the standard time of 4.63 minutes.

Question B is answered by assuming that when the average of the group is 4.63 minutes, the poorest learner will be 20 percent slower or $4.63 \times 1.20 = 5.56$ minutes. Thus, if a large percentage of the group should attain the standard in 2400 minutes, the standard should be 5.56 minutes.

Production Progress Functions

Production progress functions are a method of measuring and estimating the rate at which an active organization learns to produce a product. This type of learning has been found to follow the same negative power functions that are used for human learning. However, the learning rate is considerably faster, the most common rate being an 80 percent curve. Also, the dependent variable is usually dollars/unit or time/unit.

A number of studies have been reported in the literature concerning the percent production progress found in various industries. Reference 5 contains the most comprehensive list as summarized in Fig. 3–9.

The factors that caused the large reductions in unit times above varied from situation to situation. The following were the major reasons:

A. Organizational improvements
B. Improvement in the dimensions of the process to be assembled
C. Improvement in work methods
D. Improvements in the means of production (new machines)
E. Increase in the skill of the employee (human learning)

Production progress, thus, occurs in many industries at a rapid rate. It continues for large numbers of units. Although the examples given are unit times, many organizations experience the same percent production progress rates where dollars/unit are plotted versus number of units. In a real sense, production progress is the antithesis of standard costs, because with production progress causing costs to drop 20 percent or more every time the units are doubled, standard costs do not stay standard for very long. Knowledge of production progress functions is very important to those who are involved in bidding, cost anal-

Source	Percent Production Progress
a. Volkswagen, 1945 to 1949	40
b. Volkswagen, 1950 to 1954	20
c. Twenty Light Alloy Products	20
d. Repair of Goods Wagons	20
e. Home Construction	14–27
f. Welding of Thin Steel	30
g. Airplane Production	25–30
h. Shipbuilding	10–26
i. Vehicle Bodies	20–30
j. German Armament Industry	18–35
k. Railway Carriages	7–25

Figure 3–9. A sample of the percent production progress found in various industries. (The percent learning curve would be 100 − percent production progress.)

ysis, and the pricing of products. For example, suppose we have two competing businesses, and both companies decide to produce a very similar product. Let us further assume that both companies start production at the same time, however, company A experiences a 75 percent progress curve, and company B experiences an 80 percent curve as depicted in Fig. 3–10. It is obvious, without doing any computations, that company A will have lower costs/unit providing company A produces at the same rate as company B.

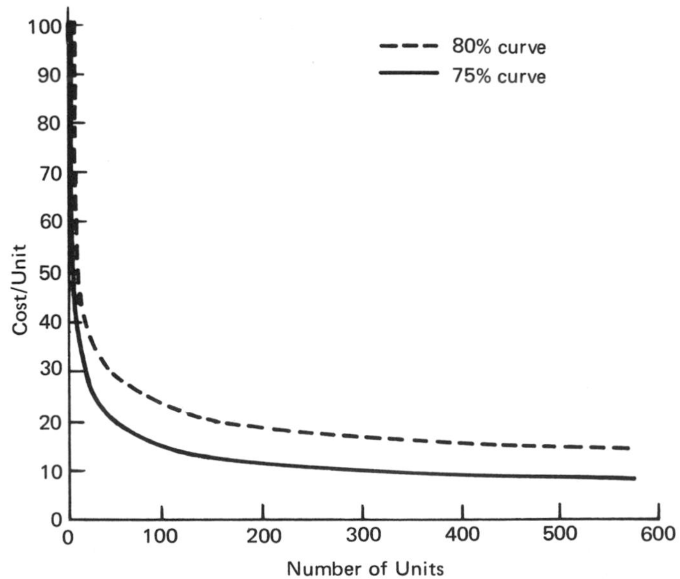

Figure 3–10. A plot of 75 percent and 80 percent production progress curves.

The equations presented earlier can be used to compute a number of factors as follows:

A. What is the unit cost of the 200th unit of company B if $K = 100$? Using equation (1) with $A = 0.3219$ for an 80 percent curve:

$$Y = 100(200)^{-0.3219} = \$18.17$$

B. What would be the total cost of the first 200 units of company B if $K = 100$? Using equation (4):

$$C = 100\left[\frac{(200 + 1/2)^{1-0.3219} - (1 - 1/2)^{1-0.3219}}{1 - 0.3219}\right]$$

$$= \$5{,}275.31 = \text{total cost of the first 200 units.}$$

C. What would be the average cost of the first 200 units of company B? Using the result in B, and dividing by 200: $\frac{\$5{,}275.31}{200} = \$26.38$

D. Production is planned for 20,000 units next year. At the beginning of the year we will have already produced 3,000 units. What should be the budget for company B for this product? Using equation (4):

$$C = 100\left[\frac{(23{,}000 + 1/2)^{1-0.3219} - (3001 - 1/2)^{1-0.3219}}{1 - 0.3219}\right]$$

$$= \$100{,}163.31.$$

E. What should be our unit cost in D above?

$$\frac{\$100{,}163.31}{20{,}000} = \$5.01 = \text{unit cost.}$$

F. There is a change in planning for company B. Instead of 20,000 units, 40,000 units are planned for the next year. What will be the unit cost? Using equation (3):

$$AV = 100\left[\frac{(43{,}000 + 1/2)^{1-0.3219} - (3001 - 1/2)^{1-0.3219}}{(1 - 0.3219)\,(43{,}000 - 3000)}\right]$$

$$= \$4.27 = \text{the unit cost.}$$

G. Company B is experiencing a 10 percent/year increase in the cost of materials and labor. If the unit cost is \$7.60 at 3000 units, how many units would have to be produced so that the production progress would enable the company to continue to sell the product at \$7.60? Solution: $X \times 1.10 = \$7.60$, $X = \$6.90$, which is average unit cost that will have to be realized over the next year. Using equation (3):

$$\$6.90 = 100\left[\frac{(N_2 + 1/2)^{1-0.3219} - (3{,}001 - 1/2)^{1-0.3219}}{(1 - 0.3219)\,(N_2 - 3{,}000)}\right]$$

$$N_2 - N_1 = 2240,$$

which is the number of units needed to be produced to counter the cost increase due to inflation. *Note:* the equation is best solved by iteration using a programmable calculator.

H. Company B is a job shop and produces 3000 units. Three months later they are asked to bid on the production of 2500 more units. What would be their estimated costs for the 2500 units?

1. Using equation (5), but using the units of dollars for S, R, and K, and assuming S is \$6.08 at $X_1 = 6000$, the value of R at $X_2 = 3001$, using equation (5) would be:

$$R = 100 - \left(\frac{100 - 6.08}{6000}\right)(3001) = \$53.02,$$

where $CS = X_1$.

2. A must be recomputed from the following calculation:

$$6.08 = 53.02(6000 - 3000)^{-A}$$
$$A = 0.2705.$$

3. Using equation (4) where $R = K$, $N_2 = 2500$, $N_1 = 1$

$$C = \frac{53.02}{0.7295}[(2500 + 1/2)^{0.7295} - (1 - 0.5)^{0.7295}]$$
$$= \$21{,}847.64 \text{ for 2500 units.}$$

References

1. Barnes, R. M. *Motion and Time Study, Design and Measurement of Work,* 5th ed., John Wiley & Sons Inc., New York, 1964, p. 324.
2. Chaffin, D. B. and Hancock, W. M., *Factors in Manual Skill Training,* Research Report 114, MTM Association for Standards and Research, Fair Lawn, N.J., August, 1966.
3. Clifford, R. R. and Hancock, W. M., "An Industrial Study of Learning," *Journal of Methods Time Measurement,* Vol. IX, No. 3, MTM Association for Standards and Research, Fair Lawn, N.J./Ann Arbor, Michigan, January–February, 1964, pp. 12–27.
4. De Jong, J. R., "Increasing Skill and Reduction of Work Time," *Time and Motion Study,* September, 1964, Chapter 1, pp. 28–41.
5. De Jong, J. R., "Increasing Skill and Reduction of Work Time—Concluded," *Time and Motion Study,* October, 1964, Chapter 3, pp. 20–33.
6. Hancock, W. M. "The Learning Curve," Chapter 5, *Industrial Engineering Handbook,* edited by H. B. Maynard, 3rd ed., McGraw-Hill, New York, 1971, pp. 7–102 to 7–114.
7. Hancock, W. M. and Åberg, U. *Design Criteria of Predetermined Time Systems with Special Reference to the MTM System,* International MTM Directorate, Stockholm, Sweden, 1968.
8. Hancock, W. M. and Sathe, P., *Learning Curve Research on Manual Operations,* Research Report 113A, MTM Association for Standards and Research, Fair Lawn, N.J., 1969.
9. Mundel, M. F., *Motion and Time Study Principles and Practices,* 4th ed., Prentice-Hall, Englewood, N.J., 1970, p. 303.
10. Welford, A. T., *Aging and Human Skill,* Oxford University Press, 1958.

Bibliography

Abramowitz, J. G., and Shattuck, G. A., Jr., *The Learning Curve, A Technique for Planning, Measurement and Control.* Report No. 31.101, 4th ed., IBM, Harrison, N.J.

Barnes, R. M., and Amrine, H. T. "The Effect on Practice of Various Elements Used in Screwdriver Work." *Journal of Applied Psychology,* April, 1942, pp. 197–209.

Karger, D. W., and Bayha, F. H. *Engineered Work Measurement,* 3rd ed., Industrial Press, New York, 1977.

Nanda, R., and Alder, G. L. *Learning Curves, Theory and Application,* American Institute of Industrial Engineers, 1977.

A Few Additional Thoughts On Organizational Learning

If time/unit produced is used as a measure of this function, it means that the focus of the study is almost solely on the factory production function. Such information is valuable, but since the whole organization seems to learn how to do "it" better, cost/unit is a more generally meaningful measure. In fact, more organizations are focusing on overall learning as measured by cost/unit.

Cost/unit is affected by material costs (suppliers also learn), marketing costs, financial costs (like interest expense), general and administrative expense, direct labor cost, indirect labor cost, and all elements of overhead—all involved costs.

Since this text is largely concerned with the factory, let us briefly focus on the production function. Workers as previously described learn how to meet the production standards in a normal and straightforward manner. However, they also conceive and/or learn how to make major and minor improvements. The large improvements usually get processed through the suggestion system, but the minor ones are put into effect without telling anyone—they do not want the standard changed. Often the supervisor is aware of the small changes but says nothing since it only amounts to 3–5 percent. Unfortunately, each year sees equivalent improvements taking place so that the effect over 5 to 10 years is 15–50+ percent. Improved quality of parts also often contribute to this creeping improvement and erosion of labor standards. How to deal with this problem is discussed in the companion text, *Engineered Work Measurement.*

If a product is redesigned, the new parts and operations start a new learning curve for their portion of the overall cost; this applies to both the primary producer and all of the involved suppliers. The parts and operations that are not changed proceed on their old learning curve, in this case a preferred term would be, in our opinion, experience curve, or organizational learning curve, or (in a restricted case) a production progress function curve.

A generalized discussion of experience curves can be found in *Perspectives on Experience* by The Boston Consulting Group.

Experience curves, when used for corporate planning or for strategy formulation by upper executives, should be plotted using log-log coordinates since the curve then appears as a straight line and inferences and conclusions are easier to arrive at by nonmathematically oriented people.

In studying the phenomenon, it was noted that product price seems to become stable (not subject to major fluctuations in rate of change) when the slope of the price curve is parallel to that of the learning curve of the lowest cost producer. This phenomenon is illustrated in Fig. 3–11, which also illustrates the presence of a competitor and how company positions reversed over time.

If a firm is interested in trying to determine a competitor's underlying rate of learning, it is suggested that the reader try plotting the deflated price curves for one or two typical and somewhat mature products on log-log coordinate paper. If the price curves have assumed the typical shape illustrated in Fig. 3–11, then it is likely that the slope of the lower portion of the curve is a close approximation of the company's rate of learning—although be aware that some subsidiaries of large corporations may exhibit characteristics different from other subsidiaries or divisions. Such curves need to be deflated by an appropriate factor if the underlying inflation rate is very high in order for the appearance of the typical shape to occur. Also, remember that experience curves often change slope when a new CEO arrives in a company with a slow rate of learning and/or when a product experiences a major redesign.

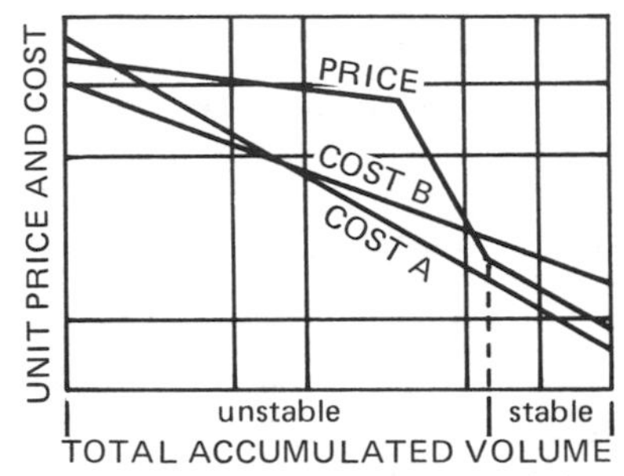

Figure 3–11. An illustration showing the cost curves and the deflated price curve of a common product. Company "A" eventually becomes the lowest cost producer. The price of the product remains unstable until its slope becomes parallel to the cost curve of the lowest cost producer.

The MTM-1 Learning Curve Methodology

Earlier in the chapter, it was mentioned that typical learning curves are between 88 and 92 percent. For most of the calculations in the first section a 90 percent curve was used. If one is desirous of getting more precise predictions of conditioned learning rates, there is a methodology developed for use with the MTM-1 system,[1] which will be helpful in certain instances when using the methodologies presented earlier in this chapter.

The methodology consists of a series of linear equations, which, when properly applied, enable the user of the MTM-1 system to predict the number of cycles to standard, coefficients of the learning curve $Y = KX^{-A}$, and the percent learning. The following is an outline of how to use the methodology:

1. Write the MTM-1 standard for the operation involved. We will call this the Total Analysis Standard.
2. Identify subcycles where a subcycle is a suboperation that is done more than once. Separate the MTM analysis of the subcycles from the total analysis. What is left of the total analysis we will call the Main Analysis.
3. Figure 3–12 is a series of linear equations which are to be used in conjuction with

[1]Hancock, W. M. and Sathé, P., *Learning Curve Research On Manual Operations. Phase II: Industrial Studies,* Report 113A, MTM Association for Standards and Research, Fair Lawn, N.J., 1969.

MTM Element	MTM Std.	0–500 Cycles One Hand	0–500 Cycles Simo	500–1000 Cycles One Hand	500–1000 Cycles Simo	Comments
R-A	8.7	1.142 − 0.000065(N)	1.403 − 0.000144(N)	1.109 − 0.000193(M)	1.331 − 0.000193(M)	
R-B		0.776 − 0.000075(N)	0.964 − 0.000155(N)	0.739 − 0.000094(M)	0.887 − 0.000094(M)	R-C, R-E
R-C	12.9	0.776 − 0.000075(N)	0.964 − 0.000155(N)	0.739 − 0.000094(M)	0.877 − 0.000094(M)	
M-A	11.3	0.916 − 0.000022(N)	1.115 − 0.000058(N)	0.905 + 0.000058(M)	1.086 + 0.000058(M)	
M-B		0.580 + 0.000068(N)	1.090 − 0.000336(N)	0.614 − 0.000128(M)	0.922 + 0.000093(M)	M-C
M-C	11.8	0.580 + 0.000068(N)	1.090 − 0.000336(N)	0.614 − 0.000128(M)	0.922 + 0.000093(M)	
G1A	2.0	2.365 − 0.001270(N)	2.927 − 0.001702(N)	1.730 − 0.000430(M)	2.076 − 0.000430(M)	
G1B		0.765 + 0.000048(N)	1.010 − 0.000126(N)	0.789 − 0.000149(M)	0.947 − 0.000149(M)	G1C2
G1C2	8.7	0.765 + 0.000048(N)	1.010 − 0.000126(N)	0.789 − 0.000149(M)	0.947 − 0.000149(M)	{G1C1 G1C3
G2		0.765 + 0.000048(N)	1.010 − 0.000126(N)	0.789 − 0.000149(M)	0.947 − 0.000149(M)	G1C2
G3		0.765 + 0.000048(N)	1.010 − 0.000126(N)	0.789 − 0.000149(M)	0.947 − 0.000149(M)	G1C2
G4A	7.3	1.663 − 0.000318(N)	2.247 − 0.000412(N)	1.504 − 0.000001(M)	2.041 − 0.000086(M)	G4B
P1SE	5.6	1.995 − 0.000182(N)	2.568 − 0.000608(N)	1.904 + 0.000029(M)	2.264 − 0.000018(M)	
P1SSE	9.1	1.309 + 0.000574(N)	2.208 − 0.000583(N)	1.596 − 0.000346(M)	1.917 − 0.000209(M)	P1NSE Simo
P1NSE	10.4	1.929 − 0.000737(N)	2.208 − 0.000583(N)	1.561 + 0.000155(M)	1.917 − 0.000209(M)	
P21S		0.940 − 0.000091(N)	1.208 − 0.000284(N)	0.895 + 0.000013(M)	1.066 − 0.000008(M)	P22S2
P22S-	11.9	0.940 − 0.000091(N)	1.208 − 0.000284(N)	0.895 + 0.000013(M)	1.066 − 0.000008(M)	
Remainder Equation		1.148 − 0.000345(N)	1.581 − 0.000667(N)	0.976 − 0.000189(M)	1.247 − 0.000100(M)	

Figure 3–12. Learning equations for 0–500 and 500–1000 cycles for 120 percent performance. Reproduced with modifications by permission of the MTM Association For Standards and Research.

an MTM-1 analysis. These equations are used to determine the number of cycles to standard for a worker at 120 percent performance. Two sets of equations are given because it was not possible to use linear expressions without breaking at 500 cycles. The left-most column gives the motions that the equations in the same row apply to. The "Comments" column gives other motions to which the equations can be applied. The equations are normalized so that the number of TMU for each motion involved can be multiplied with the intercept and coefficient of the equations listed. For example, if there were 150 TMU of R – A single handed motions, then

$$150(1.142) - (150)0.000065(N) = 171.3 - 0.0098(N)$$

is the equation for the R – A motion. The remainder equation at the bottom of Fig. 3–12 is used for all elements not listed.

4. If there are subcycles involved, the equations are by the TMU element content as indicated in Step 3. Then the linear equations are summed to get an overall equation for the subelement. Since the subelement occurs more frequently than the main cycles, the learning of the subcycles goes at a much faster rate. To account for this, the coefficient of N or M is multiplied by the frequency of the subelement. For example, if the subelement occurs five times every time the total element is performed, the coefficient of N or M would be multiplied by 5.
5. The TMUs are determined by motion category for the main analysis. The equations are determined as described in Step 3.
6. The equations are summed for the subcycles (Step 4) and the main analysis (Step 5). *Note:* Linear equations are summed by summing their respective times. For example, suppose the main analysis equation was 972 – 0.04178(N) and the subanalysis was 155 – 0.00462(N), then the summed equation would be 1127 – 0.04640(N).
7. A decision now has to be made about the pace of the workers during learning. The equations are for 120 percent pace (see Chapter 2 for a discussion of pace if you are at all confused as to the specific meaning of the term). If any other pace is to be achieved, then the constant value of the total equation has to be changed accordingly. For example, if 100 percent pace is desired, the intercept of the *summed* equation in Step 6 would be modified as follows:

$$1.20\,(1127) - 0.04640(N) = 1352 - 0.04640(N).$$

If 85 percent of normal is acceptable, then modify the intercept of the summed equation as follows:

$$1.35\,(1127) - 0.04640(N) = 1521.45 - 0.04640(N).$$

8. We must now determine the number of cycles to standard. At this point, we should have two total cycle equations: One for 0–500 cycles and one for 500–1000 cycles. In order to determine which equation to use, we compare the intercepts with the total analysis standard. For example, if the standard were 130.8 TMU, the 0–500 cycle equation were 181.620 – 0.042872(N), and the 500–1000 cycle standard were 155.911 – 0.009815(M), the standard is less than the intercept of the 500–1000 cycle equation (155.911). Therefore, the 500–1000 cycle equation would be used. Solving for M:

$$155.911 - 0.009805(M) = 130.8$$
$$M = \frac{155.911 - 130.8}{0.009805} = 2561$$

Since N = M + 500,

$$N = 2561 + 500 = 3061 \text{ cycles.} \quad (1)$$

This is the number of cycles to standard for an operator to obtain the MTM standard.

9. We must now determine the learning curve equation by first determining the value of A in the equation $Y = KX^{-A}$:

$$A = \frac{\log_{10} K'}{\log_{10} X_s} \quad (2)$$

where K' is the constant found experimentally to be 2.5 for conditioned learning and X_s is the number of cycles to standard. Substituting using the example results of Step 8:

$$A = \frac{\log_{10} 2.5}{\log_{10} 3061} = 0.1142$$

The learning curve is of the form:

$$Y = KX^{-A} \quad (3)$$

where:

$K = K'(T_s)$,
T_s = Standard Time
X = the number of cycles
Y = the cycle time.

Thus the learning equation is:

$$Y = 2.5(130.8)\, X^{-0.1142} \text{ or}$$
$$Y = 327\, X^{-0.1142}.$$

The percent learning in this case is obtained by using the fact that every time the number of cycles is doubled, the cycle time decreases by a constant amount. Thus

$$P = \frac{Y_i}{Y_{i/2}} \times 100 \quad (4)$$

where P is the percent learning and Y_i is the cycle time at cycle X_i.
Using the example equation and solving for Y_i in two instances where the number of cycles are doubled (say $X = 2$ and $X = 8$), we get

$$Y_2 = 327(2)^{-0.1142} = 302.114$$
$$Y_4 = 327(4)^{-0.1142} = 279.121$$
$$P = \frac{279.121 \times 100}{302.114} = 92.4 \text{ percent.}$$

We are now in a position to use the methodologies of "The Learning Curve Methodologies" section with more precision.

Fatigue Allowances

What Is Fatigue?

Fatigue—as applied to people working—has many meanings, varying from feeling tired, not feeling like working, the actual tiring of the muscles of the body (physiological fatigue), or a reduction in output due to becoming overheated (heat stress).

Feeling Tired

It is common experience to start to feel tired after working for a period of time, but the feeling of tiredness is not a very reliable indicator of the ability to work. In many cases, output can be sustained with no injurious effects to health. One can feel tired before work is ever started. Lack of sleep, personal discord, or physical activity not associated with work can cause the feeling of tiredness to occur. Furthermore, there is a wide variation in how tired people feel even after doing the same activity.

Not Feeling Like Working

Not feeling like working can be due to many causes. It can be due to tiredness or to the desire to do something else or to be somewhere else. Generally not feeling like working is called lack of motivation, but lack of motivation may not be due to fatigue. The fatigue aspects will be discussed in this chapter under physiological fatigue and heat stress. Other reasons for lack of motivation are outside the scope of this chapter.

Physiological Muscle Fatigue

Physiological fatigue occurs when the work produced by the human body consumes more of the available energy than the body can sustain. The human machine consumes food and oxygen and through the metabolic process produces heat and physical work. In practically all cases, the human has plenty of food reserves to draw on, but he/she may reach the limits of being able to absorb enough oxygen into the blood stream to enable the muscles to produce the motions involved in work. When this happens, a phenomenon known as oxygen debt occurs in which the muscles continue to function by using the energy available within the muscles themselves. After a period of time the muscles' capacity to act will diminish. When the person goes into oxygen debt, he/she may or may not feel tired.

Physiological Heat Stress

When a person is using his/her muscles, part of the metabolic output is consumed by the muscle groups involved. The other part results in heat. If the human's exercise is rigorous enough, he/she will start to perspire. As the perspiration evaporates from the surface of the body or from clothing next to the body, the body heat is removed. Thus the perspiration can be thought of as the body's air-conditioning system. When the perspiration provides insufficient cooling, the body will start to heat up and the body temperature will rise. If the body temperature stays within certain limits, no problems are likely to occur, but if the body temperature gets too high, then rest and removal from the environment must occur or the body can go into heat shock.

Body temperature is also affected by the environment in which the work occurs. If the person is working and being exposed to the sun at the same time, or if the work is near any other source of radiant energy such as a furnace, then the body must get rid of additional amounts of heat. Of course, if the radiant source is colder than the human, then the source will aid in the cooling of the body rather than adding to the heat load.

The perspiration system has varying abilities to cool the body due to temperature and humidity of the air surrounding the worker. Generally, low humidity will cause more perspiration to evaporate resulting in a greater amount of cooling. Also, the higher the atmospheric temperature, the more evaporation. On the other hand, if the humidity is high, evaporation will occur at a much slower rate, and the body will become stressed much faster.

Present-Day Use of Fatigue Allowances

Now that we know a little about fatigue, we can deal with how fatigue allowances should be developed and used. In order to discuss this, it is important to first consider the typical work measurement approach to fatigue allowances.

The fatigue allowances as applied to MTM systems or to any other work measurement system are usually multipliers of the standard time to allow for "fatigue." The following two types of multipliers are used:

The Fixed Percentage Multipliers—Typically, the figure is about 8 percent of the normal standard time. For example, if an MTM standard were 1500 TMU, then 1500 × 1.08 = 1620 TMU would be the standard time factored for inclusion of the fatigue allowance. Often, the 8 percent is applied to all jobs regardless of the work or the environment of the work.

The Fatigue Factor Set by Environment and/or Work Factors—Typically a range of points are assigned to a number of characteristics of a job and the job environment. Then each job is graded with respect to each of the characteristics, the resultant points added and the sum converted to a percent fatigue allowance.[1]

[1]Niebel, B., *Motion and Time Study*, 5th ed., Irwin, 1972, pp. 364–365.

The problem with these systems is that neither one has any technological basis. To apply a fixed percentage independent of all work factors or environment implies that all work conditions and work are the same. To attempt to differentiate between different types of work and work environments is a much better approach except that the range of the points for each job and environmental characteristics are arbitrarily determined. The word arbitrary in this context means that little or no data were gathered and analyzed and that there may be little or no relationship between the allowance given and the fatigue experienced.

Fatigue allowances as set by the above systems are being used presently because:

A. Until recently they were the only methods available to set fatigue allowances.
B. They have been made part of the bargaining process. To provide for fatigue, at least on the surface, seems to be reasonable, and since until recently no one knew how to measure fatigue, or define it, one value is as good as another.
C. Many industrial engineers have been educated to automatically apply a fatigue allowance as part of the PF&D (Personal Fatigue and Delay) Allowance without questioning whether or not it is needed.
D. There is confusion on the difference between fatigue and reasonable fatigue. Some proponents feel that any feeling of tiredness should result in a fatigue allowance whereas others feel it is perfectly natural to get tired over the period of a working day and that the fatigue allowance should only be used to enable the employee to recover from unreasonable fatigue. Since this issue has not been resolved in many cases, fatigue allowances are given to everyone.

Fatigue Allowances as Applied to the MTM Systems

Based on the assumption that a fatigue allowance is to provide for an unreasonable level of fatigue so that the worker can take a rest, no fatigue allowance is needed in the vast majority of MTM applications. The MTM values are based on a work rate that can be sustained for 8 hours, 5 days/week for the working life of the employee providing he/she stays healthy. Inasmuch as the movement of objects requires more effort than the movement of an empty hand, it can be stated that move motions have some provision for fatigue built into the time values. Thus, no fatigue allowance should be given unless unreasonable circumstances exist. Presently, the unreasonable circumstances where fatigue allowances should be given are:

A. The energy output of the total body exceeds certain limits. This is physiological muscle fatigue.
B. An adverse environment exists that results in unhealthy physiological changes to the body. The most frequent case is physiological heat stress.
C. The work place requires the same muscle groups to be used repeatedly without the muscles being able to relax. This is called local muscle fatigue. The most obvious example is a static hold, where the worker must hold an object without moving for long periods of time.

Fatigue allowances rarely exist for local muscle fatigue because the job can be redesigned to reduce it to reasonable levels. If this cannot be done, then special studies must be made to determine how to enable the operator to get the proper rest.

Physiological Muscle Fatigue

Rapid progress has been made in the field of work physiology in recent years. Questions such as how much a worker can do in 8 hours, the difference in the physiological capabilities of workers (male and female) as individuals and as working populations, and the use of physiological measurements to improve workplace design have all been attacked. Available data are at a point now where the results can be used to predict fatigue allowances; however, the methodologies require that the work measurement analyst becomes versed in the use of oxygen uptake equipment and the mathematical formula necessary to make the fatigue allowance calculations.

The steps necessary to determine whether a particular job (worker) should receive a rest allowance are as follows:

A. An estimate must be made of the energy consumption required to do the job.
B. The aerobic capacity of the individual must be estimated or working population figures appropriate to the work force must be used.

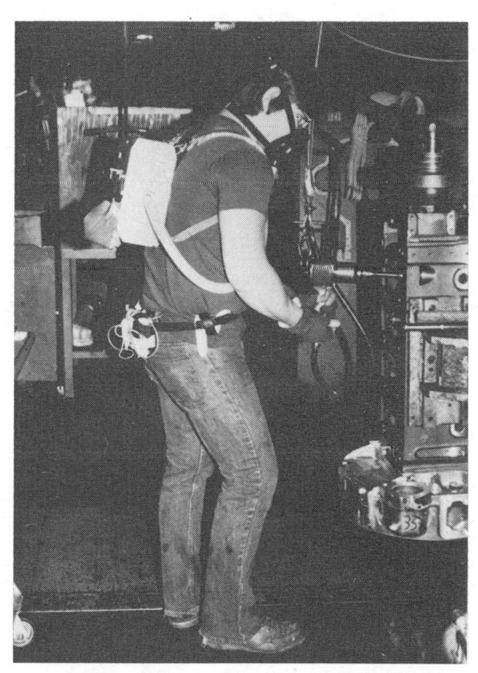

Figure 4–1. An example of the use of a portable respirator.

C. The percent of the aerobic capacity that has been sustained for 8 hours (or any other period) must be determined.

D. Calculations necessary to determine whether or not a fatigue allowance is appropriate and the magnitude of the allowance must be made.

A discussion of these steps follows.

A. *An estimate of the job's energy requirements*. There is a high correlation between the amount of oxygen a person uses and the amount of energy he/she is using to do the work. Thus, if one can sample the expired air of the worker while working, and compare it with the oxygen content of the surrounding air, then the amount of energy being consumed can be computed. The unit of measurement is kilogram calories/minute. The most common method of obtaining oxygen consumption figures is to use a portable oxygen sampler such as the Max Planck respirator. This equipment is specifically designed to be used in industrial environments. Figure 4–1 shows a person using a respirator.

Figures 4–2, 4–3, and 4–4 are lists of values determined by this method for a number of jobs. Of course, the values listed are for the work being done and

	kcal/min/145 lb man	
	Range	Mean
Woodwork factory		
carpenter—assembling	2.9–5.0	3.9
carpenter—finishing		2.9
cabinetmaker	5.5–5.7	5.6
laminating machine operator		4.0
milling machine operator		3.8
sanding machine operator		4.3
spray painter		3.9
wood stainer		4.7
packaging		4.1
Electrical industry		
armature winding		2.2
battery plate casting	3.6–4.2	3.9
battery plate punching and cutting	2.5–5.0	3.4
coil assembly	3.1–4.9	4.0
dipper	5.2–5.5	5.4
ebonite moulding	2.9–4.4	3.4
galvanizing	4.5–4.9	4.7
materials handler	2.9–3.7	3.3
punch press operator	3.3–5.0	4.2
relay and part adjuster	1.9–3.0	2.3

Figure 4–2. The energy expenditures for various jobs for a 145 lb man.[2]

[2]Chaffin, D. B., "Some Effects of Physical Exertion," Dept. Of Industrial Engineering, University of Michigan, 1972, pp. 132–133. Taken from Durnin, J.V.G.A. and Passmore, R., *Energy, Work and Leisure,* Heinemann Books, Ltd., London, 1967, pp. 53–54. Reproduced by permission.

	kcal/min/145 lb man	
	Range	Mean
radio mechanics		2.7
rolling machine operator	1.9–4.3	2.7
stock room work	3.8–4.5	4.2
trimming	3.4–5.0	4.2
wire-drawing machine operator	2.7–5.5	4.1
Machine-tool, metal, and motor industries		
bending and straightening	2.3–4.1	3.2
drilling and tapping	2.4–4.5	3.5
fitting	2.7–4.0	3.4
lathe operator	3.2–3.6	3.4
machinist	2.2–4.3	3.1
metal polishing	1.7–4.4	3.1
punch press operator	5.6–5.8	5.7
sheet metal worker	1.5–4.3	3.1
spray painting	3.2–3.5	3.4
tapping and drilling	3.3–5.2	4.2
toolmaking	1.4–4.7	3.2
tool setting	3.3–4.8	4.1
turning	2.2–3.7	3.0
welding	2.7–4.1	3.4
Chemical industry—paint production		
machine operator—oil refining		3.6
semi-skilled work—dispatch	3.4–3.7	3.6
grinding	4.7–5.0	4.9
stirring machine operator	5.7–6.1	5.9
stock room work	6.0–6.5	6.3
General laboratory work	1.8–3.2	2.3
gas analysis (by Haldane's method)	1.7–2.2	1.9
setting up equipment		3.0
chromatography		2.5
examining slides		2.0
repairing oscilloscope	1.8–2.8	2.2
screening a motor	2.2–3.4	2.8
glassblowing	1.4–2.3	1.9
setting up camera, taking photograph, developing and printing	2.2–2.5	2.4
Workshop		
working at lathe	1.2–2.5	1.9
sawing and filing metal	3.3–4.2	3.7
bending metal by machine		2.4
repairing machinery		2.1
tidying up	2.4–3.8	3.3

Figure 4–2. Continued

	kcal/min/121 lb woman	
	Range	Mean
Electrical industry		
assembly	1.4–2.1	1.9
coil winding		1.4
packing		2.5
punch press	1.4–4.2	2.6
Laundry		
checking overalls		3.6
ironing		3.4
marking		3.1
pressing		2.9
shaking the linen		4.0
Machine-tool, metal and general factory work		
assembly work in car factory	1.8–2.8	2.3
forging	2.1–4.2	3.0
gasket inspector	1.5–2.3	1.9
gasket packing	1.9–3.0	2.5
gear assembly	1.7–2.0	1.9
inspector	1.0–1.6	1.3
packer of small parts	1.5–2.3	1.8
packing and labeling boxes	2.2–2.9	2.6
polishing machine worker	2.8–3.0	2.9
pressing, both hand and foot presses	1.4–3.7	2.2
tool setting	1.8–5.3	3.1
metal turning	1.5–4.3	2.5
turning	1.3–4.6	2.4
turning and finishing	1.1–4.9	2.9
finishing	1.3–3.0	2.1
walking, carrying loads	2.9–5.4	3.9
laboring	3.6–5.3	4.4

Figure 4–3. The energy expenditures for various jobs from a 121 lb woman.[3]

the basal (output at rest) rate of workers involved. Other workers would give different rates depending upon their body weight. Please note that Fig. 4–2 is for a 145 lb man and that Fig. 4–3 is for a 121 lb woman. Figure 4–4 is for men with no specified weight. These tables may be used to get an idea of the energy expenditures of different kinds of jobs. Of course nothing is specified regarding the weights, the pace of the workers, or the distances involved so that the values should be only used as guidelines.

B. Determine the aerobic capacity of the individual or use working popula-

[3]Chaffin, D.B., "Some Effects of Physical Exertion," Dept. of Industrial Engineering, University of Michigan, 1972, pp. 132–134. Taken from Durnin, J.V.G.A. and Passmore, R., *Energy, Work and Leisure,* Heinemann Books, Ltd., London, 1967, pp 55–57. Reproduced by permission.

Activity	Metabolic Rate (kcal/min)
Sitting: Light or Moderate Work	
a. Sitting at a desk, writing, calculating, etc.; driving a car; lying down under a car as by service mechanic	1.5
b. Using hand tools; light assembly work; radio repair; typing; just driving a truck	1.8
c. Working heavy levers; dredge	2.1
d. Riding mower as in individual work	2.5
e. Driving heavy truck or trailer rig (includes getting off and on frequently and miscellaneous work like this)	3.0
Standing: Moderate Work	
a. Standing and assembling light or medium machine parts where speed is not a factor; working at own pace or a moderate rate—can be using light hand tools	2.5
b. Using hand tools (as by a gas station operator) and other jobs that do not involve assembly work all day, weights 10 lb occasionally	2.7
c. Scrubbing, waxing, polishing (floors, walls, cars, windows	2.7
d. Assembling or repairing heavy machine parts such as farm machinery; plumbing airplane motors; light welding; washing machine repairman	3.5
e. Stocking shelves; packing or unpacking small or medium objects	3.0
f. Sanding floors with a power sander	3.0
g. Assembling light or medium machine parts on assembly line or working with tools on line when objects appear at an approximate rate of 500 times a day or more and when their weight is 45 lb or less (as working with compressors)	3.5
h. Same as (g) above when weight of objects is more than 45 lb	4.0
i. Cranking up dollies; hitching trailers; operating large levers or jacks	3.5
j. Pulling on wires; twisting cables, jerking on ropes, cables, etc., as in rewiring houses	3.5
k. Masonry; painting; paperhanging	4.0
Occupations: "janitor" (unspecified)	3.1

(continued)

Figure 4–4. Energy expenditures for industrial jobs.[4]

[4]Reiff, G. G., Montoye, H. J., Remington, R. D., Napier, J. A., Metzner, H. L., and Epstein, F. H., "Assessment of Physical Activity by Questionnaire and Interview," *The Journal of Sports Medicine and Physical Fitness*, Vol. 7 (3), 135–142, September, 1967.

Activity	Metabolic Rate (kcal/min)
Walking: Moderate Work	
a. Walking	4.0
b. Carrying trays, dishes	4.2
c. Walking involved in gas station work including some tire changing, wrecker work	4.5
Standing and/or Walking: Heavy Arm Work	
a. Heavy tools	
1. Pneumatic tools (jackhammers, drills, spades, tampers)	6.0
2. Shovel, pick tunnel bar	8.0
b. Moving, pushing heavy objects 75 lb or more (desks, file cabinets, heavy stock furniture such as moving van work); also includes pushing against heavy spring tension as in boiler room, etc.	8.0
c. Pushing a cart or dolly loaded with objects weighing less than 75 lb	4.4
d. Laying railroad track	7.0
e. Cutting trees, chopping wood—automatically	3.0
—hand axe or saw	5.5
f. Carpentry—hammering, sawing, planing	6.0

Figure 4–4. Continued

tion figures. To determine whether a particular worker can do the job with or without rest allowances, we must determine his/her aerobic capacity. One way to do this is to ask the person to walk on a treadmill at increasing rates until his/her oxygen consumption no longer increases. The energy being expended at the point where oxygen consumption no longer increases is the aerobic capacity of the individual. This method, however, is not recommended because of the health risks involved. Less strenuous methods have been developed to determine aerobic capacity, one of which is the following[5]:

1. Have the subject to be tested perform a minimum of three grades of work intensity that utilize the muscles most closely associated with his/her occupational work. (The work intensities should vary from a standing rest to above 50 percent of the average aerobic capacity predicted for a normal individual of the same age, weight, and sex.)
2. During the performance of the tasks, measure the steady-state oxygen uptake and heart rates.
3. Plot the heart rate on the ordinate and the energy expenditure rate, as predicted from the oxygen uptake, on the abscissa.
4. With reference to the following relationship of heart rate decrement as a function of age, compute and plot the maximum heart rate which is attainable by the person:

[5]Martiz, J.S., Morrison, J. F., Peters, J., Strydon, N. B., and Wyndam, C. H., "A Practical Method of Estimating an Individual's Maximum Oxygen Uptake," *Ergonomics*, Vol. 4(2), 1961.

$$\begin{bmatrix} \text{Maximum} \\ \text{Heart Rate} \\ \text{(beats/min)} \end{bmatrix} = 190 - \left[\begin{bmatrix} \text{age in} \\ \text{years} \end{bmatrix} - 25 \right] \times 0.62. \qquad (1)$$

5. The intersection of the maximum heart rate line and the line fit to the measured task values defines the aerobic capacity of the individual.

The following is an example of the method of determining the aerobic capacity just described. The maximum heart rate for a 52 year old was obtained using equation (1):

$$\text{Max Heart Rate} = 190 - (52 - 25) \times 0.62 = 173 \text{ beats/min}$$

The intersection of the maximum heart rate line with the line drawn through the three points determined by the operator's tasks is the estimate of the aerobic capacity of the individual. Figure 4–5 is an example of the procedure. The aerobic capacity is 13.2 kcal/min.

C. *The percent of the aerobic capacity that can be sustained.* Authorities disagree on the percentage of the aerobic capacity that a person can sustain for 8 hours of continuous work. The figures vary from 33 percent to 50 percent of the aerobic capacity.[6,7,8] The preferred figure seems to be 33 percent and will be used in this example, 13.2 × 0.33 = 4.4 kcal/min. This is the number that

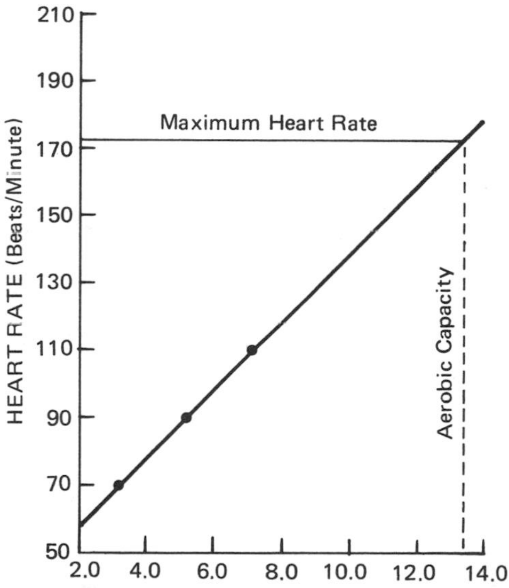

Figure 4–5. Energy expenditure rates kcal/min.

[6]Christensen, E. H., "Physical Working Capacity of Old Wokers and Physiological Background for Work Tests and Work Evaluations," *Bulletin of World Health Organization,* 1955.

[7]Bink, B., "The Physical Working Capacity in Relation to Working Time and Age," *Ergonomics,* Vol. 5(1), 1962.

[8]Snook, S. H. and Irvin, G. H.: "Psycholophysical Studies of Physiological Fatigue Criteria," *Human Factors,* Vol. 11(3), 1969.

should be compared with the energy expenditure obtained through oxygen consumption measurements obtained from the worker doing the job under study. If energy consumption is greater than 4.4 kcal/min, then a rest (fatigue allowance) should be given for this employee.

If population figures are to be used, Fig. 4–6[9] is a series of curves that gives the average and the 95 percent confidence intervals for males with an average weight of 170 lb and females with an average weight of 136 lb. These weights are probably representative of the 18–65 year old population of the United States. These curves are helpful where one has to consider the capabilities of the working population rather than an individual. See Chapter 2 for a discussion of population capabilities and their impact on the establishment of standards.

D. Compute the fatigue allowance. The basic idea here is that if the energy output required to do the job over 8 hours exceeds the acceptable energy level, then a fatigue allowance should be given. If the job takes less energy than the acceptable level, no allowance should be granted. Assuming the former, the steps are as follows:

1. Determine the energy required by the job.
2. Determine the allowable energy expenditure. Two ways are presented.

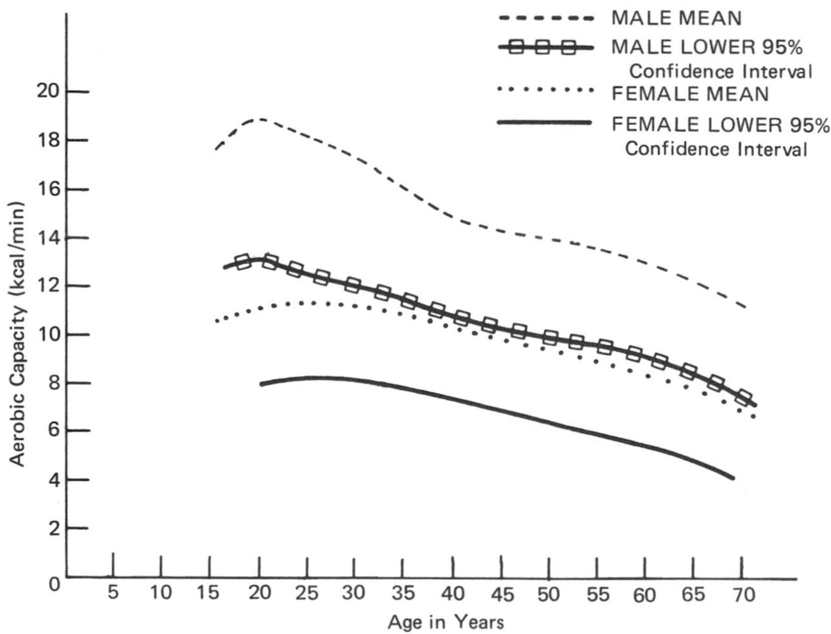

Figure 4–6. Estimated population aerobic capacities for U.S. men and women.

[9]Chaffin, D. B., "Some Effects of Physical Exertion," Department of Industrial Engineering, University of Michigan, July, 1972.

a. For a specific worker, estimate the aerobic capacity of the person and take a percent of the capacity.
b. Use curves like Fig. 4–6 to get an estimate of the population's aerobic capacity. Several alternatives exist:
    1. If only males are involved and it is desired that 95 percent of the workers equal or exceed the work rates, then the "Male Lower 95 Percent" curve should be used. To use the curve one has to obtain a distribution of the ages of the workers involved, determine the age of the 95 percent percentile worker, and obtain the aerobic capacity using Fig. 4–6.
    2. If the job requirements are already established and females are allowed to bid for the job if they can meet its demands, use the procedure in b.1. A comparison of the "Female Mean" and the "Male Lower 95 percent" of Fig. 4–6 reveals that approximately 50 percent of the females should be able to do the work.
    3. If it is desirous that 95 percent of the men and women can do the jobs, follow procedure b.1, except obtain the age distributions of all of the workers, then, using the 90th percentile from the lower age of the females, determine the age to be used in conjunction with the "Female Lower 95 percent" curves. This procedure assumes that one-half of the workers would be female.
    4. If high task or incentive operations are involved, follow procedures b.1 or b.2, depending on the sex considerations, but use the mean curves rather than the 95 percent curves. The reasoning here is that high task standards do not require that 95 percent of the population involved can meet the standards (the average incentive pace), but that the average of the population can meet them.

3. Using the following equation, compute the fatigue allowance in minutes (F):

$$F = \frac{\Delta\,(480 + MT) - M(W) - RM(RB + MT + PFD + CU)}{RM} \qquad (2)$$

where

M is the energy expenditure of the job in kcal/min.
W is the length of work time in minutes.
RM is the metabolic rate in kcal/min while resting and doing tasks other than working.
RB is the rest break time in minutes during the shift.
MT is the meal time in minutes.
PFD is the personal, fatigue, and delay time in minutes.
CU is the clean up time in minutes.
Δ is the acceptable energy expenditure in minutes.

For example, suppose:

Δ = 4.2 kcal/min.
M = 5.5 kcal/min.
RM = 1.6 kcal/min, a figure typically used for a seated person using the fingers and lower arms.
RB = 30 minutes during the shift.
MT = 30 minutes.
PFD = 33 minutes.
CU = 5 minutes.

Then, using equation (2),

$$F = \frac{4.2(480 + 30) - 5.5(W) - 1.6(30 + 30 + 33 + 5)}{1.6}$$

also

$$480 = W + RB + PFD + CU + F \qquad (3)$$

substituting and solving for W:

$$W = 412 - F$$

substituting for W in equation 1

$$F = \frac{4.2(510) - 5.5(412 - F) - 1.6(98)}{1.6}$$

$$F = 72 \text{ minutes}$$

also, $W = 412 - 72 = 340$ minutes, so that if the fatigue allowance is expressed as a percentage of time worked $72/340 \times 100 = 21.18$ percent.

Heat Stress Fatigue*

Fatigue may be increased over the physical aspects if the work is done under conditions where the body cannot get rid of the heat generated through the metabolic process and radiation sources. The following paragraphs contain a method of determining the amount of allowance to be given for heat stress. Please note that M, the metabolic load, is the same as M in equation (2) on page 65.

*References appropriate to this section are given at the end of the section.

A Practical Method for Industrial Heat Stress Allowance Determination†

Walton M. Hancock, Senior Member, AIIE

Don B. Chaffin, Senior Member, AIIE
The University of Michigan, Ann Arbor, MI 48109

The purpose of this paper is to present a practical methodology to determine the heat stress allowances to be given individuals working on production operations in a given geographical area of the United States and for given seasons of the year. This methodology has been developed for use by industrial engineers as part of their production assignment and scheduling duties. In so far as possible, attempts have been made to develop the methodology so that both exposure and recovery allowances for individual jobs can be predicted easily and the differences in allowances can be understood readily.

The basic theory that has been used was developed by Haines and Hatch.[13] Since modifications have been made in the methods of collecting data and in certain basic assumptions, the heat stress equations will be briefly presented first. This will be followed by an applied example from which the methodology is developed.

Basic Heat Stress Relationships

In order to function properly, the temperature of the human body must remain relatively stable. Body temperature stability can be maintained only as long as the metabolically produced heat load plus the environment heat imposed on the body is less than or equal to the ability of the body to lose heat. This condition, which is called thermal equilibrium, has been expressed mathematically as follows:

$$\frac{H(\text{Required})}{H(\text{Maximum})} = HSI \quad (1)$$

where

$H(\text{Required})$ = Heat load developed by the body and imposed by the environment.
$H(\text{Maximum})$ = The maximum ability of the body to rid itself of heat through evaporation.
HSI = Heat Stress Index.

Whenever the condition exists where HSI is less than 1.0, the body could be under some heat stress but not in a condition of critical stress. Heat storage occurs whenever HSI is greater than 1.0. In this situation, the body will not stay at thermal equilibrium, but will show a steadily increasing temperature.[3] This in itself is not harmful, providing the person is allowed to recover before the storage is excessive. A 2°F rise in the body's core temperature has been

†Reprinted with permission from *AIIE Transactions,* Vol. 9, No. 2.

assumed as safe in the past.[13] Recently, the NIOSH Criteria Document on Heat Stress has indicated that the body's core temperature should not exceed 100.4°F.[20] Haines and Hatch[13] have stated that:

$$H(\text{REQ}) = M \pm R \pm C \pm K \tag{2}$$

where:

M = Metabolic Load (kcal/hr),
R = Radiant Load (kcal/hr),
C = Convection Load (kcal/hr),
K = Conductive Load (kcal/hr). Note: This latter term is assumed to be negligible in industrial work, i.e., the person is insulated against significant conductive heat transfer by shoes, clothing, gloves, or low conductive seating material.

Metabolic Load in this formulation is the heat load resulting from a combination of basal (vegetative) and overt musculoskeletal activity required by the task the person is performing. The most commonly used methods of determining the metabolic rate of an individual are measuring the oxygen uptake rate and using tables of similar operations. To measure oxygen uptake, a mask is fitted to the person's face and his expired air is sampled and analyzed for the amount of oxygen used. Approximately 5 kilogram-calories of energy are expended of which 75 percent to 100 percent is realized as heat. In this latter regard, Kamon[15] suggested that 80 percent be utilized for the proportion of metabolic energy expended that results in heat energy (as opposed to mechanical energy) for common industrial tasks. Clearly, if the job involves a great deal of static, postural activity, then a value closer to 100 percent would be necessary. It is these authors' opinion that body and load motions are often involved and Kamon's 80 percent value is reasonable for general industry.

The use of metabolic energy expenditure tables of common industrial operations also has been used. Unfortunately, such tables contain little or no indication of the work methods, pace, tools, or environment and hence, their accuracy is often questionable.[11] More accurate assessments rely on specific activity analyses, as developed by Åberg *et al.*[1] and more recently by Garg.[12]

In U.S. industry, estimating the amount of energy a person develops is not done often. This is probably because the equipment required to be worn by the worker using the oxygen method is cumbersome and frequently interferes with the ability of the operator to perform his routine duties. The obvious errors incurred in using tables of common operations also deter from such estimates.

Perhaps the most compelling argument why the metabolic energy expenditure rates are not used is that stopwatch work measurement, as practiced in the United States, incorporates a subjectively established fatigue factor which is meant to compensate for energy expenditures above the "normal" rates. "Normal" is often defined for these purposes as equivalent to walking at the rate of 3.0 miles per hour.[4] Any operation that subjectively appears to be above this norm is given a time allowance for the operator to rest. Such a rest allowance, if established correctly, reduces the average pace over the working day to the equivalent of 3.0 miles per hour. In other words, the 3.0 miles per hour work rate is assumed in many industries to be the rate that people have empirically stated as an acceptable normal rate of work for long periods of time. "Long periods" in the work measurement field are construed to mean 8 hours/day, 5 days/week, for a person's working life span. When walking at 3.0 miles per hour, an average man (154 lb, 5′8″) will metabolically produce approximately 4.2 kcal/min.[8] Since body weight has an effect on this, a survey of any given population's body weight should be made before assuming a given value. As an example, a sample of 200 employees from an organization in which this particular methodology was evaluated determined that the workers were slightly heavier and taller (170 lb, 5′9″) than the hypothetical average man weighing 154 pounds, thus increasing the expected energy expenditure rate to about 4.95 kcal/min.[11] Assuming that about 80 percent of the metabolic energy expenditure is realized as a heat load to the body, as suggested by Kamon,[15] then such a hypothetical task produces about 3.96 kcal/min. This energy expenditure rate has been classified as a representative of jobs requiring "moderate work" by the OHSA Standards Advisory Committee on Heat Stress.[21]

It should also be noted that there are a number of people working in potentially heat stressing jobs who are not asked to perform much physical activity. As an example, in a glass plant studied by the authors, there were a number of operators who were essentially sedentary while monitoring the operations. These people were required to exert significant physical effort when something went wrong. For these types of jobs, a more sedentary energy expenditure rate of 1.75 kcal/min should be assumed for a 170 lb man.[8,21,23]

The Radiant Load, R, imposed on an individual in an industrial plant is often due to hot workpieces and furnaces used in certain operations. The original Haines and Hatch equations assumed a nude, average sized man for the calculation of radiant heat load. In the methodology presented here, the equations were modified according to recommendations by Hertig and Belding[7,14] for the khaki type of work clothing often worn by an employee in industrial operations, and for the specific average height and weight of the employees involved in the plants in which the evaluations took place (i.e., for 5′9″ and 170 lb workers).*

The determination of the radiant load R in the Haines and Hatch equations as modified by Hertig and Belding[7,14] is as follows:

$$R = 6.49\, B_c(t_w - t_s)\, \frac{A_{DU}}{20.0}\, F_{ex} \quad (3)$$

where:

R = Radiant Load Estimate (kcal/hr),
6.49 = Heat coefficient,
B_c = Clothing correction factor proposed by Hertig and Belding,[14] since the original transfer equation was based on a seminude, average size man. A value of 0.7 as suggested by Hertig and Belding is used,

*The specific clothing worn was cotton khaki shirt and trousers, long cotton underdrawers, tee shirt, cotton cap, cotton workgloves, cotton socks and leather shoes, and dark glasses. Such a clothing ensemble would have a clo value of about 0.8.

though one study by Kamon[15] has suggested this be reduced to 0.6.

t_s = Skin temperature (assumed to be 95°F),

t_w = Wall temperature (°F)

$= 1.8V^{0.5} (t_g - t_a)$

where:

V = Air velocity (ft/min),

t_g = Globe temperature (°F),

t_a = Air temperature (°F),

A_{DU} = Skin area by DuBois formula is square ft,

20.0 = Skin area by DuBois formula of the standard physiological man (5′8″, 154 lb),

F_{ex} = The exposed effective area of the skin surface which is assumed to be 0.975.

For the average male in our study, the equation reduces to:

$$R = 6.49(0.7)(t_w - t_s)\frac{20.78}{20.0}0.975$$

$$= 4.60(t_w - t_s).$$

The measurement of radiant loads should be done with instrumentation that has a low time constant (10 seconds). The rationale is that the worker is usually moving around while he is working causing radiant loads to vary during the work cycle. The load used in the methodology should be the time-weighted load over the work cycle. The instrumentation used in the measurement of the air velocity should be capable of measuring the air flow from all directions thus averaging the flow regardless of direction. This requirement excludes the vane-type anemometer commonly used in the heating and air conditioning industry.

Convective Load, C, is due to the heat exchanged by the air moving past the individual. When the temperature of the air is less than approximately 95°F, the air will take heat away from the body, and therefore act as a cooling medium. When the temperature is above 95°F, the air will be warmer than the skin temperature, and therefore it will impose a heat load on the body. The convective load estimating equation from Haines and Hatch[13] is:

$$C = 0.273\, F_{ex} A_{DU} B_c V^{0.6} (t_a - t_s)$$

where:

C = Convective load estimate (kcal/hr),

F_{ex} = Exposed skin area correction (assumed 0.975),

A_{DU} = DuBois estimate of skin area, square ft,

B_c = Clothing correction factor (assumed 0.7),

V = Air velocity, ft/min,

t_a = Dry bulb air temperature (°F),

t_s = Skin temperature (assumed 95°F).

For the slightly larger average male in this plant study, whose weight is 170 pounds and stature is 69 inches, the equation reduces to:

$$C = 0.273(0.975)(20.78)(0.7)\ V^{0.6}\ (t_a - 95)$$
$$= 3.87V^{0.6}\ (t_a - 95).$$

H(Max), represents the maximum cooling ability of the human body. The major method of cooling the body is by sweat evaporation. A number of studies have been made concerning the maximum amount of water an acclimatized person can perspire over an 8 hour period. The maximum amount of heat loss through the process of evaporation has been set at about 605.45 kcal/hr based on a maximum physiological sweating rate of 1.0 liter per hour sustained over 8 hours.[3] This physiological limit is not often reached, however, since this will only occur with dry and/or relatively high velocity air. Thus the environmental conditions for H(Max) are computed as follows:

$$H(\text{Max}) = 0.05174F_{ex}A_{DU}B_cV^{0.6}\ (VP_{skin} - VP_{air}) \quad (5)$$

where:

H(Max) = Maximum heat loss by sweating, kcal/hr,
0.05174 = Heat coefficient,
F_{ex} = Exposed skin area correction factor (assumed 0.975),
A_{DU} = DuBois estimate of skin area, square ft,
B_c = Clothing constant (assumed 0.7),
V = Air velocity, ft/min,
VP_{skin} = Vapor pressure of saturated air at the skin when at 95°F, mm Hg (assumed = 42),
VP_{air} = Water vapor of air at t_{air}, mm Hg.

For the employees in this study, the equation reduces to:

$$H(\text{Max}) = 0.05174\ (0.975)\ (20.78)\ (0.7)\ V^{0.6}\ (42 - VP_{air})$$
$$= 0.7339\ (V^{0.6})\ (42 - VP_{air}).$$

Development and Application of the Heat Stress Methodology

The foregoing discussion contains the existing technology necessary to determine the heat stress index for many different working conditions. With the exception of the special instrumentation and the corrections for population differences from the "standard" physiological man, the basis for the equations can be found in the references. In practice, the industrial engineer frequently is concerned with the need to establish work and rest time allowances for many different "hot" operations wherein many employees are exposed such as in a foundry or in a glass plant. In order to do this on a consistent basis, it was found necessary to examine those aspects of the working environment that were common to all employees, to establish work and rest policies for these, and then to determine individual heat stress allowances.

The method used to determine the general working conditions were based on an analysis of United States Weather Bureau data. Since primary concern is for the summer months, the data in one study were obtained from the Weather Bureau for periods April 15 to October 14 for the three years. The outside ambient temperatures for every three hours were compiled into a computer file.

Table 1. Daily ambient temperatures (°F) according to percentile ranking, specific summers, and times of the day.

Summer	Per-centile	Time of Day							
		1 AM	4 AM	7 AM	10 AM	1 PM	4 PM	7 PM	10 PM
April 15 to October 14, Year 1	80	71	69	70	78	82	83	80	74
	90	74	72	73	83	88	88	85	78
	95	77	74	75	85	90	90	86	81
	100	81	79	80	90	93	93	91	88
April 15 to October 14, Year 2	80	69	67	69	77	83	83	80	73
	90	72	70	72	81	86	87	82	77
	95	75	72	74	83	88	89	85	79
	100	81	78	79	89	93	93	89	85
May 1 to October 14, Year 3	80	70	62	68	76	81	91	79	72
	90	73	70	72	79	85	85	80	75
	95	74	72	73	82	86	88	83	77
	100	78	76	76	88	91	92	90	81

The variations were examined between times of the day, between months, and between seasons. Table 1 summarizes some of the data as an illustration of variations in the Great Lakes region. From this analysis it was concluded that:

1. The 100 percentile temperature (highest daily ambient temperature) occurred during the summer of years 1 and 2, at 4:00 PM, and was 93°F.
2. The 100 percentile temperatures (maximum daily temperatures) for the summer periods studies were as follows:

   Year 1 – 93°F
   Year 2 – 93°F
   Year 3 – 93°F
3. Over the three summers the hottest period of the day was consistently between 10:00 AM and 7:00 PM.

Similar data for the principal cities in the United States and Canada are contained in Ref. 2.

These data were summarized in such a way that various policies could be made regarding the temperatures to be used in computing heat stress allowances. As suggested by the OHSA Standards Advisory Committee on Heat Stress,[21] the worst possible conditions should be used to determine if any heat stress allowances are needed (i.e., do not evaluate a job in the winter). Heat stress allowances could then be applied in the expected hot periods based on the seasonal data. In some operations, however, the allowances may be economical over the year. For instance, of the total operations in a plant, there are a very small number of jobs that are heat stressed (support is further given for this assumption by Dinman *et al.*[9]) and many operations are paced operations where changing the heat stress allowance for a few individuals and jobs would make it necessary to rebalance the work assignments of a large number of employees.

Whatever the policy, it is necessary to relate the Weather Bureau data to the conditions in the plants of interest. This can most easily be accomplished by using automatic temperature recording equipment to enable continual monitoring of the inside and outside temperatures in the plants. These data are then used to develop regression equations which relate the temperature inside the plants to the conditions in the metropolitan area, as supplied by the Weather Bureau. A similar concept has been used by Mutchler *et al.*[19] in a recent paper.

It may be found that there is substantial variation between plants regarding the temperature outside versus the temperature inside. For example, assuming the worst possible ambient outside temperature of 93°F, as shown in Table 1, the average maximum temperatures inside three different plants at six randomly selected work stations were found to be:

Table 2. Maximum temperatures at six work stations for days yielding 93°F outside maximum air temperatures.

Plant	Average Temperature in Plant
1	110°F
2	114°F
3	108°F

The work station plant temperatures did not reach a maximum until approximately 1 hour after the maximum outside was observed. This analysis of the temperature inside different plants in the same geographical location indicates that a wide difference exists in the individual plant temperature regulation and cooling. Ideally, in the authors' opinion, sufficient air should be circulated so that the inside temperature will not rise more than a stated amount above the outside temperature. Information of this type is often useful to the facility design staff in detecting inadequate air ventilation systems and in improving the design of new buildings to provide more adequate air flow.

The dew point temperature is the air temperature where water vapor starts to condense. Because the Weather Bureau obtains data on the dew point temperature at the same time it obtains information on the ambient air temperature, VP_{air} (Eq. 5) can be determined with the use of a psychrometric chart. The data on dew point temperatures for the summers of years 1, 2 and 3 in the study discussed earlier are presented in Table 3. The following is a summary of the data:

1. The highest dew point occurred in the summer of year 1 and was 77°F (100th percentile 7:00 PM).
2. The maximum dew points (100th percentile) for the summers studied were as follows:

   Year 1 – 77°F
   Year 2 – 73°F
   Year 3 – 73°F

Since it is rare to have significant vapor pressure gradients within a single geographic area, it is reasonable to assume that the vapor pressure in the plant

Table 3. Distribution of dew point temperature (°F) according to percentile ranking, specific summers, and times of the day.

Summer	Per-centile	Time of Day 1 AM	4 AM	7 AM	10 AM	1 PM	4 PM	7 PM	10 PM
April 15 to October 14, Year 1	80	62	63	63	66	64	64	62	63
	90	66	66	66	68	68	68	67	67
	95	68	68	68	70	70	70	70	69
	100	74	73	73	75	76	75	77	73
April 15 to October 14, Year 2	80	60	60	61	61	59	59	59	60
	90	64	64	65	65	65	64	64	65
	95	66	66	67	69	67	67	66	67
	100	69	69	70	73	70	70	68	68
May 1 to October 14, Year 3	80	62	62	62	63	64	63	62	62
	90	64	64	65	65	66	66	65	64
	95	65	65	66	67	69	68	68	66
	100	68	68	69	73	72	71	71	71

in the summertime with the windows open is the same as in the surrounding area, providing large quantities of water are not given off by the manufacturing or air circulation processes.

Allowances Based upon the Heat Stress Index

As discussed previously, heat storage occurs when $HSI > 1.0$. Under these conditions, the core temperature of the body will begin to rise and a person can be permitted to stay in the environment for only a limited length of time. The best information available indicates that the core temperature can safely rise only 2°F without any adverse effects on the individual. This figure was formerly proposed by Belding and Hatch[6] but may be lowered by NIOSH.[20] The term "Maximum Exposure Time" is used to express the amount of time a person can stay in an environment before his core temperature will rise to an unsafe level. The number of kcal's, to raise the body temperature, is given by the following equation:

$$S = 0.253\ WC_p\ \Delta T \tag{6}$$

where:

S = Storage of heat in the human body, kcal/hr,
W = Body weight, lb,
C_p = Specific heat of the human body (assumed = 0.83),
ΔT = Rate of change in mean body temperature, °F/hr.

For a 170 lb person and a 2°F/hr rise:

$$S = 0.253(170)\ (0.83)\ (2) = 71.4 \text{ kcal/hr}$$

Maximum exposure time is:

$$TMAX = \frac{60\ S}{H(\text{REQ}) - H(\text{Max})} \tag{7}$$

where:

TMAX = Maximum exposure time, min.

If a person's core temperature has reached a 2°F rise above normal, a second important question could be: How long will it take him to recover? This is dependent, of course, on the recovery environment. The following equation is used for a 170 lb man:

$$TRECM = \frac{60\ S}{H(\text{Max}) - H(\text{REQ})} \tag{8}$$

where:

TRECM = Maximum recovery time, min.

Assuming a 460 min work day and constant environmental conditions throughout the working day, the daily heat stress allowance can be expressed as follows:

Heat Stress Allowance (in minutes per day)

$$HSA = \left(\frac{TRECM}{TMAX + TRECM}\right) 460 \tag{9}$$

The equation presumes that the allowances must be given at uniform times even though the dry bulb and dew point temperatures are varying throughout the day. The use of uniform allowances appears to be in contradiction to the variation in dry bulb and dew point temperatures shown in Tables 2 and 3. Of course it is technically feasible to adjust the heat stress allowances throughout the work day, but this would require real time detection and computation of the variables involved. Also, there are many production processes that cannot be adjusted to varying rest times on an hourly, weekly, or even monthly basis. Another complicating factor is the use of fatigue factors mentioned earlier in the report, which are usually negotiated independently of real time changes in climatic conditions. Under these conditions, a seasonal allowance might be the best solution.

Equation (9) could be further modified to take into account lunch breaks. Lunches were found to be eaten, in most cases, in the same work environment except the radiant load from the process was not present. Equation (9) would then become, assuming that the employee can fully recover during lunch time:

Heat Stress Allowance (in minutes per day)

$$HSA = \frac{(\text{No. of Recovery Times} - 1)(\text{Time/Recovery})}{TMAX + (\text{No. of Recovery Times} - 1)(\text{Time/Recovery})} 460 \tag{10}$$

Equation (10) also assumes that the recovery time coincides with lunch time. This situation may not exist unless lunch can be taken at the discretion of the worker. This is unusual in the midwest where the studies were conducted, so Eq. (9) will be used in the examples presented in this paper.

A Simplified Methodology

Because heat stress allowances are part of the fatigue allowances that are normally negotiated between workers and employers, it is the authors' opinion that it is necessary to provide the industrial engineer with simple to use tables that can be readily understood and consistently applied by managers, union officials, and industrial engineers. The tables, of course, would reflect the agreements made by the parties involved concerning the temperatures, dew points, metabolic expenditures and the heights and weights of the working force involved. The authors found, in the situation described in the paper, that without the availability of such tables, the technology of Heat Stress remained the province of the Industrial Hygiene Department and, because of the complexity of the calculations, was not being used often to determine allowances. In other words, the allowances were being negotiated without any technological input.

The methodology, which reduces the computations of maximum exposure time to one table lookup, requires the following information about the operator.

1. The time-weighted mean radiant load for the various work sites of a job.
2. The ambient temperature in the plant, established as being representative of the conditions to which the Heat Stress Allowance is to be applied (chosen to be maximum meteorological and plant conditions as discussed earlier). This, of course, is a "worst case" approach.
3. Whether the individual is performing at the standard physical activity level (3.96 kcal/min), or at a monitoring activity level (1.75 kcal/min).

An optimal ventilation rate has been assumed based on the following. For a maximum dew point of 77°F (as found in the three plants described earlier), Eq. (4) was used to determine the air velocity necessary for the worker to cool himself at the maximum rate of 605.45 kcal/hr. (It is assumed that the individual is capable of sweating 1 liter/hr.)

Since VP_{air} = 24 mm Hg for a dew point of 77°F,

$$E(\text{Max}) = 0.7339V^{0.6}(42 - VP_{air})$$

$$605.45 = 0.7339V^{0.6}(42 - 24)$$

therefore:

$$V = 587 \text{ feet/minute.}$$

Since 587 ft/min is 6.67 miles/hr, an assumption that the work places can be designed or redesigned to provide this level of air flow is practical and reasonable. 6.67 miles/hr is a high air flow, but it should be kept in mind that it will only be needed in work situations infrequently occurring. At other times the air flow can be reduced.

ble 4. Maximum exposure times for monitoring activity (3.96 kcal/min) where H(Max) = 10.1 kcal/min. The radiant load is the average load at the work place and the dry bulb temperature is the temperature at the work place. A blank indicates that the exposure time exceeds 460 minutes.

diant oad al/hr)	Maximum Exposure Times																		
66	17	16	15	14	13	13	12	11	11	10	10	10	9	9	8	8	8	8	7
43	18	17	16	15	14	13	12	12	11	11	10	10	9	9	9	8	8	8	8
31	19	18	16	15	14	14	13	12	12	11	11	10	10	9	9	9	8	8	8
18	20	19	17	16	15	14	13	13	12	11	11	10	10	10	9	9	9	8	8
05	22	20	18	17	16	15	14	13	12	12	11	11	10	10	9	9	9	8	8
93	23	21	19	18	17	15	14	14	13	12	12	11	11	10	10	9	9	9	8
82	25	22	20	19	17	16	15	14	13	13	12	11	11	10	10	10	9	9	9
68	27	24	22	20	18	17	16	15	14	13	12	12	11	11	10	10	9	9	9
55	29	26	23	21	19	18	17	16	15	14	13	12	12	11	11	10	10	9	9
42	32	28	25	23	21	19	17	16	15	14	13	13	12	12	11	10	10	10	9
30	35	30	27	24	22	20	18	17	16	15	14	13	13	12	11	11	10	10	10
17	39	33	29	26	23	21	20	18	17	16	15	14	13	12	12	11	11	10	10
05	44	37	32	28	25	23	21	19	18	16	15	14	14	13	12	12	11	11	10
92	50	41	35	31	27	24	22	20	19	17	16	15	14	13	13	12	11	11	10
79	59	47	39	34	29	26	24	21	20	18	17	16	15	14	13	12	12	11	11
67	72	55	45	37	32	28	25	23	21	9	18	16	15	14	14	13	12	12	11
54	91	66	51	42	36	31	27	25	22	20	19	17	16	15	14	13	13	12	11
41	124	81	60	48	40	34	30	26	24	22	20	18	17	16	15	14	13	12	12
29	196	107	74	56	45	38	33	29	26	23	21	19	18	17	15	15	14	13	12
16		156	94	67	52	43	36	31	28	25	22	20	19	17	16	15	14	13	13
04		291	130	84	62	49	41	35	30	27	24	22	20	18	17	16	15	14	13
91			211	111	76	57	46	39	33	29	26	23	21	19	18	17	16	15	14
78				166	97	69	53	43	37	32	28	25	23	21	19	18	16	15	14
66				326	137	86	63	50	41	35	30	27	24	22	20	19	17	16	15
53					229	116	78	58	47	39	34	29	26	23	21	20	18	17	16
41						177	101	71	54	44	37	32	28	25	23	21	19	18	16
28						371	144	89	67	51	42	35	31	27	24	22	20	19	17
15							251	121	80	60	48	40	34	30	26	24	22	20	18
03								189	105	73	55	45	38	33	29	25	23	21	19
90								430	152	92	66	52	42	36	31	28	25	22	20
78									276	127	83	61	49	40	34	30	27	24	22
65										204	109	75	57	46	38	33	29	26	23
52											161	96	68	53	43	36	32	28	25
40											308	133	85	63	49	41	35	30	27
27												220	114	77	58	46	39	33	29
	90	92	94	96	98	100	102	104	106	108	110	112	114	116	118	120	122	124	126
	Dry Bulb Temperature (°F)																		

Tables 4 and 5 contain the data on maximum exposure times for the normal physical activity level of 3.96 kcal/min, and for a monitoring level of activity (1.76 kcal/min).

Table 6 is a recovery time table based on the earlier recovery allowance equation (Eq. 8). The information necessary to use this table is as follows:

1. The ventilation velocity of the recovery area.
2. The dry bulb temperature of the recovery area.

The following assumptions were necessary in developing this particular table:

1. The employee will be at seated rest while in the recovery area. This, therefore,

Table 5. Maximum exposure times for monitoring activity (1.75 kcal/min) where H(M = 10.1 kcal/min. The radiant load is the average load at the work place and the dr bulb temperature is the temperature at the work place. A blank indicates that the exposure time exceeds 460 minutes.

Radiant Load (kcal/hr)	Maximum Exposure Times																	
666	38	33	29	26	23	21	19	18	17	15	15	14	13	12	12	11	11	10
643	43	36	31	28	25	22	20	19	17	16	15	14	13	13	12	11	11	10
631	49	41	25	30	27	24	22	20	18	17	16	15	14	13	13	12	11	11
618	57	46	39	33	29	26	23	21	19	18	17	16	15	14	13	12	12	11
605	69	53	43	37	32	28	25	23	21	19	18	16	15	14	14	13	12	12
503	87	63	50	41	35	31	27	24	22	20	19	17	16	15	14	13	13	12
582	116	78	59	47	39	34	29	26	23	21	20	18	17	16	15	14	13	12
568	177	101	71	54	44	37	32	28	25	23	21	19	18	16	15	14	14	13
555	363	144	89	65	51	42	36	31	27	24	22	20	19	17	16	15	14	13
542		252	122	80	60	48	40	34	30	26	24	22	20	18	17	16	15	14
530			190	105	73	56	45	38	33	29	25	23	21	19	18	17	15	14
517			433	153	93	66	52	42	36	31	28	25	22	20	19	17	16	15
505				278	127	83	61	49	40	34	30	27	24	22	20	18	17	16
492					204	109	75	57	46	38	33	29	26	23	21	19	18	17
479						162	96	68	53	43	36	32	28	25	23	21	19	17
467						310	134	85	63	49	41	25	30	27	24	22	20	18
454							221	114	77	58	46	29	33	29	26	23	21	19
441								172	99	70	54	44	37	32	28	25	23	21
429								350	141	88	64	50	42	35	31	57	24	22
416									241	119	79	59	47	39	34	30	26	24
404										184	103	72	55	45	37	32	28	25
391										402	148	91	66	51	42	36	31	27
378											265	125	81	61	48	40	34	30
366												197	107	74	56	45	38	33
353													157	94	67	52	43	36
341													293	131	84	62	49	41
328														213	112	76	57	46
315															167	98	69	53
303															329	137	87	63
290																231	114	78
278																	178	101
265																	375	145
252																		253
240																		
227																		
	90	92	94	96	98	100	102	104	106	108	110	112	114	116	118	120	122	124
	Dry Bulb Temperature (°F)																	

allows the assumption that the worker is metabolically expending only 1.59 kcal/min on the average.

2. The employee recovery area will not be subjected to any significant radiant load. In other words, the worker will not be asked to rest in the presence of hot castings or furnaces.

The heat stress allowance for the shift is then computed by simply adding the recovery and exposure times, given in the tables, and dividing it into the exposure times, as described earlier (Eq. 9).

A Comparison with Traditional Heat Stress Allowances

To aid in evaluating the impact of the proposed procedure, Table 7 contains information on 19 jobs from three different plants that were studied. The jobs

Table 6. Recovery time for a dew point temperature of 77° F and an A_{DU} of 20.78 sq ft (170 lb man).

Ventilation Rate(ft/min)	Recovery Time (min)												
587	5	5	6	6	7	7	8	8	9	10	11	13	15
575	5	6	6	6	7	7	8	8	9	10	12	13	15
550	5	6	6	7	7	7	8	9	10	11	12	14	16
525	6	6	6	7	7	8	8	9	10	11	12	14	16
500	6	6	6	7	7	8	9	9	10	11	13	15	17
475	6	6	7	7	8	8	9	10	11	12	13	15	18
450	6	7	7	7	8	9	9	10	11	12	14	16	19
425	6	7	7	8	8	9	10	11	12	13	15	17	20
400	7	7	8	8	9	9	10	11	12	14	15	18	21
375	7	7	8	8	9	10	11	12	13	14	16	19	22
350	7	8	8	9	10	10	11	12	14	15	17	20	23
325	8	8	9	9	10	11	12	13	14	16	18	21	25
300	8	9	9	10	11	12	13	14	15	17	20	23	27
275	9	9	10	11	11	12	13	15	16	18	21	25	29
250	9	10	11	11	12	13	14	16	18	20	23	27	32
225	10	11	11	12	13	14	16	17	19	22	25	30	36
200	11	12	13	13	15	16	17	19	22	24	28	34	41
175	12	13	14	15	16	18	19	22	24	28	32	39	49
150	14	15	16	17	18	20	22	25	28	32	38	47	60
125	16	17	18	20	22	24	26	30	34	39	47	59	78
100	19	21	22	24	27	29	33	37	43	52	64	83	120
75	25	27	29	32	35	49	45	53	63	78	103	152	287
50	37	41	45	51	58	67	80	100	132	193	363		
25	98	116	143	185	263	453							
	60	65	70	75	80	85	90	95	100	105	110	115	120
	Dry Bulb Temperature (°F)												

were selected because heat allowances had already been established by more subjective methods.

An examination of Table 7 reveals that there are many cases where no heat stress allowances should be granted under the proposed system, whether or not conditions are changed. In one case, Job 12, the present allowance granted is not sufficient under the "proposed" or "recommended" systems. Since there was very little specific information available as to how the "present" heat stress allowances were determined, only a few general comments can be made regarding the practical validity of this methodology:

1. The "present" heat stress allowances appeared to have no rational basis. This could have serious consequences in labor-relations negotiations.
2. It is well known that a person's ability to withstand heat stress is dependent on the general fitness and acclimatization to heat, both of which vary greatly in the population. The proposed system is based on the assumption that a person's health is good. Perhaps some of the apparent inconsistencies in the present allowances are

Table 7. Data from 19 heat stressed jobs.[a]

1. Operation Number	2. Maximum Temperature, (°F)	3. Actual Ventilation Rate (ft/min)	4. Radiant Load (kcal/hr)	Recovery Minutes		
				Present 5.	Recommended 6.	Proposed 7.
1	100	400	0	30	0	0
2	100	90	0	30	128	0
3	100	350	53	80	0	0
4	100	170	0	70	0	0
5	100	570	106	138	0	0
6	114	200	119	72	274	0
7	114	690	119	48	0	0
8	114	400	0	48	0	0
9	114	520	71	10	0	0
10	114	570	18	48	0	0
11	114	900	0	48	0	0
12	108	340	406	120	346	151
13	108	200	162	80	267	0
14	108	450	212	250	109	0
15	108	280	0	58	0	0
16	108	470	0	120	0	0
17	108	370	62	120	0	0
18	108	150	49	80	0	0
19	108	300	35	212	0	0

[a]Column

1. Operation Number—This is a coded number for a particular operation.
2. Maximum Temperature—This is the maximum dry bulb temperature that we would predict would occur in the work station according to the Weather Bureau data for the summers of Years 1–3 and according to the relationship developed between the temperature inside the respective plants and the Weather Bureau data.
3. Actual Ventilation Rate—This is the average air flow that was measured at the work place. Please note that many of the operations do not have the recommended velocity of 587 ft/min.
4. Radiant Load (kcal/hr)—This is the average radiant load on the individual at the work station from external sources. This was determined by the use of a hot wire thermometer enclosed in a globe where readings were taken at the upper and lower trunk each time the operator changed position. A "0" in the column indicates that there was no radiant load from external sources.
5. Recovery Minutes Allowed: Present—This column contains the present allowances granted by the company for heat stress operations in minutes.
6. Recovery Minutes Allowed: Recommended—This column contains the heat stress allowances that would be given under the proposed system if the conditions at the work place were not changed, i.e., ventilation rate not changed. In order to make these computations, a recovery ventilation of 150 ft/min was assumed. This is a necessary assumption because in many cases there was not a definite place where the employees took their rest. Consequently, 150 ft/min was assumed, which is a minimum for the type of forced circulation used.
7. Recovery Minutes Allowed: Proposed—This column contains the proposed heat allowances that were developed assuming an optimal ventilation rate of 587 ft/min in the recovery area.

due to people complaining who are in relatively poor health. As now strongly recommended by the OSHA Standards Advisory Committee on Heat Stress,[21] the medical staff must play a more prominent role in examining people who complain of heat stress and who are to be working on hot jobs.

3. The ventilation systems were often not designed with regard to heat balance. In some cases, the ventilation rates were higher than necessary and in others too low. In a very few cases, the employees could control the ventilation. Since about 587 feet per minute was determined as the criterion for maximum cooling of most people for the worst conditions, it would be needed only on a few days. On other days it could produce chilling unless the operators could control the flow and/or direction.
4. The geographical area studied was in the Great Lakes area. If the same jobs were located, for instance, in the Southeastern United States, the dry bulb and dew point temperatures would be considerably higher, and thus substantially higher allowances or ventilations would be necessary.
5. Since the present allowances were established in the absence of a general policy regarding heat stress control, the allowances probably reflected the perceptions of specific individuals regarding the degree of heat stress. The methodology presented in this paper is primarily based on physiological and quantitative criteria which consider what a group of healthy men can safely tolerate, once acclimated to the heat.
6. Similarly, the metabolic load assumed to develop the recommendations in Tables 4, 5, and 6 are gross estimates of the metabolic loads for various activities. Certainly if fatigue allowances are estimated based on metabolic rate monitoring of workers, then these data could be used directly in the heat stress computations described earlier.

In the example in this paper, a heat stress index of 100 was used. It would be clear, however, that recommendations could be developed for other levels if desired. A heat stress index of 100 was used, in this case, because of the relatively few days during thê year and the few hours during these days wherein maximum conditions were expected.

A Comparison with Traditional Stop Watch Time Study Allowances

It is a common practice in many industrial situations to grant an across-the-board fatigue allowance on a company plant or departmental basis. These allowances developed for heat stress, therefore, could be part of the general fatigue allowances. Under these conditions, an additional allowance for heat stress should not be given unless:

$$\text{Heat Stress Allowance in minutes/working day} > \text{Fatigue Allowance in minutes/working day.}$$

In many cases 8% fatigue allowance is given for a 480 minute day. No additional allowance would be given unless the heat stress allowance exceeds 38 minutes/day. If the heat stress allowance exceeds 38 minutes/day, then the heat stress allowance minus 38 should be given.

Summary

In summary, tabular values can be developed that will enable the industrial engineer to carry out management policy regarding heat stress allowances consistently and economically. The industrial engineer can do this as a routine part of his production assignment duties with the use of inexpensive equipment and the tables provided him. The methodology that is described is consistent with stop watch, work measurement practices in wide use in the United States today. Sample tables are presented, which are based on assumptions which were reasonable for the areas of application. Other situations may require different assumptions which could substantially change the numerical values.

Evidence is given that there may be many situations where heat stress allowances need not be given at all if the work places are correctly designed.

References

1. Åberg, U., Elgstrand, R., Margnus, P., Lindholm, A., "Analysis of Components and Prediction of Energy Expenditure in Manual Tasks," *International Journal of Production Research,* **6**(3), pp. 189–196 (1968).
2. *ASHRAE Handbook of Fundamentals,* The American Society of Heating, Refrigeration and Air Conditioning Engineers, Inc., New York, N.Y., Chapter 33, pp. 667–689 (1972).
3. American Industrial Hygiene Association, *Heating and Cooling for Man in Industry,* monograph, available from George Clayton Associates, Southfield, Michigan, pp. 2–10 (1970). (Also available from AIHA.)
4. Barnes, R. M., *Motion and Time Study, Design and Measurement of Work,* Sixth Edition, John Wiley and Sons, Inc., p. 386.
5. Belding, H. S., Narwood, S., Hertig, B. A., and Riedesel, M. L., "Laboratory Simulation of a Hot Industrial Job to Find Effective Heat Stress and Resulting Physiologic Strain," *American Industrial Hygiene Association Journal,* **21**(1), pp. 25–31 (Feb. 1969).
6. Belding, H. S. and Hatch, T., "Index for Evaluation of Heat Stress in Terms of Resulting Physiological Strains," *Heating, Piping and Air Conditioning,* August, 1955.
7. Belding, H. S., Hertig, B. A., and Riedesel, M. L., "Laboratory Simulation of a Hot Industrial Job to Find Effective Heat Stress and Physiological Strain," *American Industrial Hygiene Association Journal,* **21**(1), pp. 33–39 (Feb. 1960).
8. Consolazio, C. F., Johnson, R. E., and Pecora, L. J., *Physiological Measurements of Metabolic Functions in Man,* McGraw-Hill Book Co., Inc., New York, p. 332 (1963).
9. Dinman, B., Stephenson, S., Horvath, S., and Wolwell, M., "Work in Hot Environments: 1. Field Studies of Work Load, Thermal Stress and Physiologic Response," *Journal of Occupational Medicine,* **16**(12), pp. 785–795 (1974).
10. DuBois, D., and DuBois, E. F., "A Formula to Estimate the Approximate Surface Area if Height and Weight Be Known," *The Archives of Internal Medicine,* **VII,** pp. 863–871 (1916).
11. Dumin, J. V. G. A. and Passmore, R., *Energy, Work, and Leisure,* Heinemann Educational Books Limited, London, pp. 47–82 (1967).
12. Garg, A., "A Metabolic Rate Prediction Model for Manual Material Handling Jobs," PhD Dissertation, University of Michigan (1976).
13. Haines, G. F. and Hatch, T., "Industrial Heat Exposures Evaluation and Control," *Heating and Ventilation,* pp. 11–18 (November 1952).
14. Hertig, M. P. H. and Belding, H. S., "Evaluation and Control of Heat Hazards," from *Temperature, Its Measurement and Control in Science and Industry,* Vol. 3, Pt. 3, Reinhold Publishing Co. (1963).
15. Kamon, E., "Ergonomics of Heat and Cold," *Texas Reports of Biology and Medicine,* **33**(1) (1975).

16. Kuehn, L. A., "Response of the Globe Thermometer," *DCIEM Report 859,* Department of National Defense, Canada, p. 3 (Jan. 1973).
17. Kuehn, L. A., MacHattie, L. E., and Kroon, J., "An Improved Direct Reading Instrument of WBGT Index," AIHA Conference, Miami (May 1974).
18. MacHattie, L. E. and Kuehn, L. A., "An Improved Direct Reading Meter for WBGT Index," *DCIEM Report 73-R-947,* Defense Research Board, Canada, p. 3 (Jan. 1973).
19. Mutchler, J., Malzahn, D., Vecchio, J., and Saule, R., "An Improved Method of Monitoring Heat Stress Levels in the Workplace," *American Industrial Hygiene Association Journal,* pp. 151–164 (March 1976).
20. NIOSH, criteria document, *Occupational Exposure to Hot Environments,* HSM72-10269, pp. I-2 and V-4 (1972).
21. OSHA Standards Advisory Committee on Heat Stress, "Recommendation for a Standard for Work in Hot Environments," Draft 5, *BNA Reporter,* **3**(33), p. 1055 (January 17, 1974).
22. vaderWalt, W. H. and Wyndham, C. H., "An Equation for Prediction of Energy Expenditure of Walking and Running," *Journal of Applied Physiology,* **34**(5), pp. 559–563 (May 1973).
23. Webb, P., *Bioastronautics Data Book,* National Aeronautics and Space Administration, No. SP-3006, Washington, D.C., pp. 159–200 (1964).

CHAPTER 5

Decision Times

The words "decision times" or "human decision making" have many connotations. In the context of this chapter, we are talking about the decisions necessary to perform a given task or job once the worker understands what needs to be done.

Types of Decision Processes

Basically there are three types of decisions that are made in doing a task where members of the body and the sensory functions (eyes, ears), kinesthetic (position), and tactile (touch) are used. These are:

1. The decisions necessary to decide what to do. For example, in doing work assembly, one may have to decide what parts are to be assembled first and then the most appropriate tools to use. These decisions are typically made while the worker is learning how to do the work. After he/she has learned the job, decisions are no longer necessary because the operator has "learned" what parts are to be assembled, their sequence, and what tools to use.
2. The decisions necessary by an experienced worker to perform the basic motions. Using the MTM-1 system as an example, each of the motion categories has a certain amount of time for decisions to be made. As an example, an R5B has a value of 7.8 TMU whereas an R5C has a value of 9.4 TMU. The difference in the two values is primarily due to the increased content of the R5C. Reaching to an object that is jumbled with other objects (R__C) requires more decisions than reaching to a single object (R__B). Using the concepts of information theory, reaching for a jumbled object has more uncertainty than reaching for the single object. If we use Gilbreth's Therblig concept, more searching and selecting occurs with the R__C than with the R__B.

   It is important to point out that the difference in TMU between, for example, an R5C and an R5B of 1.6 TMU is not the total decision time differential between the two motions. It is the additional amount of decision time that is not overlapped with movement time. This means that the act of collecting information and making decisions (information processing) is occurring in parallel with the movement of the members of the body. If the information processing can occur while the body member performs at the same rate as if the information processing were not

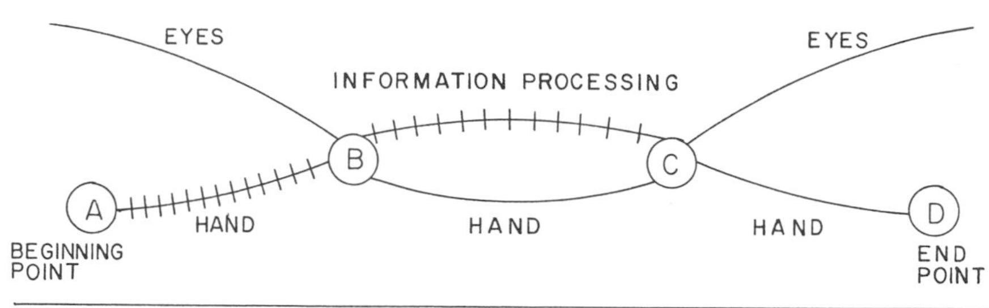

Figure 5–1. An example of the parallel nature of the hands, eyes, and information processing during a reach motion where the hashed line is the critical time path.

occurring, no time need be added to the motion involved. Figure 5–1 is a diagram of an R_B reach.

The critical path, which is denoted by the hashed line, is the path that takes the most time. Note that the information processing system is on the critical path from (B) to (C). At other times, the eye movement and focus and the information processing occur, but are not restricting the hands and thus cost nothing with respect to time. The difference between an R_B and an R_C can be thought of as the increased time from (B) to (C) for the R_C.

3. Decisions that are required where the information processing is on the critical path take longer than the MTM-1 motions have allowed. In Item 2 above, the eyes are continually leading the hands so that the eye transport focus and information processing can occur with minimum time spent on the critical path. This sequence of the human attempting to conserve time by doing as much of the eye/information processing in parallel with the hands is part of the "rhythm" of doing work. However, this "rhythm" can be interrupted by something abnormal happening. Typical abnormalities are the dropping of a part or tool, the failure of a component to go together as expected, or an inspection situation where the worker cannot move his/her hands until the information is processed.

All of these situations have in common insufficient time for the eye travel, eye focus, and information processing to be done without extending the time distance between (B) and (C) of Fig. 5–1. This lack of time is called "lack of preview" because there is insufficient time for the eyes to move and focus and the information to be processed without extending the overall cycle time.

Why would there be a "lack of preview" if the operator drops a tool? As soon as the operator drops the tool, he must process more information in order to restore the tool to the proper position. The eyes must return to the part if the drop occurs after the eyes have left it to get information for the next motion, and additional information must be processed. The return eye movement, eye focus, if necessary, the information processing, and the part repositioning motions will be on the critical time path. Furthermore, the eyes will have to return to their original position in order to resume the operation at the "steady-state" pace.

A similar situation exists if a component does not fit as expected. The worker's eyes will leave the part as soon as he/she perceives that the position will occur as expected.The eyes have to return to the part in order to accomplish the position.

The return of the eyes, focus, if necessary, information processing, and the return of the eyes to the original position will all be on the critical time path.

In the case of an inspection example, the operator must process information before deciding which path the hands must take. Frequently, the quality of the part will be unknown until a gauge is read or a measurement taken. In these cases, the hands will stop or hesitate until a decision can be made as to the appropriate path.

Decide Action

The MTM-1 system (and all of the other MTM systems, since they are based on MTM-1) is a system where the values are given for the motions of the human including the hands, feet, body, and eyes where steady-state conditions are assumed for the information processing. This is a very reasonable assumption the vast majority of the time because humans strive to work under steady-state conditions where the eyes lead the hands so that eye travel, eye focus, and information processing are occurring in parallel with the movements of members of the body as much as possible. However, there are cases where "steady-state" conditions do not exist. Under these conditions the hands hesitate or slow down until the necessary eye motions and information processing can occur. The information processing is not allowed for, except for a simple (binary) decision that is part of the eye focus (EF). If a more complicated decision has to occur, such as may happen with the part that does not position correctly or during an inspection operation, then the time that is taken is called process time. Time study is usually suggested as the method of determining the process time. However, information processing usually only takes a few seconds and is almost impossible to time with conventional time study equipment such as a stopwatch.

Many studies have been made by psychologists and human engineers to determine the reaction time and the information processing capabilities of humans. In recent years a number of studies have been specifically directed toward predicting the time it takes for people to process information. As a result of these studies, a Decide Action Table has been proposed for the MTM-1 system. The table and numerous examples of its use are described in the cited reference.[1] The purpose here is to present the table, briefly describe how it is to be used, and then give several examples. Figure 5–2 is the proposed Decide Action data, including the proposed symbols for Decide Action.

Variables Involved In Decide Action Data

The Degree of Preview Before the Information is Needed. Three cases are provided. *Full* preview is the steady-state condition that requires no additional time. This condition exists if the signal is available to be used by the information processing system at least 50 TMU before the particular motion requiring the information processing takes place. For example, in an R6C the eyes can travel

[1]Bayha, F. H. and Hancock, W. M., *Application Guidelines for Decide Action Research.* MTM Association for Standards and Research, Fairlawn, N.J., 1973.

Degree of Preview for Action Signal		Decision Time TMU for Choice of Action	
Case	Degree	Limited—L (2–4 Options)	Multiple—M (5–16 Options)
–	Full	Use manual time	
1	Partial	6.6*	9.8*
2	None	13.1*	16.4*

*Add 7.3 TMU if an eye focus does not immediately precede the decide action. If an eye focus does not occur, add the symbol E to the designation.

Figure 5–2. The Decide Action Table. Values are in TMUs.[2]

to the target whenever it is convenient, i.e., without any restrictions, due to the availability of the target area to the eyes, so that at least 50 TMU are always available. Thus the steady-state condition exists and no additional time is needed.

Partial preview exists when there is from 7 to 50 TMU of preview time available prior to the motion. Additional time is needed for information processing because more time will be taken for the information processing than in the steady-state situation. This class is for those cases where the target is available for only a short period before the motion has to occur. A typical example could be an inspection operation where a meter indication occurs within the range of 7 to 50 TMU before the operator has to take action as a result of the reading.

None preview occurs when there is no preview. This was the situation in a previous example when the part was dropped with no advanced warning. Once the part is dropped, the hands must immediately stop their motion and all of the eye travel, eye focus, and decide action are on the critical path.

The "none" situation also occurs in the example where the part will not fit as anticipated. Here again all motion must cease and eye travel, focus, and information processing to the original location must occur on the critical path in order to rectify the situation.

The Number of Alternatives Available

From information theory experiments we know that the time it takes to process information (make a decision) is dependent on the number of alternatives (options) that the worker has available to him/her. For example, if the inspector reads a meter, two options are available if the operator is to either reject or accept the part. Suppose, however, the worker must make an adjustment as a result of reading the meter. There may be three possibilities: adjust up, adjust down, or leave alone. This is a three-option decision.

In the previous example of dropping the part, how many options are there?

[2]Bayha, F. H. and Hancock, W. M., *Application Guidelines for Decide Action Research*. MTM Association for Standards and Research, Fairlawn, N.J., 1973, p. 2.

The answer depends on the shape of the part and the number of ways an operator would *consider* grasping it. It is important to point out that *all* parts can probably be grasped in many different ways, but an experienced operator will decide on only a few different ways that are practical for the operation at hand. For example, if the part is a hex headed bolt, he/she would probably consider grasping it with the thumb and forefinger on three of the six flat surfaces or on the end. Thus a four-option decision has occurred.

In the example of the part that will not fit, the operator has not released the part so that there are probably no options concerning how to grasp the part, but the operator must decide on what to do in order to get the part to position properly. Generally, he/she will attempt to correct the alignment by a series of short moves. For most align adjustments the initial attempt will usually consist of two options: to move the part this way or that way (two options). More difficult aligns may also include a rotate and increased pressure (four options).

As an example of a larger number of options, consider the decisions necessary to put the cards from a deck of cards into piles by letter and number. Here, a decision has to be made each time a card is examined and a decision made about placement. This is a 13-option decision.

It is important to note that it is difficult to find options higher than four where operators with experience are being observed so that the default case, if there is any question, is to use the limited (L) category.

Eye Focus (EF) Effects. The table values are calculated on the assumption that an EF occurs immediately before the Decide Action.

Figure 5–3 is a decision diagram for the Decide Action Table.

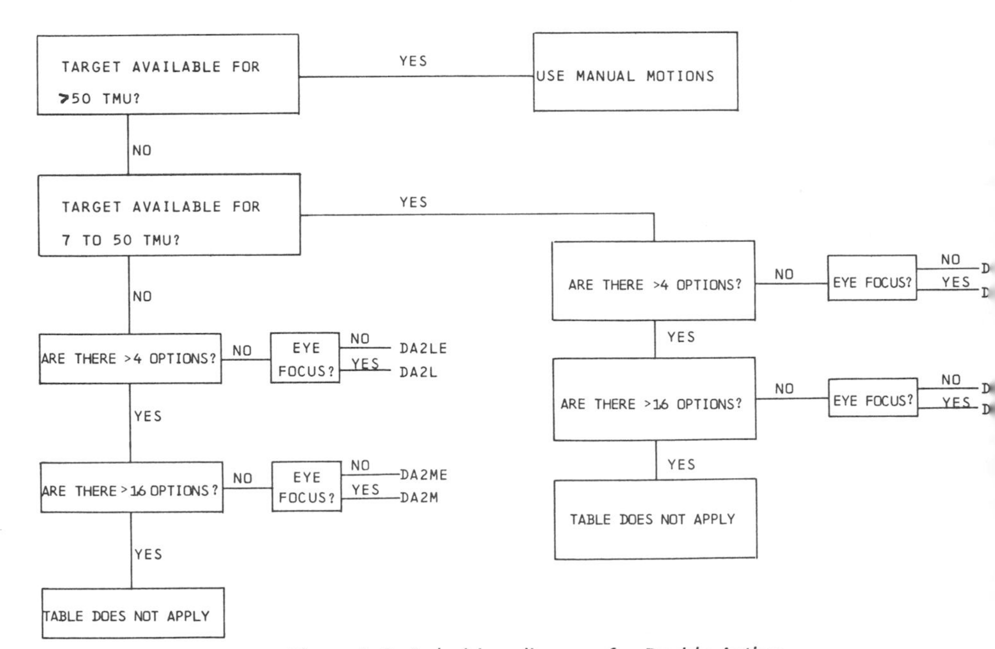

Figure 5–3. A decision diagram for Decide Action.

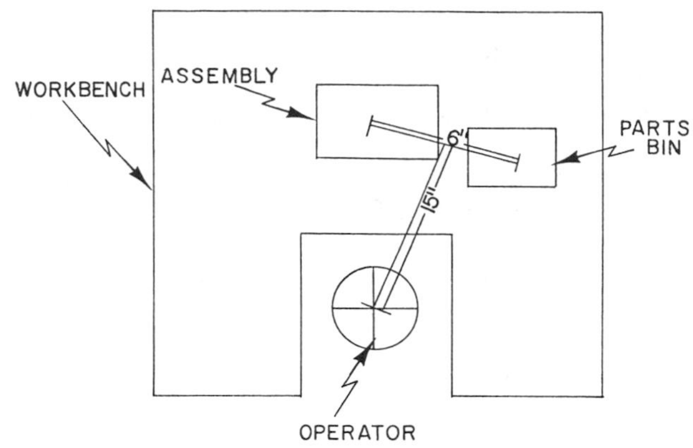

Figure 5–4. The work place layout for example 1.

Examples of the Use of Decide Action

To aid in the understanding of the use of the Decide Action, the MTM-1 analysis for the examples used in this chapter are as follows:

Example 1. An operator is performing an assembly motion which consists of reaching for a part, grasping it, and transporting it to be positioned in an assembly. Figure 5–4 is the workplace layout.

The "steady-state" MTM-1 pattern for this operation would be:

10.1	R6C	To parts bin from assembly
9.1	G4B	
10.3	M6C	
16.2	P2SE	
45.7	TMU Total	

If the operator were to drop the part during the M6C, then the motion pattern could be as follows:

10.1	R6C	To parts bin from assembly
9.1	G4B	
5.7	M3B	Part dropped after moving 3 inches
3.0	ET 3/15	Eyes would have to return for the assembly 15.2 × 3/15 = 3.0
7.3	EF	
13.1	DA2L	No preview, assume two or three options
8.7	G1C2	
3.0	ET (3/15 × 15.2)	
5.2	M2C	5 inches to assembly
16.2	P2SE	
81.4	TMU Total	

Many MTM analysts will wonder why an example of dropping a part is used. They will argue that the assumption is made that parts are not dropped. The example was used to help show the difference between a typical steady-state operation where Decide Action is not necessary and an interrupted operation where Decide Action is needed. Even experienced operators will drop a part, and this example may help the analyst to analyze these cases if it is necessary.

The EF may not be necessary in both analyses if the parts bin and assembly are within the depth of focus of the operator. If this is the case, then the DA2L would be a DA2LE because an eye focus would not immediately precede the DA.

Example 2. This example is similar to Example 1. We can use the same layout (Fig. 5–4). The difference is that the part will not position correctly in the assembly and the operator's eyes will have left the part in anticipation of the reach before he/she realizes that the position has not gone as expected. The motion pattern would be as follows starting with the position:

16.2	P2SE	The original position is attempted
2.0	ET 2/15	Assume eye traveled 2 inches 15.2 × 2/15 = 2.0
20.4	DA2LE	Decision to realign part—assumption: No eye focus precedes motion
2.0	MfC	After move the rest of the position occurs
2.0	ET 2/15	Eyes travel to original location 15.2 × 2/15 = 2.0
10.1	R6C	
52.7	TMU Total	

Note that in both examples 1 and 2, two ET are involved; one to get the eyes back to the problem area and the other to allow the eyes to return to their original position. The second ET is to compensate for the additional time required to get the human system back to "steady state" after the interruption.

Example 3. In this example we want to deal a deck of cards into 13 piles. The cards are lying face up in a pile. Figure 5–5 is the layout.

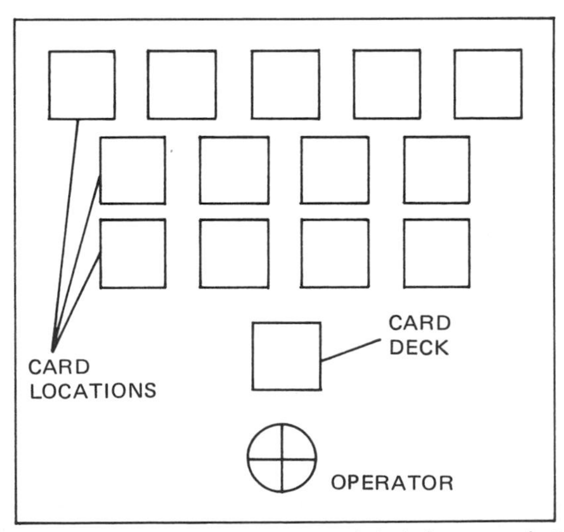

Figure 5–5. The layout of the card deck for example 3.

For purposes of simplicity, let us assume the average distance to the card locations is 8 inches. If the cards are face up, then the motion pattern is:

~~EF~~	10.1	RB	EF occurs during reach
	9.8	DA1M	
	3.5	G1B	
	11.8	M8C	
	5.6	PISE	
	40.8	TMU Total	

In this example the next card is available after the grasp until the end of the reach. This is 27.5 TMU of preview time. Thus limited preview is involved. Thirteen options require a multiple options category. The EF occurs but is limited out by the reach.

Example 4. This example is the same as example 3 except that the cards are laying face down instead of face up. The motion pattern would then be as follows:

~~EF~~	10.1	R8B
	3.5	G1B
	9.4	T180S
	23.7	DA2ME
	11.8	M8C
	5.6	P1SE
	64.1	TMU Total

The DA2ME is needed because the EF, which is limited out by the R8A, does not occur immediately before the DA and there is no preview of the cards.

Example 5.[3] An electronics division of a major chemical company tried the Decide Action data on one of their commonest operations—100 percent manual inspection of small axial-lead components, such as capacitors, following the rejection of the lot during sampling inspection. Lots arrive and depart on slotted trays, with good parts left on the arrival tray and bad parts sorted to two reject trays for the two defects of bad printing and solder break(s). There are on the average 12 percent defects: 6 percent solder and 6 percent printing. In order to inspect, the parts are kept on the arrival tray by passive restraint of the right hand while the left hand twirls the part by its lead. With only two types of defects, the reject options are limited; and, since the parts lie close together in regular array with the upper half exposed to easy scan, the preview is Partial for the top. However, the preview for the bottom is None because it can only be seen after the part is rotated. Thus there are two DA, one for the top half and one for the bottom half.

The analysis for the inspection is:

to next part	RfA	2.0	RfA	to next part
obtain lead	G1A	2.0	G1A	close fingers
		5.7	EF (0.78 × 7.3)(0.12EF occurs during return from trays)	
	DA1L	6.6		inspect top of part

[3]This example is a modification of an example in the volume cited in footnote 1.

twirl part	~~MfB~~		~~MfB~~	(allow slippage)
		7.3	EF	
		13.1	DA2L	
	RL1	2.0	RL1	
		38.7	TMU Total	

The analysis for parts disposal which occurs 12 percent of the time is:

part to reject tray	M9B	11.5	~~RL1~~	open fingers
drop part in slot	RL1	2.0	wait	
hand return	R9D	12.2	wait	
		25.7	TMU	

The standard would be (38.7) + 0.12 (25.7) = 41.8 TMU.

The Verification of Decide Action

MTM analysts may want to determine if the use of DA provides them with superior standards. The proper procedure is to use computer-aided timing where the computer clock has accuracy at least to 0.1 TMU. The standard approach for other motions has been to compare the results achieved with time study. This is considered a very gross method because the DA is usually a very small percentage of the overall operations and the DA values are not capable of being timed accurately by stopwatches.

If time study is the only method available, then there is a tendency to choose cycles where DA is a major fraction of the cycle time. This usually results in short cycles such as the examples above and means that we have to worry about the prediction accuracy of the MTM-1 system. If the NRT is greater than 100 TMU, then the $\%A$ can be obtained from Fig. 12–2, Chapter 12. If the NRT is less than 100 TMU as in Example 4, then percentage can be computed using footnote 6, Chapter 12, where s_j is derived by solving the equation for s_j for a point on the graph. For example, if we choose the 100 TMU NRT point on the graph, then $\%A = 20$. Thus,

$$20 = \frac{1.96(s_j) \times 100}{\sqrt{100}} \quad \text{or} \quad s_j = 1.0204$$

For example 4, the NRT is 38.7 + 25.7 = 64.4 (all frequencies are set to 1), then

$$\%A = \frac{1.96(1.0204)}{\sqrt{64.4}} \times 100 = 24.9$$

The 95 percent confidence interval is 41.8 TMU ± 0.249 (64.4) or 41.8 ± 16.0 TMU. Therefore, if the time study produced a standard between 41.8 − 16.0 TMU and 41.8 + 16.0 TMU or 25.8 to 57.8 TMU, the standards are not significantly different from each other.

Summary

Decide Action, although it occurs infrequently, is an important area for the MTM analyst to understand because it aids in the comprehension of how humans make decisions and the time consumed in the process. The concepts of critical path, steady-state, and information processing are particularly important concepts for the analyst to understand, particularly where short cycle, inspection, or delays are part of the standard.

CHAPTER 6

MTM-2: A High-Level, Less-Detailed System than MTM-1[1]

Until the mid-1950s, the application of predetermined times in work measurement was accomplished by employing a basic level system. MTM was such a system, long composed of very small observable motions. Since a large majority of industrial operations are composed of common recurring motion patterns, application of the small elements of MTM required considerable work and time for the analyst. In a sense, with regard to the use of small elements, it was slow and repetitious. In order to make the measurement of longer cycles of work more economically feasible, it was deemed necessary to develop a system which had fewer but longer motion aggregates. Such a system is MTM-2. It is called "MTM-2" because it contains motion aggregates, rather than single motions, and it is therefore known as a "higher-level" system. At the time of its introduction to the work measurement world, the basic, original MTM system became known as MTM-1.

Discussions leading to the system's development were actually started in 1962 by the Swedish MTM Association (Svenska MTM Foreningen AB). In part the motivation for MTM-2's development was provided by the development of Master Standard Data (MSD), a proprietory system of the Serge A Birn Co., and by the MTM Association's[2] General Purpose Data's (MTM-GPD) beginning development. Moreover, account was taken of the research work carried out by the Svenska MTM-Gruppen AB into basic motion frequency and motion sequence. Active development by IMD[3] actually started in 1963 with the project under the control of Svenska MTM-Gruppen AB and supported by Svenska MTM Foreningen through the action of its board.

The initial work included a computerized determination of sequences and frequencies of MTM basic motions from analyses of manual operations carried out in a variety of industries. These analyses contained 14,000 basic motions representing 30,000 man hours. A data card representing these motion sequences was even developed.

[1]While this discussion of MTM-2 is in considerable detail and the system appears simple, do not be misled. There are many problems that will arise in its application for which there is a standard solution leading to correct results. Other solutions which may appear satisfactory will lead to errors. Do secure the proper training in an approved MTM-2 Application Training Course before applying the system.

[2]The MTM Association for Standards & Research, 16–01 Broadway, Fair Lawn, N.J.

[3]IMD (International MTM Directorate) is an overall managing board elected by the various National MTM Associations.

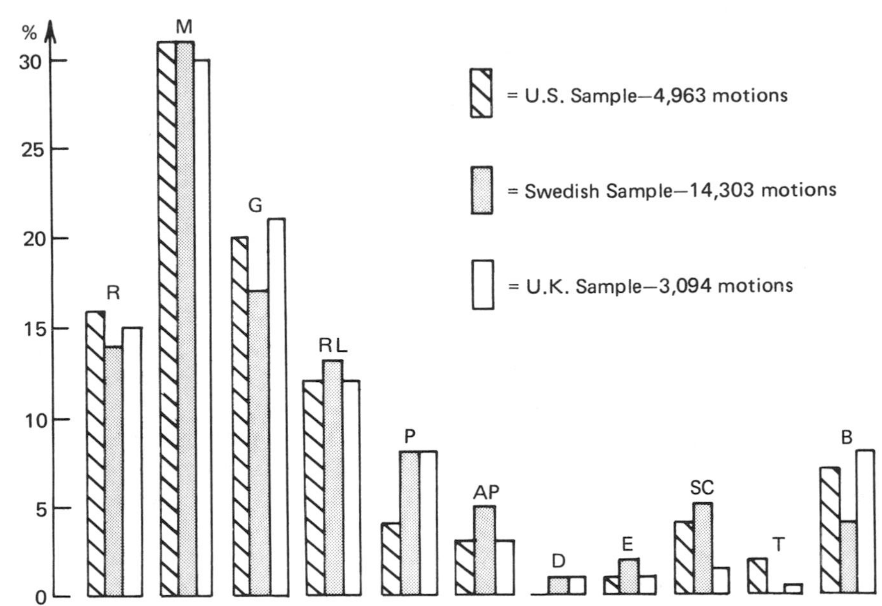

Figure 6–1. The distribution of MTM motions. © MTM Association for Standards & Research, Fair Lawn, NJ.

Almost simultaneously, discussions were taking place within the IMD on the standardization of MTM data at the second level. In the autumn of 1964 the IMD's standing Applied Research Committee (ARC) was given the task of examining existing systems of combined MTM with a view of proposing one system.[4] From the beginning it was agreed that the project should follow the main thrust of the Swedish effort. To the Swedish project were added thoughts from the entire IMD membership and especially those of the United States (developers of MTM-1) and the UK Association who now joined in the efforts of the Swedish Association by directly participating in the MTM-2 development. To the Swedish basic motion frequency study were added data from the United Kingdom and the United States. These data are summarized in Fig. 6–1. Ultimately, this led to the 11 June 1965 approval of MTM-2 by the Managing Board of the International MTM Directorate in Munich, West Germany.

Definition[5]:

MTM-2 is a synthesized system of MTM data on the second generic level based exclusively on MTM-1 single motions and motion sequence combinations.

[4]Dr. Walton Hancock, one of the authors of this text, was a member of the ARC when it examined existing systems of combined MTM and made its report to the IMD. He also actively worked on the development of both MTM-2 and MTM-3 with Svenska MTM-Gruppen AB.

[5]This definition and the other definitions of the various systems and system elements are all officially approved definitions and are given through the courtesy of the MTM Association for Standards & Research, Fair Lawn, N.J. Subsequent reference to these facts will be annotated (Courtesy of U.S./Canada MTM Association. See footnote 5, Chapter 6).

Some aspects of the definition should be clarified and elaborated. MTM-2 is a synthesized system in that the data elements are statistically combined single MTM-1 data elements or they are elements that include two or more MTM-1 elements. While more information on this is given later in this chapter, the reader may also want to study the material referenced in this chapter.

System Accuracy and General System Details

In the case of MTM-2, the Swedish MTM Association first set desired system characteristics in terms of total system error, application error, and speed of application. The MTM-2 system concept was endorsed by IMD in 1964. MTM-3's definition and development (discussed in the next chapter) followed after MTM-2.

MTM-2 is a *synthetic system* since it is essentially a statistical composite of MTM-1 elements and combinations of elements. Simplicity was achieved by both technical simplification and statistical simplification (making use of probability theory).

As mentioned in Chapter 1, *generic data* or *generic systems* can be applied almost anywhere due to their generality. Generic data have a broad application spectrum, or generality of application, because they are oriented to human behavior. Its elements (in our case statistically derived elements of MTM-1 system elements) are recognizable as distinct human actions by the practitioners and by the workers.

It is suggested that the reader review at this point the discussion of generic, functional, and specific data systems, including the discussion of data levels, as found in Chapter 1.

The MTM-2 system satisfies all of the major requirements imposed upon its design. These system characteristics include:

1. Consistency between analysts *and* between areas of application.
2. Speed of application almost twice as fast as MTM-1, and with practice it is even significantly faster than "twice as fast."
3. Specifications for system accuracy.
    (a) MTM-2 analyses, if produced by adequately trained and experienced practitioners, are within ± 5 percent (4.7 percent) with 95 percent confidence from MTM-1 analyses at approximately 2000 TMU of nonrepeated elements.

    Total system variance for both MTM-1 and MTM-2 are shown in Figs. 6–2 and 6–3. The difference between the two curves at any given TMU value is the MTM-2 variance from MTM-1. The data are plotted in log-log coordinates so that the reader can work with "straight" lines.

    The difference in percentage of MTM-2 from MTM-1 decreases as the cycle length increases, for example, at 6 minute cycles (10,000 TMU) the variance from MTM-1 is only about ± 2.1 percent.

    The variance of MTM-1 is about ± 2.1 percent and that of MTM-2 is ± 4.2 percent. This means, for example, with regard to MTM-1 at 10,000

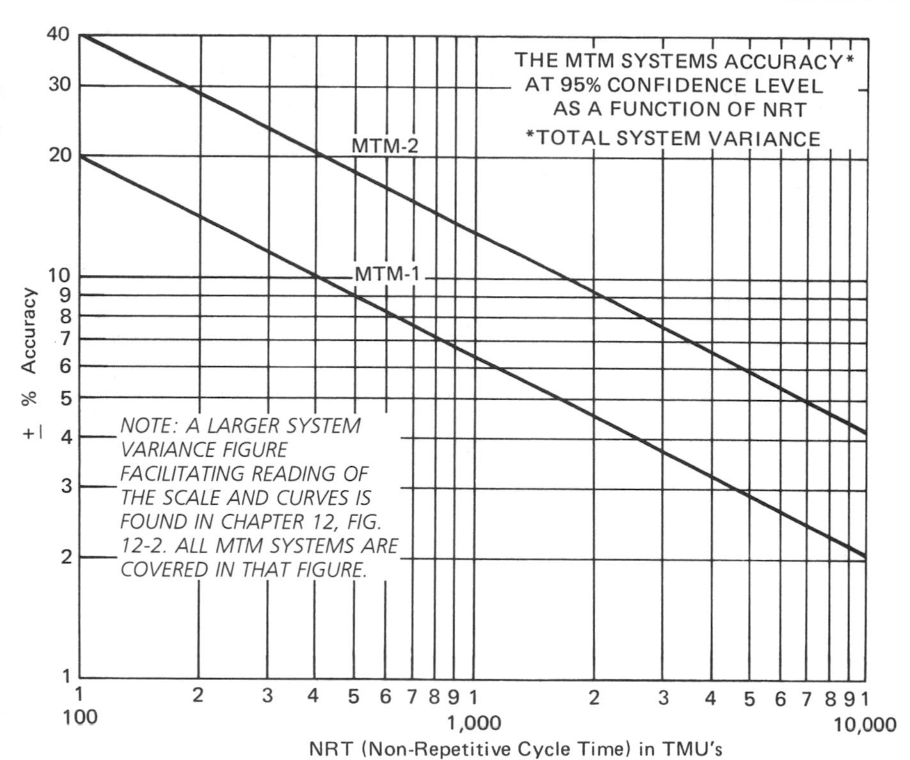

Figure 6–2. The MTM-1 curve shows the total system variance (applicator + system error). The MTM-2 curve shows the sum of the system variations from MTM-1 and the random applicator variations. © MTM Association for Standards & Research, Fair Lawn, NJ.

TMU nonrepeated cycles, that 95 out of 100 analysts would be within ± 2.1 percent of each other (assuming we have "read" the curve correctly), while 5 of the analysts would fall outside of these limits.

The same information is given *on the curves* illustrated in Fig. 6–3 for cycles of 200 TMU, 2000 TMU, and 7000 TMU.

(b) Average bias is 0.

4. MTM-2 is easy to understand.
5. The coding and, therefore, the motion patterns are descriptive of the method.
6. MTM-2 is compatible with other MTM data in that MTM-2 motion patterns can be directly combined with other varieties of MTM data.
7. MTM-2 is derived from MTM-1.

Since industrial engineers and work analysts should always be concerned with accuracy, a few statements on this subject need to be made.

Conventional wisdom says that the average time study is within ± 5 percent of the correct value. While this is possible at a 95 percent confidence level from a *timing* view, it cannot be done by timing only 5 to 20 continuous operations

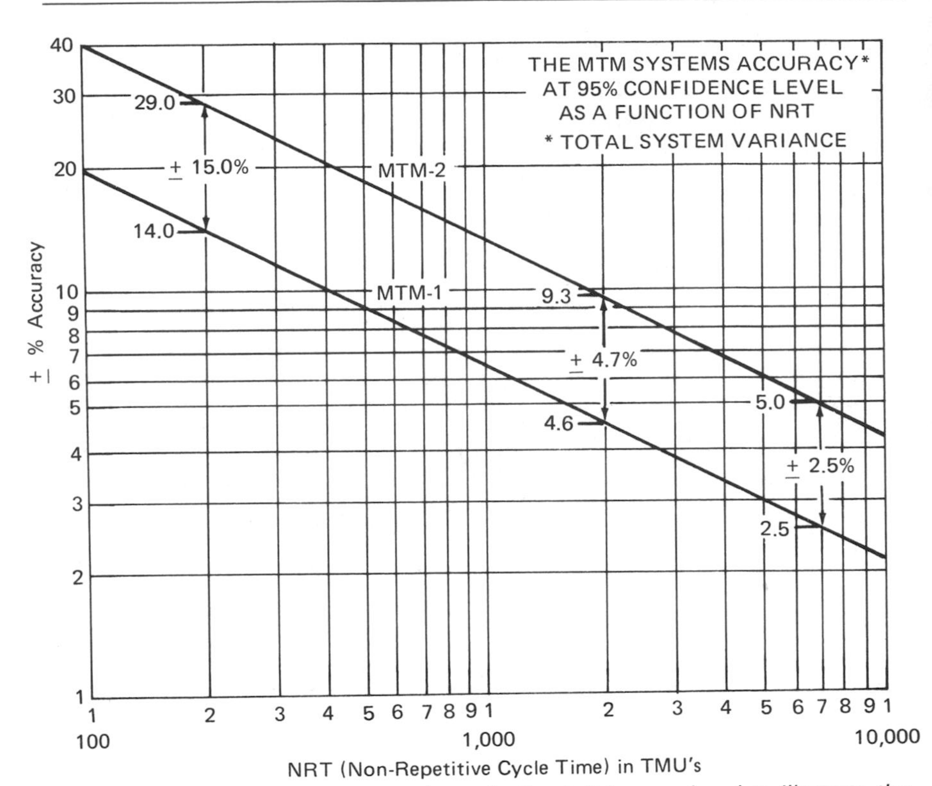

Figure 6–3. A portion of the curve shown in Fig. 6–2 is reproduced to illustrate the calculations referred to. © MTM Association for Standards & Research, Fair Lawn, NJ.

(see pages 151 through 166 of *Engineered Work Measurement*, EWM). It requires random observations over a period of days (usually), and the total number of studies is far more than 20. Also, when the time study analyst finishes the observations, he/she must still rate the operator(s). Any reader who has rated operators in rating films knows that this, too, is an inaccurate procedure and will likely further degrade, to a significant degree, the accuracy of the time study.

As explained in EWM, all these problems are eliminated in MTM, and MTM-1 is approximately the equivalent of a properly designed and executed statistically valid time study supported by multiple ratings (by several experienced and fully trained raters) of an adequately trained operator. When practitioners understand how MTM-1 was designed, was developed, and is used,[6] they will understand how MTM-1 comes far closer to a system accuracy of ± 5 percent or less than does the average time study.

In the statement of MTM-2 accuracy it was emphasized that this was for nonrepeated elements. Also, when one adds the facts of how MTM-1 elements

[6]See EWM (footnote 1, Chapter 1).

were used to produce MTM-2, it should be easy to understand that MTM-2 should not be used in place of MTM-1 when any of the following are characteristic of the work to be studied:

1. cycles are less than 1 minute (about 1600 TMU);
2. the motion pattern is composed mostly of complex, simo[7] finger motions;
3. the operation consists of a large number of repetitive motion patterns or elements.

It was mentioned that statistical analysis was used to reduce the large amount of data available on the MTM-1 data card. Actually, the establishment of the system elements was accomplished by (1) combining, (2) averaging, (3) substituting, or (4) eliminating basic MTM-1 motions. Examples of these actions are as follows:

1. Reach, Grasp, and Release were combined to give the MTM-2 category GET. Frequency of usage was involved in determining the way the MTM-1 elements were combined.
2. APA and APB were averaged to give the MTM-2 category APPLY PRESSURE, but, again, it was a statistically weighted average that took into account the frequency of occurrence of each.
3. Turn was substituted by Reach in GET and by Move in PUT.
4. Sit/Stand were eliminated and analyzed as BEND AND ARISE plus other relevant motions. Again frequency of element usage entered into the design of this element.

The MTM-2 System Details

MTM-2 Data Categories	*Identification Codes*
GET	GA
	GB
	GC
GET WEIGHT (MTM-1 Static Weight Factor)	GW
PUT	PA
	PB
	PC
PUT WEIGHT (MTM-1 Dynamic Weight Factor)	PW
REGRASP	R
APPLY PRESSURE	A
EYE ACTION	E
FOOT MOTION	F
STEP	S
BEND AND ARISE	B
CRANK	C

[7] Simo stands for simultaneous motions, the condition when two motions are performed simultaneously by *different* body members.

MTM-2

RANGE	Code	G A	G B	G C	P A	P B	P C
Up to 2"	-2	3	7	14	3	10	21
Over 2" - 6"	-6	6	10	19	6	15	26
Over 6" - 12"	-12	9	14	23	11	19	30
Over 12"-18"	-18	13	18	27	15	24	36
Over 18"	-32	17	23	32	20	30	41
MTM ASSOCIATION ®	GW 1 - per 2 lb.				PW 1 - per 10 lb.		
	A	R	E	C	S	F	B
	14	6	7	15	18	9	61

Figure 6–4. MTM-2 data card. The MTM-2 system provides 39 time standards ranging from 1 to 61 TMU. These are shown on the data card. © MTM Association for Standards & Research, Fair Lawn, NJ.

The same categories are used throughout the world and licensed practitioners are not to delete a category or change a time value. The same holds true for the entire MTM family of systems.

The MTM-2 system consists of only the 9 categories and 15 codes listed above. This makes the data card, illustrated in Fig. 6–4, extremely simple. It provides 39 time standards ranging from 1 TMU to 61 TMU.

The Learning Effect

MTM-2, like MTM-1 and the other MTM systems, indicates the time required by an experienced operator to perform the motions at normal performance level. An untrained worker will slow down his or her motions and will need to use the eyes excessively. Sometimes such workers will also perform incorrect and/or additional motions.

Standards are to be developed for the experienced operator, hence it is not accepted practice to record motions of an untrained operator—rather analyze what should occur with a fully trained and experienced worker (EWM contains information on how to conduct a work study).

Learning and associated research is discussed at length in Chapter 2 for the general case and for MTM-1. Learning principles are essentially the same for MTM-2 and the other MTM systems, but the method shown in Chapter 2 for computing the learning time of MTM-1 does not apply *directly* to the higher level MTM systems.

GET—G

Definition[8]:

GET is an action with the predominant purpose of reaching with the hand or fingers to an object, grasping the object, and subsequently releasing it.

[8]Courtesy of U.S./Canada MTM Association. See footnote 5, Chapter 6.

Like all MTM motions and elements it has identifiable beginning and ending points which coincide from system to system. It starts with a reach to the object, includes the grasping or gaining control of the object, and ends when the object is released—which means that the eventual release (usually after a PUT, later described) is included in GET or "G." The following diagram will be helpful in understanding the relationship between "GET" and "PUT."

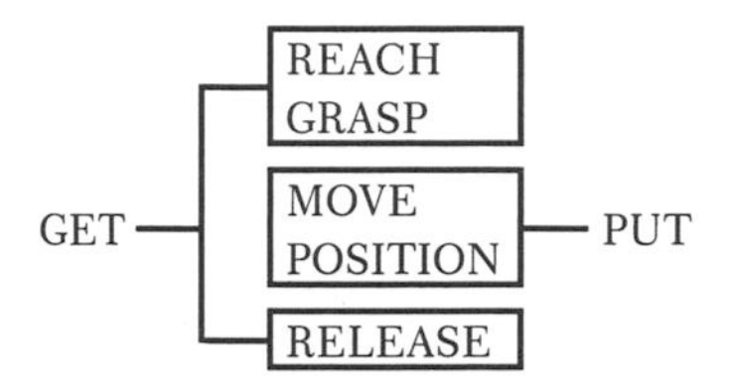

Variables

There are three variables. The first variable is the CASE OF GET. The cases are related to the kind of grasping action required.

Case	*Symbol*	*Description*
A	GA	No grasping action is required—the grasp is the contact, sliding, or hook grasp of MTM-1. In this case sufficient control for lifting the object does not result. Other much less frequent GAs involve no object contact or consist of a plain reach to an indefinite location to get the hand in position for body balance or the next motion, or out of the way. This latter type of GA is rarely time limiting, which means it is usually "limited out" by one or more other motions that consume more time. More will be said later on limiting and limited motions.
B	GB	The grasp requires only one grasping motion, the simple closing of the fingers as in the MTM-1 pick-up grasp (G1A) of a small, medium, or large *easily* grasped object that is not mixed with other objects (it is by itself). It is the most common case of GET.
C	GC	The gaining of control of the object requires more than one grasping motion. This would be generally true for grasping (gaining control) of a small object (like a needle or a single piece of paper) lying close against a flat surface, or grasping an object jumbled with other objects (comparable MTM-1 GRASPs would be G1B, G1C, G4A, G4B, or G4C). GC should not normally be followed by a REGRASP.

The decision model for determining the case of GET is simple and is illustrated in Fig. 6–5. For those unfamiliar with such models, the user enters at the top and by answering the questions automatically arrives at the case in most

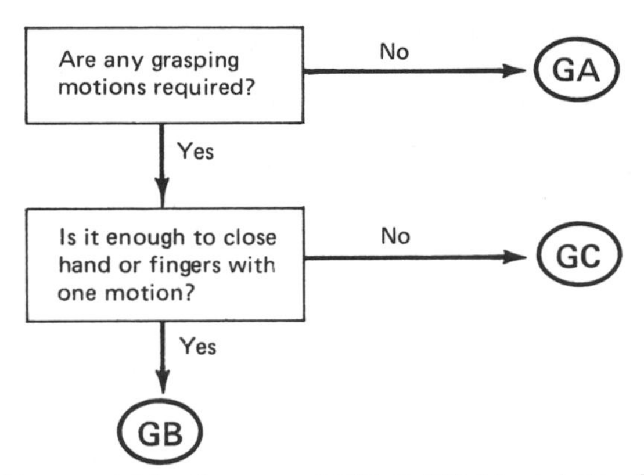

Figure 6–5. Decision model for the case of GET. © MTM Association for Standards & Research, Fair Lawn, NJ.

situations where the object characteristics are straightforward and unambiguous.

The second variable is the DISTANCE REACHED after subtracting any body assistance.[9] No interpolation or extrapolation is required or permitted because of the way the distance categories are defined. Rather than increase system accuracy, such actions will likely lower it. Estimation should be the normal approach to determining the distance category, but only after experience has been acquired.

Distances are expressed by the upper limits of the ranges which are 5, 15, 30, 45, and over 45 centimeters in the International Data and 2, 6, 12, 18, and over 18 inches in the U.S./Canada Data.

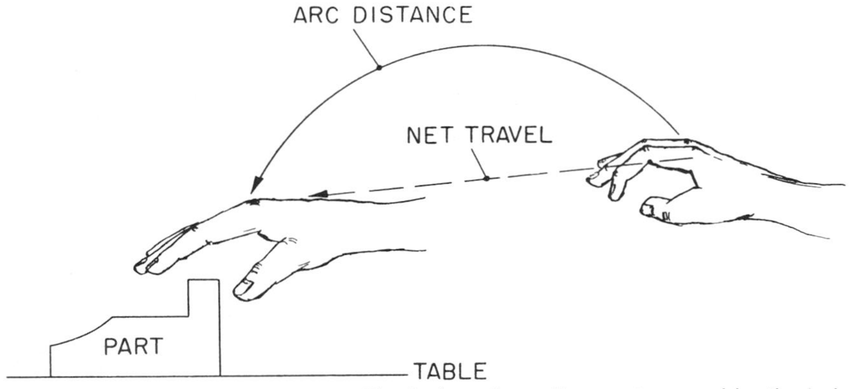

Figure 6–6. Distance measurement. The inches of arc distance traversed by the index knuckle, as shown, is the chargeable distance of the REACH. It is improper to use the straight line net travel as REACH distance. Reprinted from Engineered Work Measurement*, p. 314.*

[9]See EWM (footnote 1, Chapter 1)

As in MTM-1, the reach distance is *not* a straight line from where the hand started to where it contacted the object, but a slightly curved path as illustrated in Fig. 6–6. Measuring the distance in order to develop skill in estimating correctly should be done using a flexible rule. A practical reference point for measurement or estimation is the knuckle of the index finger on the hand being moved. The average person can reach a maximum of only about 24 inches directly away from the body without making assisting body movements. If a radially forward reach (GET in this case) exceeds this distance, some body movement will be employed. Simple body assistance is illustrated in Fig. 6–7. In order to determine the effective length of the GET it is suggested that one first put the arm–hand in the original position. Then, holding the arm rigidly in this position relative to the body, move the body to its final location. Only then make the reach or GET motion to determine the effective length of the motion. The short forward motion of the trunk will not be limiting and need not be recorded—unless it is a true BEND as later defined. If a BEND is necessary, this motion would be limiting and only a very short GET would be needed to complete the sequence (as later described).

Radial body assistance by turning the torso also occurs. It is handled in a similar manner. Again the body motion is ordinarily not limiting and need not be recorded.

One other often occurring assistance to a GET involves wrist assistance. It

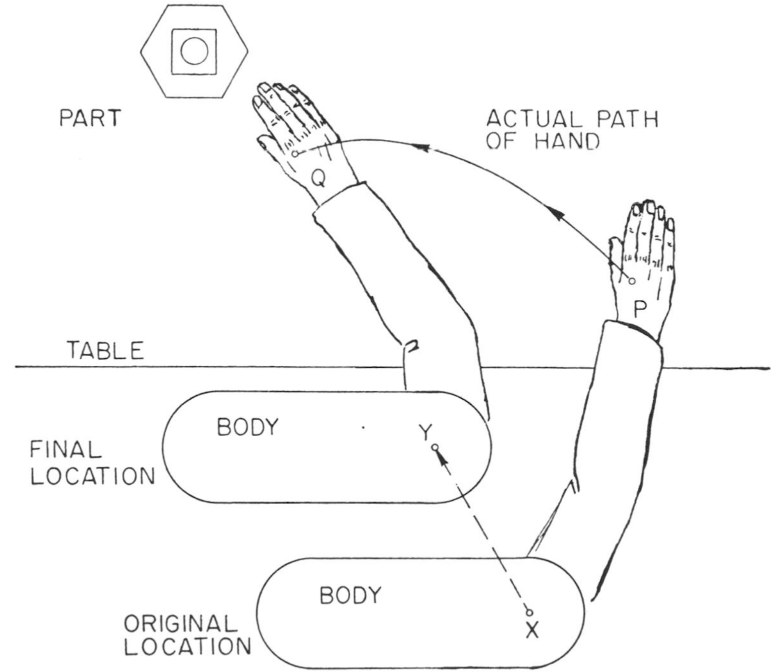

Figure 6–7. Simple body assistance to REACH. Reprinted from Engineered Work Measurement, *p. 317.*

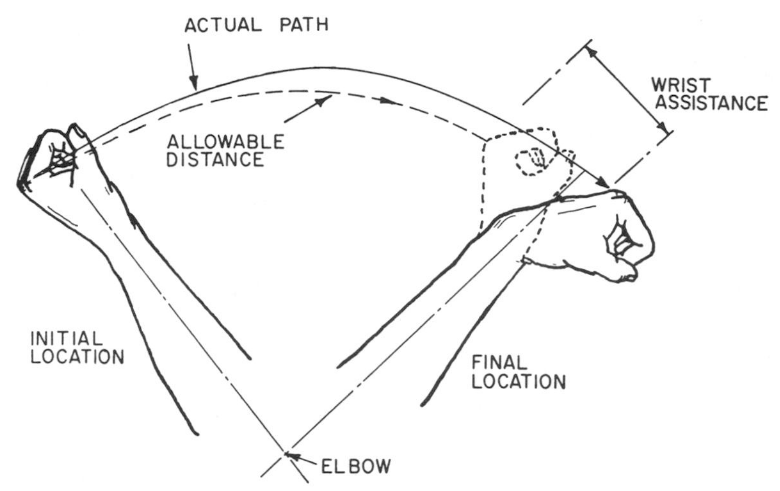

Figure 6–8. Wrist assistance to arm motion. Reprinted from Engineered Work Measurement, *p. 319.*

is illustrated in Fig. 6–8 (occurring during the movement of the arm and hand in GET) and is not limiting and should be subtracted from the distance of the *total* path traveled by the hand.

In a few rare cases, the analyst will find that a required body movement adds to the arm–hand movement.

The table in Fig. 6–9 shows the TMU for both the inch and metric categories of GET.

The third and final variable is the WEIGHT OF THE OBJECT or its RESISTANCE TO MOTION. This is, for GET, the static weight factor of MTM-1. While more on this subject is given in EWM, the factor is briefly discussed below under the topic of GET WEIGHT—GW.

Weight as a variable must also be considered in the MTM-2 action called PUT. In this case the weight is equivalent to the dynamic weight factor of MTM-1.[10]

	Inches US/Canada Data		Centimeters International Data		
Code	Over	Not over	Over	Not over	Code
2	0	2	0	5.0	5
6	2	6	5.0	15.0	15
12	6	12	15.0	30.0	30
18	12	18	30.0	45.0	45
32	18	upward	45.0	upward	80

Figure 6–9. Metric and inch categories of GET. © MTM Association for Standards & Research, Fair Lawn, N.J.

[10] *Ibid.*

Rules of Thumb

As with all systems of this sort, some "rules of thumb" were either originally conceived or eventually generated because of need. EWM fully explains these application rules and presents them for each MTM-1 motion category. Those often applying to GET follow.

If an object is transferred to the other hand, the actual *transfer* (not the preceding or following motions) should normally be analyzed as a GB2 (GET class GB with the 0–2″ or GB5, 0–5 cm distance range).

The distance notation in MTM-2 (where distance is involved) is always recorded as the highest number in the range and after the notation for the type of motion and its class.

An example illustrating the above concepts follows:

	LH	TMU	RH
		18	GB18 Get object
Toward object	(G-) (circled)	11	PA12 to other hand
Get object	GB2	7	
Place or put aside	PA12	11	
Total TMU		47	

Note that the limiting symbolic conventions used in MTM-1 also apply in MTM-2. In the above example, the GET toward the object is limited out by the PA12—it does not take as long as the simultaneous PA12 motion. The "circle" around the G- is one of the correct ways of showing this.

Another rule that possibly falls in the "rule of thumb" category is that a GC is assigned if an operator GETs a handful of small parts out of a supply box or pan.

A GA or GB can be performed when the object is out of the operator's vision when adequate practice opportunity has made the action automatic.

If the hand reaches to and grasps an object while holding and palming (in MTM-1 parlance) a tool, like a scissors, the action is a GET. The predominant purpose was not to move the scissors. This and other similar comments will clearly show that a knowledge of MTM-1 is necessary, or the analyst will need special training by a qualified instructor who is capable of imparting the missing knowledge. Approved MTM Association training courses are available for personnel, both trained and untrained in MTM-1.

GET WEIGHT—GW

Definition[11]:

GET WEIGHT is the action[12] required for the muscles of the hand and arm to take up the weight of the object.[13]

[11]Courtesy of U.S./Canada MTM Association. See footnote 5, Chapter 6.

[12]Predominant purpose.

[13]The previously referenced static weight factor of MTM-1.

This action starts upon the completion of the grasp after the fingers have closed upon the object in the preceding GET, and ends when the muscular forces have acquired sufficient control to permit moving the object. GW must be accomplished before any object movement can take place.

Up to and including 2½ pounds of weight to be sustained (or overcome) by the hand requires *no* GW assignment, i.e., no additional time allowance or recording is required. When the object weight is over 2½ pounds but not over 4 pounds, the code in the MTM-2 motion pattern recorded on the work study sheet is GW4. Since 1 TMU must be assigned for every 2 pounds of total weight (but assignment of *time* does not begin until weight exceeds 2½ pounds), 4 pounds requires an allowance or assignment of 2 TMU. Above four pounds per hand, 1 TMU must be assigned for *every* 2 pounds of total weight rounded to the next highest 2-pound increment. For example, 10 pounds requires a time allowance of 5 TMU. GW10 would also be shown and allocated if the weight were 8.1 pounds, or any value above eight pounds up to and including 10 pounds, remembering that each 2 pounds requires 1 TMU, since we are dealing in 2-pound units. Always show the upper weight limit of the category involved, just as for the distance indication in GET. Show *nothing* if the weight is 2½ pounds or less.

Use exact weights to determine what weight category to show, if practical; estimate only when necessary. The weight category in GW, if one hand is to control the weight in any following spatial action, is the weight of the object in pounds expressed in the analysis as the higher figure of the appropriate 2-pound increment as indicated above. If two hands are to control the object during subsequent spatial PUTs (or nonspatial moves where there is resistance to movement), assign a GW that represents one-half of the actual weight (if this half weight exceeds 2½ pounds). Again show it as the upper limit of the appropriate 2-pound category.

Weight as used here can be RESISTANCE TO MOVEMENT, instead of actual weight—like sliding an object resting on the floor or on a bench. One can use actual resistance as measured with a spring scale, just as one can use the scale in measuring weight.

A second method is to calculate the resistance using the object's weight and the coefficient of friction. Average coefficients of friction as taken from EWM are shown in Table 6–1. The formula for the force required to slide an object is W (weight) $\times$ F_c (coefficient of friction). If two hands are used equally to slide the object, divide the weight by 2 in the formula or divide the force exerted by 2. This force is the EFFECTIVE NET WEIGHT (ENW). If two hands support

Table 6–1. Coefficients of Friction (Average Values)

Surface	Friction Coefficient	Average Value
Wood on Wood	0.3 to 0.5	0.4
Wood on Metal	0.2 to 0.6	0.4
Metal on Metal	0.3	0.3

Source: *Engineered Work Measurement*, p. 344.

or lift an object spatially, the ENW is the weight of the object divided by 2. The ENW is the weight used in calculating GW and PW (PUT WEIGHT, explained later).

A third method is to use the following rule of thumb: Take GW as 40 percent of the actual weight if only one hand is used and 20 percent when two hands are used. This, too, is a practical compromise (of the F_c data).

In MTM-2 it is usual to consider 40 pounds/hand as the maximum permissible ENW. Again, this is a compromise figure since about 35 pounds per hand (ENW) is considered to be the maximum for women and 50 pounds for men.

A GW can occur without a GET. For example, if a wrench is used to unscrew a rusted-on nut or a lag screw into a hard wood, extra time will sometimes be needed in acquiring control of the resistance to turning, although no GW was needed in acquiring control of (or PW in transporting the wrench to the nut or screw) the wrench weight.

In the above case, leverage is also involved. If a spring scale is used to measure resistance, be sure to connect it to the wrench where the hand grasps it. Any other point will give either too high or too low a reading—the reading will depend upon the arm length of the lever.

PUT—P

Definition[14]:

PUT is an action with the predominant purpose of moving an object to a destination with the hand or fingers.

The action starts immediately after the object has been grasped and control of it has been secured at the object's location. Securing control means the inclusion of a GW in the MTM-2 work pattern following the GET if the object has a weight of more than 4 pounds and only one hand is used. PUT ends when the object has arrived at its intended destination. All transporting and correcting motions between these end points are included in the PUT. The movement of the object may be laying it aside or placing it against a stop or moving it to an approximate location (the most common PUT).

The third variable, which really determines the case, is the number of correcting motions which occur between the end points, the "point" of beginning and the "point" (location) of the end of the PUT.

Case	*Symbol*	*Description*
A	PA	If there are no correcting motions, evidenced by a smooth motion from start to finish, it is a case A, symbolized PA.
B	PB	This case contains one correcting motion and is the most frequent kind of PUT. However, it is very hard to

[14]Courtesy of U.S./Canada MTM Association. See footnote 5, Chapter 6.

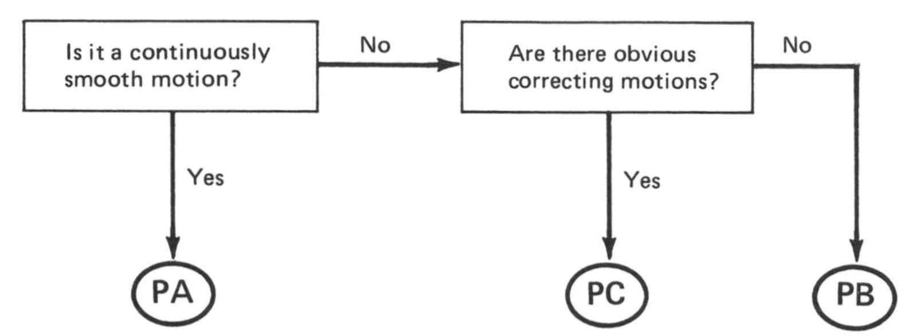

Figure 6–10. Decision model for the case of PUT. © MTM Association for Standards & Research, Fair Lawn, N.J.

Case	*Symbol*	*Description (continued)*
		see or recognize *directly*. This fact is taken into consideration in the decision model for all cases of PUT and is shown in Fig. 6–10. It indicates how to recognize this case of PUT by means of an exception. If it is not a PA or a PC, PUT probably is a case B (symbolized PB). It is generally associated with easy to handle objects where a loose fit (in positioning the object) occurs.
C	PC	This case of PUT contains more than one correcting motion. The correcting motions have to do with the *travel path,* not the manner in which the object is grasped. Usually there are several and they are therefore usually recognizable when observing the operator who should be experienced and fully trained. These correcting motions are usually caused by the object being difficult to handle, close fits, lack of symmetry of engaging parts, and/or an uncomfortable working position. These latter object characteristics are clues suggesting that the work analyst look for the usually accompanying correcting motions. A correcting motion is a short unintentional (unplanned) motion. This additional description is given in order to help the reader distinguish it from a short PA. A PA is a *purposeful* movement of a length that ordinarily is easily discernable.

The decision model in Fig. 6–10 will aid in deciding the case of PUT. If there is any doubt as to the proper case to assign, assign the higher case.

The motion distance is handled the same as in GET, showing the upper limit in the category into which the motion falls.

One rule of thumb for PUT is that if an engagement of parts occurs following a correction movement, then allow an additional PUT (PA) if the engagement distance exceeds 1 inch.

PUT WEIGHT—PW

Definition[15]:

PUT WEIGHT is an addition to a PUT motion. The amount of time depends on the weight of the object moved.

It is similar or analogous to the dynamic factor of weight in MTM-1.[16] Its beginning "end point" is when the move of the PUT commences and it ends when the move of the object ceases. The additional time to compensate for *moving* a weighted object is an additive factor *when the weight sustained per hand exceeds 2½ pounds*. Do not show the PW code unless a PW is involved.

Time for weights are calculated or determined as in GW. For weights over 2½ pounds up to 10 pounds, 1 TMU is allowed and the code is PW10. Each 10-pound increment above 10 pounds is allowed 1 TMU, and the code indications are PW20, PW30. Thirty-one pounds, for example, is coded PW 40 and requires the assignment of 4 TMUs. Again, a maximum of 40 pounds of effective net weight is usually considered to be the maximum *permissible* amount in using MTM-2. If greater effective net weights are found, the workplace should be changed so as to utilize hoists or some other form of operator assistance.

It was previously implied that GW was associated with GET, just as PW was indicated as being associated with PUT. The latter relation is absolutely true, but GW can also be associated with PUT under certain circumstances. Assume that a worker is fastening a 2 inch × 4 inch cross arm to a power pole with a square headed wood screw and that the pole and cross arm have been drilled so that a 10 pound force is required on the wrench used to turn the screw into the wood. The GET of the wrench has no associated GW since it weighs less than 4 pounds. The actions in MTM-2 coding, except for the final tightening, are as follows:

TMU	*Symbol*	*Description*
10	GB6	Get wrench at belt.
210	7PC12	Wrench to bolt 7 times. (The 7 in front of the motion code is an accepted way to indicate the frequency of the action.)
35	7GW	(Needed) To acquire *control* of the "resistance" or "strain" involved in turning the screw. (Not needed because of the 7PC12 or the GB6.)
77	7PA12	Turn screw.
7	PW10	Turn screw. (This is required because the resistance to accomplishing the 7PA12 is 10 pounds. The PW is related to the dynamic weight factor of MTM-1.)

If a special application of force is required in order to engage parts, do not consider it a part of the PUT. This also would be true for a final tightening

[15]*Ibid.*

[16]See EWM (footnote 1, Chapter 1).

action in the above motion pattern. The application of the latter force is handled or covered separately by an element called APPLY PRESSURE and is coded A. It is explained next.

APPLY PRESSURE—A

Definition[17]:

APPLY PRESSURE is an action with the purpose of exerting muscular force on an object.

APPLY PRESSURE is not really a motion or movement. Usually it is signaled by a hesitation—and any accompanying movement or displacement of the object must not exceed 1/4 inch during its execution. If the displacement after the A was, for example, 1/2 inch, a PA2 would need to be assigned. Any subsequent movement beyond 1/4 inch is a PUT.

The APPLY PRESSURE starts with the beginning of the hesitation, as in the final tightening of a screw or nut, or when the body member contacts the object, as in putting a shoulder against a large heavy box in order to cause it to move. After contact there is the controlled application of an increasing muscular force, followed by a minimum reaction time to permit the reversal of force and the subsequent releasing of the muscular force. It ends with the body member still in contact with the object, but the heavy muscular force is released. It could be followed by a PUT, as explained above. The PUT may or may not have an associated PUT WEIGHT; this would depend upon any resistance to the execution of the PUT. APPLY PRESSURE can be applied by any body member.

MTM-1 research clearly identified three components to APPLY PRESSURE.

1. The application of controlled increasing muscular force. The time required for this component is approximately proportional to the amount of force up to a maximum of 4 TMU at 9.4 pounds of pressure or force.
2. A minimum reaction time. The time required to sense that the required force has been applied and to signal the muscles to reverse the application of force. The time required is a constant 4.2 TMU.
3. The actual release of force MTM-1. Research showed that the time required for this element is essentially a constant 3 TMU.

The use of A to stop and then hold a heavy cart on an inclined passage in one position requires not only the assignment of an A but also the allowance of a time element best labeled "PROCESS TIME." It is usually symbolized PT, and the time must be determined by use of a timing device such as a stopwatch. PT is also used to cover machine process time.

APPLY PRESSURE in MTM-1 recognizes two cases of A, one consisting of the above three components and one recognizing the often accompanying "REGRASP" (explained in the next section). The time assigned to A in MTM-2 is a weighted average of the time for the two cases and is assigned a time of 14 TMU.

[17]Courtesy of U.S./Canada MTM Association. See footnote 5, Chapter 6.

Any time beyond the 14 TMU allowed for A, such as occurs during a holding action, must be timed with a watch. As stated above, such time and machine time is symbolized PT.

As noted above, REGRASP is included proportionately in MTM-2 APPLY PRESSURE, hence it should not be separately allowed in connection with AP. In some situations the first AP does not "break loose" the object to which the AP was applied and a second AP will be needed and noticed—it often is signaled by a REGRASP which is a *part* of the second AP.

REGRASP—R

Definition[18]:

REGRASP is a hand action with the purpose of changing the grasp of an object without losing control.

This action begins *after* the object is in the hand or hands and ends with the object being grasped at a different point or in a somewhat different manner. It primarily consists of digital and hand muscular readjustment on the object.

It occurs for the reasons given in EWM[19] for this action. Generally, it is to get a better grip on the object or to move the grasp on the object in order to facilitate the termination of the move action included in the PUT. *It is not the correcting motions involved in determining the case of PUT.*

An MTM-2 rule of thumb is that one to three fractional movements (essentially digital) during the move portion of a PA or PB signals a single regrasp. Difficult handling positions are included in PC, so no R is necessary for difficult handling.

If the hand ever relinquishes control of an object and then secures another grip, the grasp action is the start of a GET. Once control is relinquished, the action of again securing control is *not* a REGRASP.

A REGRASP should not follow a GC unless it is combined with another motion such as a P-.

EYE ACTION—E

Definition[20]:

EYE ACTION is an action with the purpose of either recognizing a readily distinguishable characteristic of an object or ***shifting the aim of the axis of vision to a new viewing area.***

If both of the above occur in the sequence defined in the definition, the action requires the assignment of two EYE ACTIONs.

The viewing area is considered to be a 4-inch-diameter circle at 16 inches

[18]*Ibid.*

[19]See EWM (footnote 1, Chapter 1)

[20]Courtesy of U.S./Canada MTM Association. See footnote 5, Chapter 6.

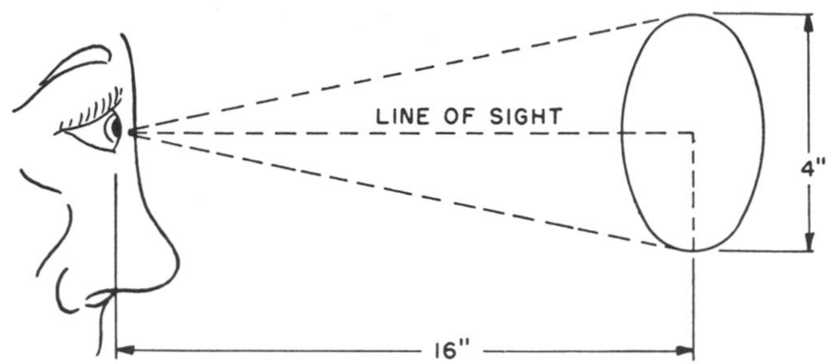

Figure 6–11. Area of normal vision (ANV). Reprinted from Engineered Work Measurement, *p. 538.*

from the eyes. This is illustrated in Fig. 6–11. The diameter of the viewing area increases as the object is a farther distance from the eyes and decreases as it is nearer to the eyes. The viewing area can be thought of as the base of a cone whose side has a constant slope regardless of the distance between the pupil of the eye and the surface being viewed.

The EYE ACTION starts when other actions cease because a characteristic of an object must be recognized—this ceasing of other actions is a clue that a limiting eye action may be taking place. EYE ACTION ends when other actions commence.

In between these occurs either (1) the readjustment of the lens of the eyes and the associated mental processes required to recognize a readily distinguishable characteristic of an object (the recognition time included in E is only sufficient for a simple binary decision, yes or no), or (2) the eye motion to shift the axis of vision to a new viewing area. More information on the basic underlying eye motion action can be found in EWM.[21]

EYE MOTIONS are always recorded on the right side of the analysis, just as are all body motions, including leg motions.

FOOT MOTION—F

Definition[22]:

FOOT MOTION is a short foot or leg motion (no more than a 12-inch leg motion) when the purpose is not to move the body.

It starts with the foot or leg at rest and ends with the foot in a new location.

The motion between the end points cannot exceed 12 inches and must be pivoted at the hip, knee, and/or instep if it is to be classed as a FOOT MOTION. The column used to indicate this on the motion analysis is always on the right.

[21]See EWM (footnote 1, Chapter 1).

[22]Courtesy of U.S./Canada MTM Association. See footnote 5, Chapter 6.

STEP—S

Definition[23]:

STEP is either a leg motion with the purpose of moving the body or a leg motion longer than 12 inches.

STEP starts with the leg at rest and ends with the leg in a new location. It is a leg motion with the purpose of achieving a displacement of the trunk, or it is a leg motion longer than 12 inches—the purpose in this latter case is to displace the leg and/or foot. It always appears in the right column.

Walking

To determine walking time it is merely necessary to count the number of times the feet hit the floor and allow an S for each occurrence.

Decision Model and Recording Convention

The decision model in Fig. 6–12 may aid the MTM-2 application engineer in deciding on F and S.

BEND AND ARISE—B

Definition[24]:

BEND AND ARISE is a lowering of the trunk followed by an arise.

The action starts with the trunk and body members in an upright position and ends with the body in the same position. In between, the trunk and other body members move to achieve such a vertical change or displacement so as to *permit* the hands to reach down to or below the knees, not whether they actually do so. This is followed by the subsequent arise. A practical rule of thumb

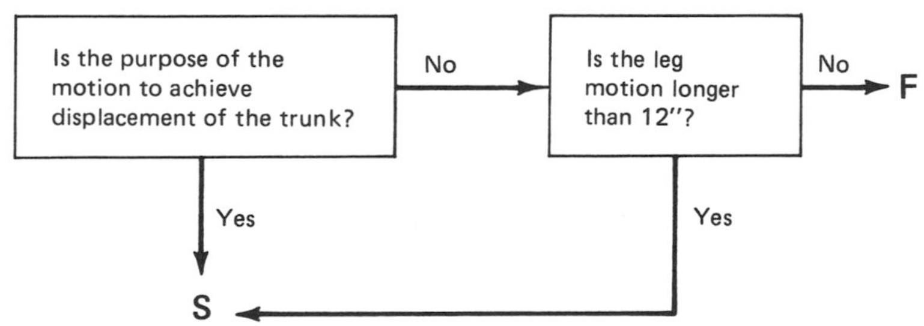

Figure 6–12. Decision model for STEP and FOOT motions. © MTM Association for Standards & Research, Fair Lawn, N.J.

[23] *Ibid.*

[24] *Ibid.*

is that kneeling on both knees should be analyzed as 2B. The right-hand column of the analysis form is used to identify BEND AND ARISE occurrences.

It should be obvious from the definition and explanation of this motion that some additional action, such as a GET, could occur while the hands were lowered in the BEND AND ARISE, and that this would not signal that another BEND AND ARISE should be shown.

CRANK—C

Definition[25]:

CRANK is a motion with the purpose of moving an object in a circular path of more than one-half a revolution with the hand or fingers.

CRANK starts with the hand on the object and ends when the object has been moved or transported in a circular motion or path of not less than one-half a revolution and not more than one revolution. The cranking action ends with the hand *on* the object when the motion ceases.

If the circular motion is more than one-half a revolution up to a full revolution, the cranking action requires the assignment of one C, even though the circular motion may continue beyond one revolution. However, this continuation starts another CRANK, and requires the assignment of another C if it continues to one-half additional revolution. Each revolution of a continuous cranking action is a C.

After the *last full revolution* of the object, any circular motion of more than one-half a revolution, but less than a full revolution, is also to be assigned a C.

CRANK includes all transporting motions required to move an object in a circular path.

Winding a string or wire around an object is as much a CRANK as turning a real crank. See Fig. 6–13 for more information on this kind of situation. Remember, the definition is moving an object in a circular path, although the actual path need not be a true circle, as in winding a string about a rectangular box. It obviously could be an ellipse or an oval or any path approximating a circle. The diameter of C is not involved in the determination of the time of C because the effect of diameter was taken into account in the design of MTM-2.

There are two variables:

1. *Number of revolutions.* Assign 15 TMU for every revolution or partial revolution over one-half, whether or not the cranking is continuous.
2. *Weight or resistance.* GW and PW rules apply to CRANK. PW applies to *each* CRANK. GW applies to each *start* from zero velocity. In other words it applies only once to a continuous cranking action, but it applies each time a CRANK motion is started in an intermittent series of cranks.

---

[25]*Ibid.*

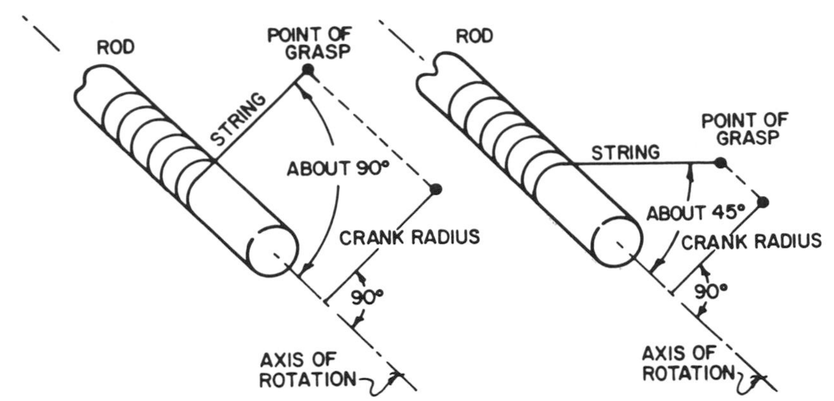

Figure 6–13. String being wrapped around an object. In MTM-2 it is not necessary to measure the crank radius. Reprinted from Engineered Work Measurement, *p. 381.*

A rule of thumb is that no correcting motions were included in CRANK. Therefore, if a correcting motion is required, it should be assigned a PUT.

The following motion pattern shows how to record the MTM-2 actions involved in turning a crank 10 inches in diameter against a resistance of 15 pounds for $5^1/_2$ continuous revolutions and for the same where there were $5^1/_2$ intermittent revolutions:

Continuous Cranking		*Intermittent Cranking*	
TMU	Action Symbol	TMU	Action Symbol
8	GW16	48	6GW16
90	6C	90	6C
12	6PW20	12	6PW20

Limited and Combined Motions

The notations for designating limiting motions are much the same as in MTM-1,[26] but there are permissible variations. The treatment of the time requirement is also much the same as in MTM-1,[27] but does differ in a few situations as to what is permissible treatment.

When two or more different motions or MTM actions are performed at the same time, the time allowed is the time for the motion with the highest time value. It is said to be the limiting motion. The other motion time is said to be limited out. This concept applies to both motions made by different body members and by motions made by the same or associated body members—as when a REGRASP is made during a PUT.

It is true that REGRASP sometimes occurs in association with a PUT, but it is not usually limiting. However, it could be limiting if it occurs as in the example below.

[26]See EWM (footnote 1, Chapter 1).
[27]*Ibid.*

LEFT	TMU	RIGHT
PB2]		
2R]	12	

The "squared" bracket signaling combined motions is generally used in Europe. The U.S./Canada MTM Association recognizes either the bracket or the parenthesis. The diagonal indicates the limited out motion.

Another MTM-2 convention (in fact, it is a general MTM convention) illustrated above is that when multiples of motions are encountered (like two REGRASPs, one after the other) they are shown by putting the number of occurrences before the symbol for the MTM-2 action involved.

Body motions may be combined with hand and/or arm motions. Also, the hand and arm may perform several motions in combination. These are essentially handled as in MTM-1.[28]

Body motions B and S are often combined with the hand and arm motions of G and P and limit out the G or P because the body motion consumes more time than the other motions. An example of this is shown later. However, in MTM-1 "Combined Motions" mean a combination of body motions in one body member—like an arm motion combined with a wrist motion. Motions involving two different body members are classed as Simo Motions, and this is really what is involved in this situation.

Earlier in this chapter it was mentioned that body motions are only shown in the right-hand column, while G or P are shown in the proper right or left column. This convention becomes clear upon studying the two examples which follow:

LEFT	TMU		RIGHT
	61	B]	To tool box on floor and get
		G-]	wrench.
		–]	Arise (included the B) and put
		P-]	tool on lathe bed.

In the same situation for a left-handed G and/or P, note that the brackets denoting combined motions appear only on the right-hand side with the body motions.

LEFT		TMU		RIGHT
	G-	61	B]	To tool box on floor and get wrench.
	P-		A]	Arise (included in the B) and put tool on lathe bed.

[28] *Ibid.*

Note that in this case (where the limited out action is on the other side of the analysis) a rectangular outline has been used to designate the limited out motion. In actual practice, an oval or circle is equally acceptable.

With regard to combined hand and arm motions, the most frequent combination is a P in combination with R, but there are others that occur less frequently:

LEFT		TMU		RIGHT
Move object to stop.	PA18 R- (R struck through)]	15		

Simultaneous Motions

The identification of simultaneous motions in the MTM-2 motion or action pattern are handled identically to such motions in the MTM-1 system[29] except that the case and distance of the limited motion is usually not recorded. The following example makes this clear:

LEFT		TMU		RIGHT
Move object #1 to fixture	PA18	15	(P-)	Move object #2 to fixture

The more the degree of control required in a set of simo motions, the more difficult it is to perform the motions simultaneously. Field of vision is also involved. For example, PBs are difficult to do simo if the target for one is out of the field of vision for the other. Recall that the field of vision is a 4-inch circle at a 16-inch distance from the eyes. If the eyes must be shifted to see the other destination, allow each PUT (more will be said about this later under motion overlap). If the targets are in the field of normal vision, two PBs can be performed simo.

Overlap

Since MTM-2 does not permit the analyst to discriminate as to the accuracy required in a PUT via a classification system and since the system does not pretend to equal MTM-1 in accuracy, some practical rules of thumb have developed. For example, if a combination right and left hand PC are to be performed simo, there will almost always be some overlap between the two PCs. This situation is covered or accounted for by allowing an additional short PC2.

LEFT		TMU		RIGHT
Screw toward threaded hole	PC12	30	[P-]	Screw toward threaded hole
		21	PC2	To hole

[29] *Ibid.*

In the case of a PB and a PC, it is difficult to perform these simo in the field of vision. One hand motion will overlap the other, so allow a PB2 or a PC2 (whichever is appropriate) for the overlapping of the motions, in addition to the full PB or PC with the other hand.

In the case of two PBs to be performed simo out of the range of vision, allow a PB2 for the overlap. If the destinations are within the field of vision, no overlap will be involved.

GC is similar to PC and can not normally be performed simo; allow a GC2 for the overlap. (Overlaps always take the same case as the limited motion.)

PA, GA, and GB require a lower degree of control and can usually be performed simo with other motions.

Difficulty with deciding on how to handle a case of simo is often a good indicator that MTM-1 instead of MTM-2 should be used in that particular segment of work.

Simo Table

The simo table shown in Fig. 6–14 provides the information MTM-2 analysts usually need to determine whether a simo motion is possible—subject to the previously mentioned application rules. A few more will also be given.

GA or GB are seldom time limiting when they are performed with S or B. However, with GC there likely will be a very short overlap, so allow an extra GC2.

GW is entirely another matter. It is never limited by a body motion, in fact it is never limited out. If two GETs are performed simo and both have associated GWs, the GWs can be performed simo. An example of GW with a body motion is now shown (examples of GW and PW with simo GETs and PUTs are given later in this chapter):

LEFT	TMU		RIGHT
	18	S]	Step toward supply box grasp a heavy lead part
		Ø-]	that has several sharp protrusions from supply box.
	14	GC2	Overlap or extra motion to avoid sharp protrusions.
	3	GW6	

PUTs were earlier mentioned in connection with body motions. PA is rarely time limiting when performed with S or B. With PB and PC there will often be an overlap motion of up to 2 inches. In this case allow an extra PB2 or PC2.

The next example illustrates how to handle simo motions involving different weights and/or distances in GET and PUT. In general, one must take both variables into account and limit on the basis of distance-class time *added to the*

	TABLE OF SIMO MOTIONS						
		GA	GB	GC	PA	PB	PC
GET	GA					X	X
	GB					X	0
	GC			0	X	0	0
PUT	PA			X		X	X
	PB	X	X	0	X	0	0
	PC	X	0	0	X	0	0

KEY:

☐ Easy to perform simo

[X] Can be performed with practice

[0] Difficult to perform even with practice

Figure 6–14. Table of simo motions. © MTM Association for Standards & Research, Fair Lawn, N.J.

weight time. The first simo example involves a GB18 combined with a GW10 vs a GB6 and a GW12. The second concerns a PB14 with a PW10 performed simo with a PB20.

LEFT			TMU		RIGHT
First Example	The sum of these two is 23TMU	GB18 GW10	18 5	(G- GW-)	The sum of these two is 16, less than 23 and therefore limited out.
Second Example	The sum of PB18 and PW10 is 25, hence it is limited out.	(PB- PW-)	30	PB32	

One final example is now presented. Assume that a worker is standing facing the workbench. Two sidesteps to the left are made while bending over to pick up an 18-pound object with both hands. Two sidesteps back to the original position are made, and the object is placed on the workbench.

Usually the first sidestep will be limiting, but the second will be limited by the bending motion (they can be performed together). However, in side stepping back to the original position, the first sidestep will be limited by the arising motion (part of the B) and the second will be limiting.

The illustrative example of the above presents the notations usually required.

LEFT	TMU		RIGHT
	18	S	First sidestep.
[G-]	61	B, $\not{S}$, $\not{G}$-	Down with second sidestep and get object.
GW10	5	GW10	These are performed simo, but nothing is limiting; they require the same time.
PA6	6	PA6	Ditto
PW10	1	PW10	Ditto

LEFT	TMU	RIGHT	
			(continued)
		-	Arise
		$	and first sidestep.
	18	S	Second sidestep.
PA6	6	PA6	Object to bench.
PW10	1	PW10	
	116 TMU		

Some Final Thoughts

As with MTM-1, the system appears simple and easy to use. This is true. However, as in MTM-1, there often are encountered unusual situations in which the action is not clear. Only through guided practice is the applicator able to proceed with confidence and produce motion patterns that do not contain significant application error. This is why the MTM Association approves training courses offered by its consulting membership for the general public. It also approves courses of its industrial members that are given to their employees. Finally, the Association also offers approved courses to the general public.

CHAPTER 7

MTM-3: The Highest-Level, Least-Detailed, and Fastest to Use General MTM System[1]

MTM-3, like its predecessor system MTM-2, was not developed by the U.S./ Canada Association.[2] Also, like MTM-2, the design parameters were established *before* the system was designed. The system is a product of the Swedish association and was produced under the general sponsorship of the International MTM Directorate.

The actual approach to the design of the system is essentially a replay of MTM-2. Because of this there will only be a limited discussion of the design aspects of MTM-3.[3]

Basically, the system was meant to be complementary to MTM-1 and MTM-2 by extending work measurement into new areas—areas somewhat impractical to cover with MTM-1 and MTM-2 in an economical manner.

MTM-3 is designed for short-run (small batch) production jobs, maintenance work, construction activity, etc. These basically involve long-cycle jobs, usually consisting of a somewhat lengthy series of tasks. Therefore, the probability is high that a positive measurement error in one element or task will be compensated for by a negative error in another, and vice versa—if the measurement system is not biased in one or the other direction. In the type of work described, there can even be the described compensation between jobs, although this does not help correct an individual job standard.

MTM-2, and especially MTM-1, when used to analyze the short-run work described above for methods improvements and the determination of work standards, produces precise and excellent results, but the result is an uneconomic use of the work study analyst's time. These systems take much more time to apply than MTM-3, and the result is excess work study costs. Moreover, enough

---

[1]This and the remaining MTM system chapters will be based upon the reader understanding Chapter 6, since the conventions, symbolization, and general parameters will be the same unless explicitly stated to be otherwise. Furthermore, when MTM-1 is referenced, the recommended nontraining course source is *Engineered Work Measurement* by D.W. Karger and F.H. Bayha, 3rd ed., Industrial Press, N.Y., 1978. These recommendations also hold for the other MTM system chapters. Finally, the chapters preceding Chapter 6 all contain related work measurement information, some of it of substantial importance to the proper application of MTM systems.

[2]Formal organization title is the MTM Association for Standards and Research, Fair Lawn, N.J.

[3]For more information on this aspect of MTM-3, review remarks in Chapter 6. Also, see *MTM-Project* by K.E. Magnusson & K. Silwerfrost, International MTM Directorate, Solna, Sweden, 1970. This latter publication contains the full technical information about MTM-3.

analyst labor hours may not be available to do the job in the time required unless MTM-3 is used.

The design objective, relative to the logical or practical application areas (from both economic and accuracy viewpoints) was to produce a system that would differ from MTM-2 in such a manner that a new MTM application area would be opened. Similarly, the area should be close enough to MTM-2's area so that no major gap would exist.

The above reasoning led to the additional objective that the new system, MTM-3, should be three times as fast to apply as MTM-2. This means that at the officially defined level the time necessary for analysis will amount to about 50 times the cycle time. However, with an increased level of expertise in working with MTM-3, the time required for analysis can be reduced to about 25 times the cycle time.

The increased speed of application achievable with MTM-3 could only be accomplished by relieving the analyst of work—thereby making necessary a reduction in the depth of analysis and the degree of methods description as compared to MTM-2. As the work can only be measured with the precision by which it is defined, the precision of time determination is lower with MTM-3 than with MTM-2. Methods analysis power is also obviously reduced.

The methods description is not all dependent upon the statement of the MTM-3 elements. Only the behavioral description of the operator is given by the MTM-3 element codes. The other portions of the job description such as the purpose of the operation, the tools used, the parts involved, and the general procedure to be followed consist of a written explanation in the analysis and can be provided in any degree of detail deemed desirable.

MTM-3 is intended to be used in cases where MTM-2 gives a higher accuracy than can be fully utilized and/or when lower cost of analysis is needed.

The demands on the design and application of MTM-3 are similar to those of MTM-2:

1. General elements
2. High speed of analysis
3. Generic elements, behaviorally oriented
4. Descriptive of methods
5. Combinable with other MTM systems
6. Easily understood and learned—derived from MTM-1
7. Defined in relation to MTM-1
8. System bias should essentially be zero

All of the stated design attributes were achieved and most of them can be understood without further elaboration—they are defined as stated. Understanding Chapter 6 will help in understanding these statements. However, a few remarks elaborating one or two of the statements may help clear up possible misunderstandings.

No system bias means that results will be over the correct (MTM-1) time about as often as under—and by amounts that approximately compensate for each other.

Total system variance is within ±5 percent at cycle times of approximately

9 minutes (95 percent confidence) where nonrepetitive elements are involved. It is ±10 percent at cycle times of approximately 1.75 minutes.

Combined system plus applicator error is such that MTM-3 analyses are within ±5 percent with 95 percent confidence from MTM-1 at cycles of 4 minutes or more.

MTM-3 can be combined with other systems based on MTM-1 since it is derived from MTM-1 and since the starting and ending points of MTM-3 elements coincide with the those in MTM-1 and MTM-2.

The system is so described that the analyst does not need to test to decide which system can be appropriately used.

MTM-3 Accuracy

This system was tested, like MTM-2, in the arena of practical work. Thirteen industrial and three consulting firms used it on a wide variety of work, including office work, foundry, sheet metal, maintenance, storeroom jobs, machine operators, and laboratory work. The test work encompassed tasks requiring 180,000 TMU, of which 21,000 TMU represented highly repetitive elements.

MTM-3 can be applied by observing work being performed or by visualization (often required in construction work) as is true for MTM systems in general.

The system variance, like for MTM-2, is presented in graphical form in Figs. 7–1A and 7–1B. The left "Y" axis or ordinate is in terms of system variance at 95 percent confidence limits and the right "Y" axis is the system variance in terms of 50 percent confidence limits.

Relative variance is found by subtracting the proper total system variances, exactly as described in Chapter 6. For example, if MTM-3 variance from MTM-1 at the 95 percent confidence level and for a cycle length of 20,000 TMU is desired, first determine the total system variance for each system:

MTM-1 variance is ± 1.5 percent
MTM-3 variance is ± 4.8 percent

In checking these "readings" on the graphs, remember that the coordinates or divisions are log-log. The difference in the above is ±3.3 percent and is the answer desired.

Figures 7–1A and 7–1B present data on system accuracy for MTM-1, MTM-2, and MTM-3 at both the 95 percent and the 50 percent confidence levels—more data than were presented in Chapter 6. Moreover, the approximate analysis time is indicated for cycles of 100, 200, 5000, 10,000, 20,000, and 50,000 TMU by the numbers appearing above and to the right of the intersection of the abscissa marker lines and the lines representing MTM-1, MTM-2, and MTM-3 overall accuracy. For example, at 1000 TMU, an MTM-1 analyst will spend 3.5 hours developing the analysis vs the MTM-3 analyst who would only spend 0.5 hour.

As a result of the above one can employ these data to select the system to use by defining total permissible system variation at the confidence level of ei-

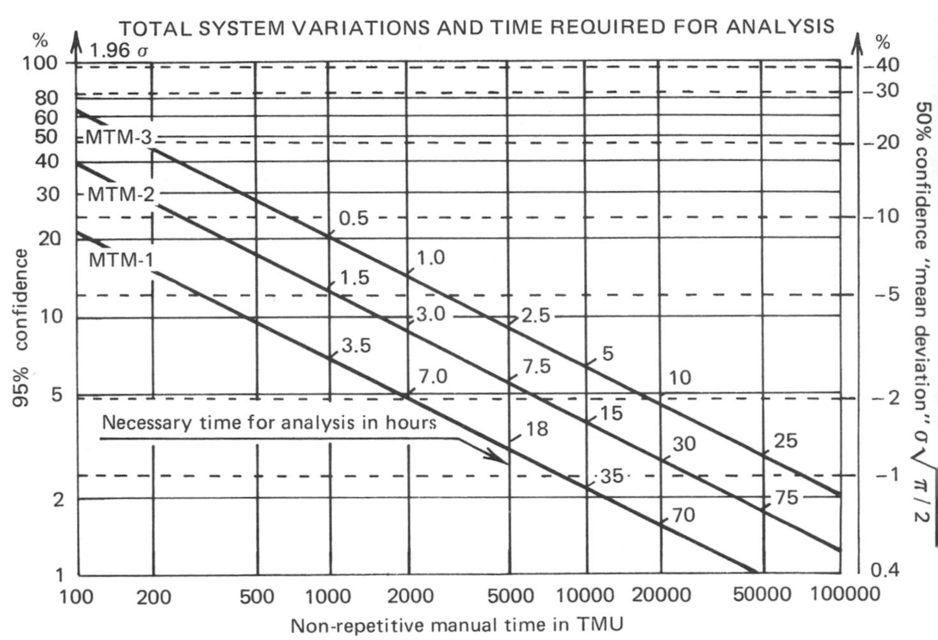

Figure 7–1A. The MTM-1 curve shows the total system variance (applicator + system error). The MTM-2 and MTM-3 curves show the sum of the system variations from MTM-1 and the random applicator variations. © MTM Association for Standards & Research, Fair Lawn, N.J.

ther 95 percent or 50 percent, and the analysis time allowable. Neglecting analysis time, one can draw a horizontal line showing the total permissible system variation and then erect a vertical line at the cycle time; under some conditions this might be average cycle time. Any system line which passes beneath the intersection can fulfill the demand on precision—and the closest one that is beneath the precision limit line will be the fastest to use.

The above system selection method ignores every other factor involved in a work study, for example, methods description, the degree of methods analysis possible, etc. The methods analysis might be of even greater importance than either of the system selection parameters (like speed of application); such a case would be likely in the event of long runs. Training and experience in application cannot be ignored, the penalty for application error is too great.

Elements and Limitations of MTM-3

There are *only* four elements in MTM-3.

H	HANDLE
T	TRANSPORT
SF	STEP
B	BEND AND ARISE

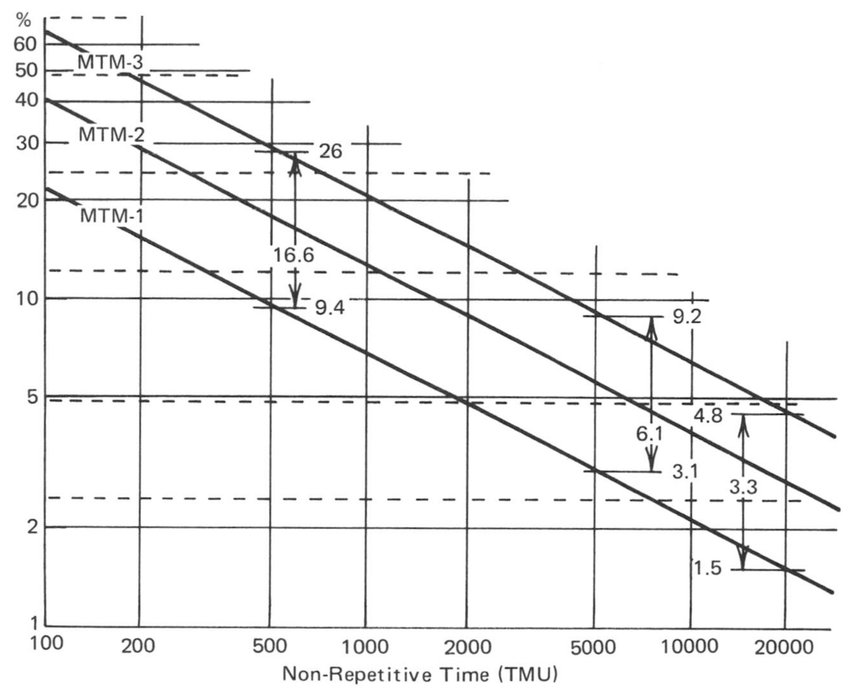

Figure 7–1B. A portion of the curve shown in Fig. 7–1A is reproduced to illustrate the calculations referred to. © MTM Association for Standards & Research, Fair Lawn, NJ.

The distance *ranges* for H and T are only two:

Up to 6 inches
Over 6 inches

The entire data card is shown in Figure 7–2.

Note that there are no elements for eye motions, apply pressure, and regrasp, and no adjustment for the effect of moving weighted objects. These are all proportionally included (based upon frequency of occurrence) in the time values for H and T.

It should be obvious from all that has been said that MTM-3 should not be applied to such work as the following:

Highly repetitive manual actions—do not use it if the frequency is above 10. However, the "rule of 10" does not apply to STEP.

Do not use where high sequences of eye motions are involved. Remember, there is no eye motion category to use in such cases, use MTM-2.

Do not use it where high accuracy is required.

If repetitive manual sequences exceed 10 for any MTM-3 element but STEP, use MTM-2 or MTM-1, whichever is the logical choice for your situation.

MTM-3

RANGE	CODE	HA	HB	TA	TB
Up to 6"	-6	18	34	7	21
Over 6"	-32	34	48	16	29
		SF	18	B	61

MTM ASSOCIATION
16-01 BROADWAY
FAIR LAWN, N.J. 07410

Figure 7–2. MTM-3 data card used by the U.S./Canada MTM Association. © MTM Association for Standards & Research, Fair Lawn, NJ.

HANDLE—H

Definition[4]:

HANDLE is a motion sequence with the purpose of getting control over an object with the hand or fingers and placing the object in a new location.

That the above definition is not as complete in itself as similar element definitions in MTM-1 will quickly become apparent. Even the starting point is not defined in the above definition since an all inclusive adjective starts the definition, handle. The element actually starts when the hands or fingers start reaching toward an object. Reach is a word with a more specific meaning; however, it could not be used as a designator of the element since other actions had to be included (some only on a statistically proportional basis). This is the result of aggregation.

HANDLE rather obviously had to include all of the necessary motions to also grasp and gain manual control over the object, move it in one direction, position it if this is required, and release the object. Weight of the object is not to be considered in MTM-3, since an allowance has been included in HANDLE for its effect, as previously mentioned.

The system design does not permit defining the end point exclusively as a release of the object, although this is a common ending. HANDLE actually ends when any one of the following occurs:

The object is released.
The transport of the object is intentionally stopped.
There is an abrupt change in direction in the motion.
The hand, if moving in a circular path (like cranking), has traveled one revolution.
An object has been inserted 1 inch (as in position).

[4]Courtesy of U.S./Canada MTM Association. See footnote 5, Chapter 6.

The sequence of motions need not be a continuous series. For example, the right hand might start a HANDLE, then stop moving while the left hand performs a TRANSPORT, and then resume motion to continue on to complete the HANDLE.

It should be obvious that the object often is not released at the end of a HANDLE. Any subsequent movements of the object are TRANSPORT motions.

If the beginning portion of HANDLE, such as a grasp and gain control sequence, occurs while the time is being limited out by an event such as machine process time or by a simo motion of the other hand, the actual remaining motion should be classed or analyzed as TRANSPORT. All HANDLES (from a sequence view) are therefore not classified as HANDLE.

Grasping an object (after reaching) and pressing firmly (applying pressure) is analyzed as a HANDLE—the application of force is viewed as a motion. In each subsequent case of application of pressure, the proper rule to follow is to assign a TA6. In fact, assign a TA6 each time the pressure is applied in a new direction and/or for each new application of pressure after the first. More on this is given under the section "Rules of Thumb for HANDLE and TRANSPORT."

HANDLE and the soon to be described TRANSPORT include the time required to position an object and insert it 1 inch (the time for this action is on a statistically based prorated basis). If the insertion is more than 1 inch, assign a TRANSPORT to the additional motion.

The object being handled may be released after HANDLE, but not necessarily. If additional movements are required, usually in another direction except for crank, such movements should be analyzed as TRANSPORT motions.

TRANSPORT—T

Definition[5]:

TRANSPORT is a motion with the purpose of placing an object in a new location with hand or fingers.

The element starts when the hand, having previously secured control of an object through an action such as HANDLE, starts moving it.

The element includes all actions necessary to move the object in *one direction,* and if required by the work cycle, to position the object and/or release it.

TRANSPORT can end, like HANDLE, in a variety of actions. These are as follows:

The object is released.
The transport of the object is intentionally stopped.
There is an abrupt change in direction in movement of the object.
The object was transported to the other hand.

[5] *Ibid.*

The hand, like in HANDLE, if moving in a circular path, has traveled one revolution (cranking).
The object has been inserted 1 inch (in a position).

Since TRANSPORT starts with the hand having "control of the object," there must have been a preceding HANDLE or TRANSPORT. It is not unusual to have several TRANSPORT motions occur in sequence. Therefore, the analyst must be careful to assign the proper number and class (later described) of TRANSPORT motions—the clues helping in this determination are the end point specifications given above.

As in HANDLE, if TRANSPORT ends in a position, and the object insertion exceeds 1 inch, the analyst should assign another TRANSPORT to cover the additional insertion distance.

Case of HANDLE and TRANSPORT

As in most MTM systems, there are different cases of some of the elements. In MTM-3 there are two cases for HANDLE and two cases of TRANSPORT, Case A and Case B—they are the same for each element.

The difference between Case A and Case B is simple in concept, but sometimes difficult to see. The concept, as stated above, applies to both HANDLE and TRANSPORT. The rule is that the element is a Case A if there are no corrective motions during *the movement of the object*. If there are such motions, it is a Case B. A good understanding of the underlying MTM-1 motions will aid in the detection of such corrective motions during the *movement of the object*.

More specifically, the corrective motions are caused by the precision demands involved in *placing* the object. These demands, if high enough, will cause such actions as unintentional stops, hesitations, or changes in direction *near the destination*. The stops are really a "slowing down." Such corrective actions as defined in MTM-3 only occur when the object is actually being transported (moved). Actions necessary to gain control over the object do not affect the case of HANDLE or TRANSPORT.

A "slowing down" or hesitation will be hard to detect if the operator is working at a very low motion speed. In fact, analyses should only be done by observing a trained operator working with approximately average skill and effort—not a super operator or not an untrained or not properly motivated operator. For these reasons follow the old time study rule: If you write the MTM-3 pattern by observing an operator, only observe a fully trained operator working somewhat close to an average pace (working with average skill and effort).

Motion Distance and Coding for HANDLE and TRANSPORT

The earlier statements regarding distance and the initial examination of the data card should have made clear that there are only two distance categories in

MTM-3 for HANDLE and TRANSPORT—up to and including 6 inches (coded −6) and distances over 6 inches (coded −32). The metric categories are 15 and 80 centimeters—coded −15 and −80.

Total coding is simple. The first position is either an H or T. The second position identifies the case of the element, either A or B. The third position is the distance category, 6 or 32 in the USA/Canada. HA32 is a Case A HANDLE at the 32 distance category (over 6 inches).

A word of *caution*—the distance *reached* in HANDLE does not affect the distance classification, only the distance the object is moved is used in deciding which distance category to use. Obviously, estimation is the normal approach to deciding what category to use, not actual measurement—except in rare cases.

Rules of Thumb for HANDLE and TRANSPORT

The following discussions cover commonly encountered work situations that are not directly addressed in the MTM-3 system. However, if the analyst utilizes the "rules of thumb" to handle the discussed motions, the system design concepts will hold.

Hammering Motions

Hammering motions are often encountered by the work analyst. It is recommended that these normally be analyzed as TA32s.

Application of Pressure

This is only partially a rule of thumb. However, it seemed that this would be a good place to discuss it. Application of pressure is a specialized action, really a sort of nonmotion. MTM-1 has an element to cover the case in detail, and MTM-2 has a simplified APPLY PRESSURE element. Application of pressure is needed, for example, in the "final tighten" of a nut or when a shaft is being inserted in a socket and it "sticks."

The decision diagram in Fig. 7–3 virtually says it all concerning this action.

Note that the first Apply Pressure is included in the HANDLE or TRANSPORT. Any additional Apply Pressures should be analyzed as a separate TRANSPORT—almost always as a TA6. Apply Pressures are signaled, in part, by pauses in the operator's actions.

Crank

This is a specialized motion covered in MTM-1 and MTM-2, but not directly in MTM-3. Hence, it must be considered by applying a "rule of thumb."

If the hand reaches to a crank and turns it five times, the first revolution is a HANDLE, successive continuing revolutions are analyzed as TRANSPORTs (four in our example). Intermittent cranks (stops between revolutions) are analyzed as HANDLEs.

Winding a string or small wire around an object should normally be analyzed as a crank; the motion does not have to be a perfect circle.

The distance class is determined in a normal manner.

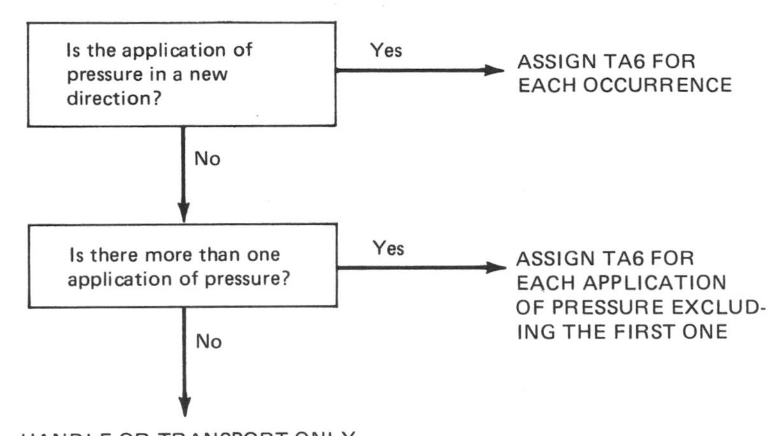

Figure 7–3. Decision model for application of pressure. © MTM Association for Standards & Research, Fair Lawn, NJ.

Disengage

Disengage is caused by the "sticking" of objects being separated. Examples could be the removal of a shaft from a tight-fitting sleeve bearing or from a socket, or the removal of a *tight-fitting* wrench from a nut (usually a nut having "faces separated by more than ½ inch"). Such disengages are signaled by a "recoil" tendency of the hand holding the disengaged object. This condition is illustrated in Figs. 7–4 and 7–5, which are taken from the often referenced *Engineered Work Measurement.*

MTM-1 has a special element category to cover Disengage, but MTM-3 does not. When using MTM-3, the following rules are to be used in handling such occurrences.

If there are no obstructions to the movement of the hand or of the disengaged object *and* if the *recoil* is in the general direction of the object being disengaged, no special considerations beyond the normal HANDLE are involved. The previously referenced figures will aid in the understanding of this discussion of Disengage.

If there are obstructions requiring care and quick control (requiring a change in direction of the movement from the recoil direction) *or* if the place the object is to be disposed requires a change of direction, the analyst is to assign a TA6 (or a TA32, depending upon the distance the object is to be moved beyond the end of the recoil), in addition to the usually required TA32 or HA32 (32 because of the distance traveled until the hand has control). In the case of the wrench, the disengage would normally be a TA6, since the recoil likely would be less than 6 inches, and the movement a TRANSPORT, since the hand already had control of the object. Also, the *need* for a change in direction is assumed. If in the case of the shaft removal the hand reached to the shaft, grasped it, and removed it, the motion would be a HANDLE, and if the distance moved was over 6 inches and no corrective motions were needed at the end of the transportation, the action would be classed as an HA32.

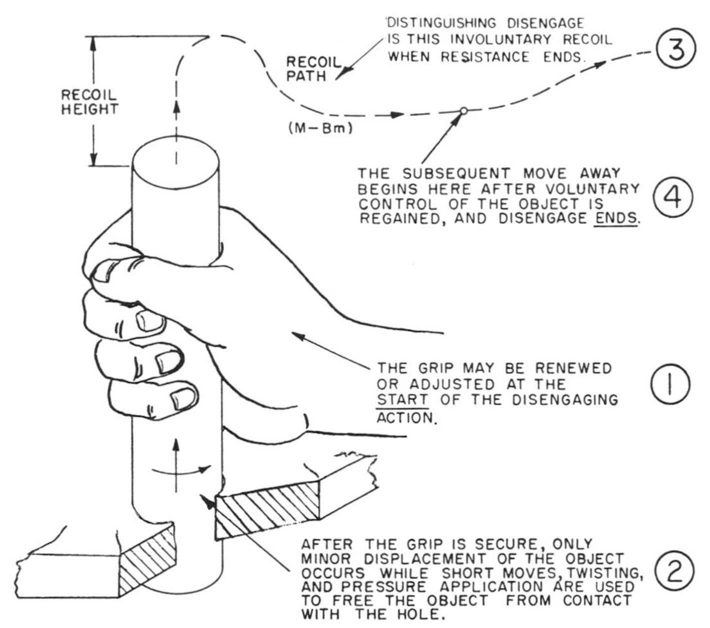

Figure 7–4. Nature of Disengage. Since recoil is an involuntary motion, the recoil path is rather indefinite and unpredictable but of the general nature here shown. Reprinted from Engineered Work Measurement, *p. 519.*

Tossing

The tossing of an object is frequently observed in industrial operations. For example, an operator reaches 20 inches with his right hand and grasps one completed part assembly. He then carries it 5 inches in a tossing motion and releases it. The object travels another 15 inches to a box. This action should be coded as an HA6; the part while in control of the hand only traveled 5 inches and there obviously were no correcting motions at the end.

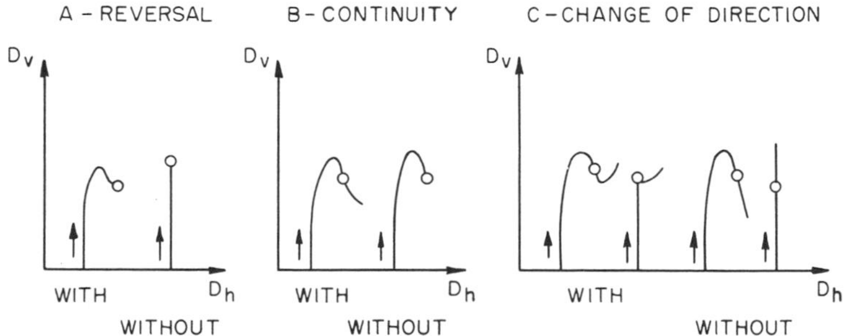

Figure 7–5. Path of recoil. D_v is the height of recoil, D_h is the lateral recoil travel. Reprinted from Engineered Work Measurement, *p. 522.*

Transfer Grasp

This again involves a case covered by a separate element of MTM-1. It concerns moving an object in one hand to the other hand. The original movement would be a HANDLE (true for most cases) if the object was not previously "in control" of the hand involved. Analysis of such a transfer to the other hand shows that the object is "in control" of the *destination* hand before the first hand releases the object, hence the movement of the object with the second hand is a TRANSPORT.

A special case of transfer grasp not covered in the above example would be where, for example, the left hand is holding a wrench (or some object) in a given fixed position and the operator next moves his/her right hand (in this example) to the wrench and holds it while the now free left hand moves aside to accomplish another task, like a HANDLE which the analyst would code. In MTM-3 there is no provision for such events as "no motion of an object has occurred." If the case is infrequent, ignore it. However, if it occurs several times in an operation, analyze it using MTM-2.

The Impact of Operation Frequency

The reader will recognize that many of the above "rules of thumb" were practical and usable compromises, in fact, MTM-3 is a compromise made in order to achieve certain objectives—speed of analysis at some loss of accuracy and precision. Therefore, when an operation is encounted that will be repeated thousands of times (like hammering a nail in construction work), it would be prudent to use MTM-2 or MTM-1.

STEP—SF

Definition[6]:

STEP is the leg motion with the purpose of moving the foot or leg in one direction.

STEP starts with the leg at rest (stopped) and ends with the leg or foot stopped in a new location. The actual possible kinds of motions included in the category are as follows:

A motion of the leg in one direction in order to move the body.
A foot motion where the foot is pivoted around the heel or instep or toe (in this latter case the toe position is a constant, but the heel changes position).
A leg motion pivoted at the knee and/or hip.

STEP, as indicated above, includes the leg actions such as sidestep as well as a step forward or backward. The category also includes the leg motions involved in body turns. Finally, it includes the leg motions not involved in moving the body.

When STEP covers walking, count the number of times the feet hit the floor in order to determine the number of step motions.

[6] *Ibid.*

BEND AND ARISE—B

Definition[7]:

BEND AND ARISE is a lowering of the trunk followed by an arise.

It involves the movement of the trunk and/or other body members to achieve a vertical change in body position so as to *permit* the hands to reach down to or below the level of the knees , and the motion is completed by an arise from this position.

The criterion is whether the operator is able to reach down to or below the level of the knees because of the new body position, not whether the operator actually does reach that low. Often work requires the shoulders to be low in relation to the knees because of an obstruction, such as a bench top, with the object to be handled being on a shelf below the bench top.

Usually B starts with a forward motion of the trunk (a "pure" bend) and ends with the trunk and body again in an upright position.

A "rule of thumb" is used to handle the situation involving the kneeling on both knees and arise. This action sequence should be analyzed as 2B, the 2 designating motion or element frequency as explained in Chapter 6.

Another Motion Problem

There is no eye action element in MTM-3. Therefore, the analyst should resort to using MTM-2 for such work. If eye actions occur rarely, they can be ignored.

Simultaneous Motions and Motion Combinations

Simultaneous Handle and Transport

Simultaneous motions are naturally and rationally made in order to simplify manual operations and thereby reduce the time required for performing them. MTM-3 rules for properly accounting for such motions do not permit the "fine tuning" of the decision on these as do the rules for MTM-1, or even those of MTM-2. The reason is simple, each motion category in MTM-3 contains or covers two to several MTM-1 motions.

In spite of the above-mentioned handicap there are available decision rules for MTM-3 motion combinations and/or simultaneous motions. These rules almost always produce satisfactory results, but sometimes some exceptions appear to have occurred. If the analyst knows MTM-1 and/or MTM-2, such situations can usually be resolved by using one of these systems for that portion of the work—but be sure to take into account properly the "end points" of the involved MTM elements.

[7] *Ibid.*

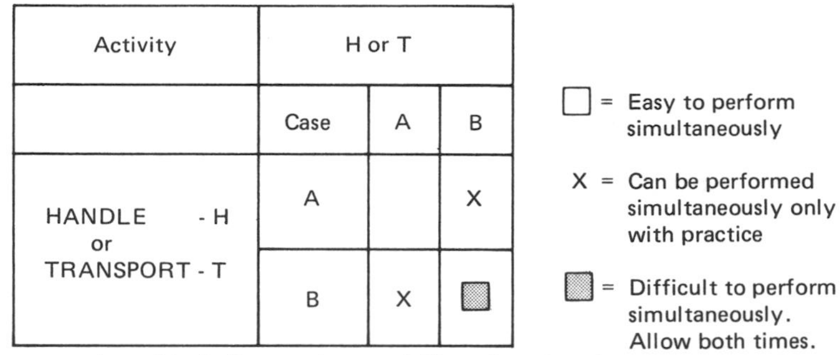

Activity	H or T		
	Case	A	B
HANDLE - H or TRANSPORT - T	A		X
	B	X	▩

Figure 7–6. The table indicates the possibility of performing HANDLE and/or TRANSPORT simultaneously. © MTM Association for Standards & Research, Fair Lawn, NJ.

The most important simultaneous motions, from a frequency of occurrence and total time view, are hand and arm motions, which in the case of MTM-3 are HANDLE and TRANSPORT. A decision table for these is presented in Fig. 7–6. Remember that the table usage and manner of indicating the limited and limiting motions is in general accordance with those given for MTM-2.

In the situation considering HANDLE WITH HANDLE, both Case B, where the objects cannot be placed simo, but can be grasped simo (MTM-1 rules help in making such decisions), the motion pattern below illustrates how to code the action and also accounts for a partial completion of the limited Case B HANDLE. First, let us assume two equal length HB6s.

LEFT	TMU	RIGHT
HB6	34	(H–)
	21	TB6

There will be some accomplishment possible in the right hand, but it, at best, will only be a partial accomplishment of the motion. Owing to the possible overlap, the H– is shown.

The reverse action could also occur:

LEFT	TMU	RIGHT
(HB–)	34	HB6
TB6	21	

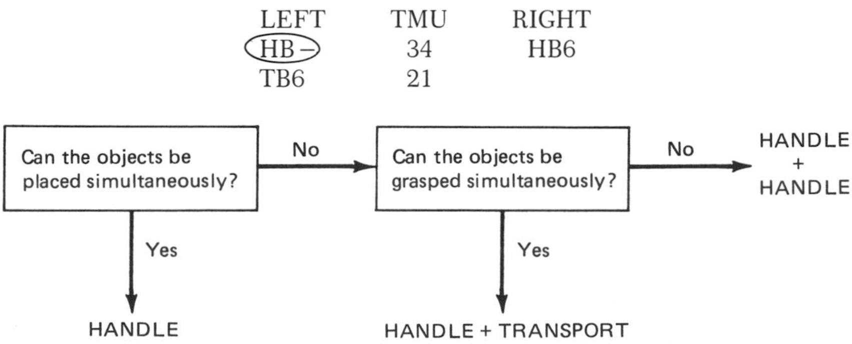

Figure 7–7. Decision model for simultaneous HANDLE where body motions are not involved. © MTM Association for Standards & Research, Fair Lawn, NJ.

If there is a required possible combination of an HB32 with an HB6 or HB32, this would be properly handled as follows:

LEFT	TMU	RIGHT
HB32	48	H–
	21	TB6

Two comments should be made at this point. The reader should again examine Fig. 7–7. Second, the TB6 is needed since only a partial HB6 could be completed; hence, the TB6 was required for completion of the action required. The TB6 is recorded *under the hand making the shorter motion* regardless of the cases. When distances are equal, but cases are different, the TA6 overlap is recorded under the limited motion.

If the involved objects cannot be grasped or placed simo, we obviously must allow separate HB6s if the distance is 6 inches. If we have a left-handed HB32 and a right-handed HB6, allow each as a separate motion—this latter case would be symbolized as follows (the reverse or opposite motion combination obviously requires the opposite symbolization):

LEFT	TMU	RIGHT
HB32	48	H–
	34	HB6

When there is an HB32 with an HA32, the allowed HB32 does not entirely limit out the limited HA32; since it can only be partially completed, there will be a motion overlap of HA6 (assume the HB32 is in the left hand):

LEFT	TMU	RIGHT
HB32	48	H–
	18	HA6

Note that when an HB32 and HB6 are performed simo, a data card value of 48 is found for both. In this case the overlap of HB32 is chosen since it provides the greater number of TMUs.

HANDLE and TRANSPORT in Combination with STEP and/or BEND

This case must be addressed since HANDLE and/or TRANSPORT usually precede and follow STEP and/or BEND. The control characteristics of the concerned HANDLE or TRANSPORT motions will determine whether they can be performed simo with the body motion—as is true for any desired simo combination. It cannot be emphasized too much that experience and training with MTM-3 and/or a good knowledge of MTM-1 will greatly help clarify questionable cases.

With MTM-3 the following rules usually yield acceptable results.

HA or TA performed with *one hand* (not simo or with a two-handed operation) will rarely be time limiting since there are no corrective motions required (control requirements are low); hence, it can usually be performed simo with STEP or BEND. The arm–hand motion will therefore be limited out.

Sometimes workplace obstructions prevent starting HA or TA at the proper time; in these cases they must be accounted for separately in the MTM-3 motion pattern. In the case of a limited one-handed HA or TA, it *can* be recorded as shown below (T would be used to show a TRANSPORT):

B ⎤
~~HA~~- ⎦

In practice the limited motion is usually not shown, only the B or SF is recorded.

HA and TA performed with both hands (simo with each other, two objects involved) will usually cause a time-limiting situation. However, only one Case A motion will be time limiting and in all normal cases the distance class 6 will be the maximum chargeable distance.

HB and TB cannot be performed unless the body is balanced and are analyzed in this situation as time limiting. Therefore, one time limiting HB or TB is allowed for each hand. If only one hand and one object is involved, the motion pattern would show an HB6 or TB6 in the proper column in addition to the B or SF.

If there are two objects and therefore two required motions and one is Case A and the other Case B (requiring corrections), allow each (show in proper right and left columns on the analysis sheet).

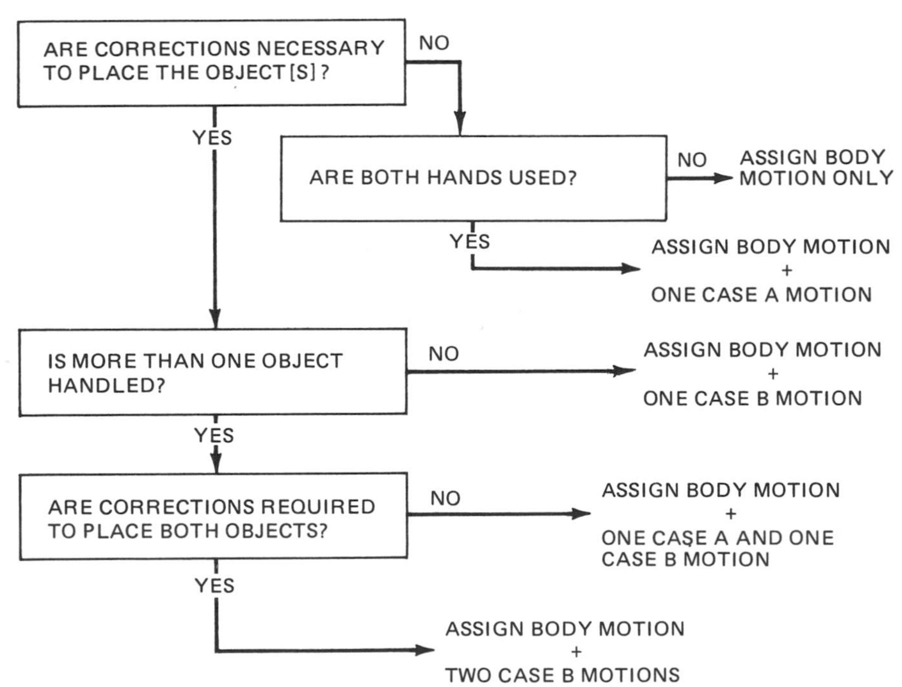

Figure 7–8. Decision model for HANDLE and TRANSPORT in conjunction with body motions. © MTM Association for Standards & Research, Fair Lawn, NJ.

If a limited Case A motion is involved, the overlap motion distance class is usually 6 because a Case A H or T can be made so as to reduce the remaining time limiting H or T to class 6.

If one object is to be handled with *both* hands, one HB or TB should be assigned.

The above rules are summarized in the decision model presented in Fig. 7–8.

Concluding Remarks

Again, remember, the way of writing patterns, the method of indicating limiting and limited simo motions, etc., are as indicated for MTM-2 in Chapter 6.

Furthermore, the same cautions about trying to apply MTM-3 without proper approved training apply. Do NOT attempt it, you are almost sure to make errors—especially if you are not trained in MTM-1 and MTM-2. Approved training courses are available for the public in North America from certain of the consulting members of the U.S./Canada Association and from the Association itself. Approved training is available, almost around the world, from consulting members and from or through the various National MTM Associations which constitute the membership of the International Directorate (explained partially in Chapter 6 and completely in *Engineered Work Measurement*, 3rd ed.).

CHAPTER 8

MTM-C: A High-Level and Fast to Use MTM System for Setting Clerical Standards—A Clerical Standard Data System for Routine Clerical Work[1]

Over one-half the nation's wage bill is spent in the services sector—and much of the increase has been in clerical work and office employment. This latter fact was caused by the organizational need for more information, the great increase in government-required documentation and data analysis, and the increasing size and complexity of business and governmental organizations.

The low productivity of office workers is legend, and often this efficiency is in the 25–50 percent range. The beneficial effects of standards and incentives can raise the productivity of office workers to the normal of the production areas of the factory on repetitive work. Perhaps a few of the related facts on the effect of standards and incentives need to be restated here.[2]

Creating work performance standards, measuring performance against them, and informing the workers of their performance significantly raises productivity, typically to the 60 to 85 percent level. This yields an approximate 50 percent labor cost reduction, neglecting the relatively minor costs of setting the standards and using them. Naturally, the use of incentives has a further beneficial effect, raising productivity to 110 to 120 percent. Also keep in mind that such increased labor efficiency reduces space requirements, number of machines, energy costs, fringe benefit costs, etc.—fringe benefits today are often in the 30+ percent range.

The usual low office worker performance is due to a number of factors. First, supervision does not "see" productivity as being a major problem or opportunity. In fact, it is usually a distasteful subject to both the supervisors and the workers. Supervisors see their role as "getting out the work" and "keeping the employees happy." They believe (and it is often true) that they win "brownie

[1]A higher-level MTM system that is really a specialized standard data system covering commonly encountered clerical work. Training and data manuals are available from the MTM Association for Standards and Research and from the approved professional consulting members of the Association. This exposition will provide the basic essential facts concerning MTM-C, but it is not as complete as the discussions of MTM-2 and MTM-3. The reason for this variation is due to space requirements vs space availability. The MTM-C manual contains all the MTM-1 backup studies, and these are necessary to any practical usage of the system. These backup studies and the size and amount of summarized data card information creates the space requirement that this text cannot accommodate. Again, the discussion of MTM-C and the exhibits are through the courtesy of the MTM Association for Standards and Research, Fair Lawn, NJ.

[2]Complete information on this subject and on the usual construction and use of standard data can be found in D. W. Karger and F. H. Bayha, *Engineered Work Measurement,* 3rd ed., Industrial Press, Inc., New York, 1977.

points" toward a larger salary and/or a promotion if they can increase the number of employees they supervise—rather than decreasing their number. Enlightened and good managements, however, will react in the opposite manner and award the "brownie points" to those supervisors who accomplish the work at a lower cost through better methods, equipment, and the management of their human assets.

It is traditional to emphasize productivity in the shop, whereas it is *not* traditional to do so in the office. There are still other reasons for low office worker productivity. For example, the office workers "see" their job as being above and far different from the factory worker's job. Historically, office workers have not been asked to perform at reasonable levels of productivity.

Worker performance will improve if they know what is expected of them and if they receive frequent feedback on their performance.

Clerical Standards

Clerical standards are like any other work standard, they can be based upon:

History
Estimates
Work Sampling
Time Study
Predetermined Time System (MTM)
Standard Data

Historical standards and estimates are only useful for estimating purposes. Such standards represent the "status quo," and their use will *not* improve performance. As to the use of historical standards for estimating, they only predict loading based on the ordinary and very substandard performance.

Work sampling can play an important role in determining kinds of work and activities involved and the amount of time spent on each category, but it does not usually play an important role in setting standards.

Time study can be and is often used in setting clerical standards, but it has handicaps such as (1) analyst time required tends to be excessive for statistically valid studies; (2) pace rating is required and it is difficult under the best of conditions and almost impossible when the observed workers are performing at 25–50 percent of standard; and (3) time study has a very bad image in the minds of office workers.

Predetermined time systems can and are being used in setting office standards, but analyst time will be excessive unless such standards are used to establish standard data. This is the motivation that produced MTM-C.

MTM-C *is* a set of specialized standard data based upon MTM-1 motion patterns. Process time, like typewriter carriage travel (or "baseball" element travel), was determined with timing devices. Its accuracy, when properly applied, is comparable to that of MTM-1. However, its speed of application is significantly greater.

For those readers who are not already familiar with the structure, coding, and usage of standard data it is suggested that they read Chapter 27 of *Engineered Work Measurement.*[3] Moreover, if the reader is not familiar with the proper procedure to follow when studying a worker so as to produce a standard, it is suggested that they read Chapter 5 of the referenced text.

General Facts and Background

MTM-C is Clerical Standard Data designed to be used for the establishment of time standards for clerical-related work tasks. Typical motion patterns appropriate for a clerical environment are synthesized from MTM-1 motions. Both Level 1 and Level 2 (there are two levels to this system) data of MTM-C are higher-level systems of MTM.

MTM-C was developed by a consortium of the U.S./Canada MTM Association members and the staff of the Association. The following design objectives were set and met:

1. Applicable in a clerical environment
2. Coded for ease of understanding
3. Faster to apply than MTM-1
4. Coding is descriptive of the operation
5. Fully documented and traceable to MTM-1
6. Compatible with other MTM systems
7. Relative to MTM-1 in accuracy and speed of application so that specific comparisons with MTM-1 are possible

The consortium members included banks, insurance companies, railroads, government agencies, manufacturers, and consultants—even a city government was a member. The following firms and organizations responded to the Association's invitation to the membership to actively participate:

American Fletcher National Bank, Indianapolis, IN
Bureau of Indian Affairs, Gallup, NM
California First Bank, San Diego, CA
Canadian National Railway, Montreal, Quebec
City of Los Angeles, Los Angeles, CA
Crocker National Bank, San Francisco, CA
First Pennsylvania Bank, Philadelphia, PA
New England Merchants National Bank, Boston, MA
Northwestern National Life, Minneapolis, MN
Rath & Strong, Inc., Lexington, MA
Republic National Bank, Dallas, TX
Stevenson & Kellogg, Ltd., Toronto, Ontario
Thom McAn, Worcester, MA

The above represented a wide diversity of organizations, clerical activities, and geographical distribution.

[3] D.W. Karger and F.H. Bayha, *Engineered Work Measurement,* 3rd ed., Industrial Press, Inc., New York, 1977.

The consortium divided the work areas covered and exchanged information and results. The major activity areas were:

Desk top activities
Filing and nonkeyboard equipment
Keyboard equipment
Coding and computer applications
Reading and writing

Each consortium member accepted assignments unique to his/her operations. The involved MTM-1 practitioners followed proper rules and recorded the work in a standardized format. The studies were exchanged among firms and groups for critique and comment. A final edit of all data used was conducted by the MTM Association staff and was also reviewed by the Training and Qualifications Committee of the U.S./Canada Association. All elements were tested to ensure that they were discrete as well as conforming to application rules.

The consortium validated the data (1) by recoding and comparing available MTM-1 clerical data (from Association and member files) to MTM-C and (2) by observing and developing standards using *both* MTM-1 and MTM-C for clerical operations of 500 to 1000 TMU and then comparing the two standards.

A total of 7366 TMU have been analyzed for accuracy validation at the time of this writing. The results indicate that MTM-C Level 1 approximates the accuracy of MTM-1 within ± 5 percent at cycle times of 1200 TMU (about 43 seconds) with 95 percent confidence. This TMU value is called the Relative Balance Time. The Relative Balance Time of MTM-C Level 2 is 4900 TMU (about 3 minutes). These figures may be slightly revised when more data are available.

The actual speed of application at the time of this writing has not yet been fully determined. It is expected that research will show it to be about the same as MTM-2 vs MTM-1. Research on the GET/PLACE category indicated the following specifics:

MTM-C1 (Level 1)	16:7.9 or twice as fast as MTM-1. Restating this in cycle time length, the above can be converted to 175 TMU/TMU when using 350 TMU/TMU for MTM-1.
MTM-C2 (Level2)	Three times as fast as MTM-1 (ratio 24:7.9) Again in cycle time length, this would be 120 TMU/TMU.

Obviously, MTM-C2 was designed to be faster than Level C1, and it has a known accuracy.

Coding of Level 2 is alphanumeric vs numeric for Level 1. Level 2 is directly traceable to Level 1 elements. It also combines GET and ASIDE (Level 1 elements) as well as using one distance range.

As explained in footnote 1, the MTM-1 backup motion patterns are often required when a somewhat different operation is encountered. Hence, in order to use this system in a practical and usable manner a substantial degree of knowledge of MTM-1 is required. As a result, approved MTM-C training courses for those not having such knowledge include an expanded MTM-1 appreciation course (an abbreviated course in MTM-1 fundamentals). For the reader not familiar with MTM-1, he/she can secure the knowledge from the related text *Engineered Work Measurement*.[4]

MTM-C1 (Level 1)

MTM-C1 has nine categories of data.

1. GET & PLACE
2. OPEN & CLOSE
3. FASTEN & UNFASTEN
4. ORGANIZE & FILE
5. READ & WRITE
6. TYPING
7. HANDLING
8. WALK & BODY MOTIONS
9. MACHINES

Distance ranges in MTM-C1 are three and resulted from an analysis of and experimentation with over 4000 data points for REACH and over 5000 for MOVE. Ninety percent of the data points fell in the 0–16-inch range. The breakpoints finally established after analysis and experimentation were:

1. 0–6″ 0 up to and including 6 inches
2. >6″–16″ Over 6 inches and up to and including 16 inches
3. >16″ Over 16 inches

The actual REACH and MOVE distance associated with each category are the following:

REACH		MOVE	
(0–6″)	= 3.35″	(0–6″)	= 3.01″
(>6″–16″)	= 10.68″	(>6″–16″)	= 10.79″
(>16″)	= 21″	(>16″)	= 22.10″

Naturally, the appropriate case of REACH and/or MOVE was used in pattern determination, but the distances used were the above.

Coding System, Level 1

A six-place numeric system was developed which has the following format:

	PURPOSE	OBJECT	AID/CONTROL
Position	1 2 3	4 5	6
Actual Notation	x x x	x x	x

[4]*Ibid.*

In general, the following statements hold:

Level 1 is completely coded in numerals.

The first place represents one of the nine previously identified categories.

The second position indicates the action being performed.

The third place denotes the appropriate distance category (1, 2, or 3). A "4" indicates additional MOVE requirements and a "0" indicates that a distance range is not applicable.

The fourth position identifies the actual object involved, "1" through "9" are used for exact objects encountered in clerical work. The "0" indicates any object or that an object category is not applicable.

The fifth position is used to further define the object. 1, 2, and 3 are used to represent small, medium, and large. A "0" indicates any size or that size is not applicable. Other specific code numbers for other category descriptors are identified for this position in the data tables.

The sixth position is used to indicate whether an aid or control is involved; details are found in the data tables. A "0" here indicates that the position category is not applicable.

Figures 8–1A, 8–1B, and 8–1C identify MTM-C1 coding categories and numeric identification.

Space available does not permit completely covering category descriptions, data cards or tables, backup data, etc., as explained earlier. Since the GET/PLACE action covered in MTM-C1 was referenced earlier with regard to speed of application, this data table is illustrated in Fig. 8–2. Figure 8–3 presents the back-up MTM-1 material (including TMU time data) for "GET ONLY—SMALL STACK" action of the GET/PLACE data.

From the title of Fig. 8–3 it is obvious that terminology such as "small stack" needs to be defined. How this is accomplished in the training manual is shown in Fig. 8–4.

The GET/PLACE coding fully utilizes the six-place system as previously described. A few examples probably provide the easiest way to convey most of the information.

1 1 2 2 1 0

GET/PLACE — 1

Get only — 1

Medium reach — 2

A stack — 2

Small in size — 1

Not applicable or needed for this task — 0

1		2		3		4		5		6	
CATEGORY		PURPOSE				OBJECT				AID/CONTROL	
1	GET/PLACE	1	Get	1	Short	1	Sheet	1	Small	1	Other hand
		2	Aside	2	Medium	2	Stack	2	Medium	2	Approximate
		3	Move	3	Long	3	Bundle	3	Large	3	Exact
				4	Addition'l	4	Pile				
						5	Single object				
						6	Jumbled object				
						7	Pad				
						8	Card				
						9	Folder	9	Contact		
2	OPEN/	1	Open	1	Short	1	Hinged cover	1	Small	1	Turn handle
	CLOSE	2	Close	2	Medium	2	Door/drawer	2	Medium	2	Turn knob
				3	Long	3	Space saver file	3	Large	3	Spring catch
						4	Binder ring			4	Latch
						5	Pouch			5	String/rope
						6	Envelope			6	Zipper
						7	Stoppered obj.			7	Ltr op'nr/knife
						8	Threaded object			8	Snap
										9	Velcro
3	FASTEN/	1	Fasten	1	Short	1	Paper clip	1	Small		
	UNFASTEN	2	Unfasten	2	Medium	2	Two-prong clip	2	Medium	2	Approximate
				3	Long	3	Clamp clip	3	Large	3	Exact
				4	Add'l.	4	Acco fastener				
						5	Staple			5	Staple remover
						6	Rubber band loop				
						7	Pin				
						8	String				
4	ORGANIZE/	1	Jostle	1	Short	1	Sheet	1	Small	1	Loose
	FILE	2	Intersort	2	Medium	2	Stack	2	Medium	2	Approximate
		3	Punch holes	3	Long	3	Bundle		-Approx.	3	Tight
		4	Identify			4	Pile		-2 holes	4	Binder rings
		5	Locate					3	Large	5	Folding
		6	Insert			6	Word(s)		-Exact	6	Envelope
		7	Tilt			7	Digit(s)		- $\geqslant$3 holes	7	Pad
		8	Remove			8	Card(s)				
		9	Replace Tilted			9	Folder				
5	READ/	1	Read	1	Short	1	Digit(s)	1	Lower Case/Small	1	Aloud
	WRITE	2	Write	2	Medium	2	Letter			2	Longhand
		3	Move eyes per inch	3	Long	3	Word(s)	2	Upper Case/Med.	3	Print
						4	Punctuation				
		4	Move to next word			5	Signs/symbols	3	Large		
						6	Draw line				

Figure 8–1 (A–C). Identifies the Level 1 coding matrices and is shown through the courtesy of the U.S./Canada Association as is true for all specific MTM system material.

1	2	3	4	5	6
CAT.	PURPOSE		OBJECT		AID/CONT.
6 Typing	1 Paper handling	1 Assemble	1 Sheet(s)	1 Small Distance	
		2 Disassemble	2 Carbon sheets	2 Medium Distance	
		3 Insert		3 Large Distance	
		4 Remove			
			5 Single object		5 After scan
		6 Straighten			
		7 Lift & scan			
	2 Key-stroke	1 Set	1 Lower case	1 Small Distance	1 Manual
		2 Release/clear	2 Upper case		2 Electric
			3 Figure	2 Medium Distance	
		4 Additional	4 Shift key		
		5 Index plates	5 Shift lock	3 Large Distance	
			6 Underline/expand		
			7 Tab(s)		
		8 Carriage return	8 Space bar		
			9 Back space		
	3 Move & locate carriage	1 To margin		1 Small Distance	1 Manual
		2 To exact space			2 Electric
		3 To exact line & space		2 Medium Distance	
				3 Large Distance	
		8 Index			
	4 Process Time	1 Continuous	1 Space		1 Manual
			2 2-space bar		2 Electric
			3 3-space bar		3 Selectric
				4 Non-expanded	
				5 Expanded	
			6 Underline, period		
			7 Tab		
			8 Carriage		
			9 Back space		
	5 Correc-tion	1 Erase	1 Sheet(s)		
		2 Separate			
		3 Insert	3 First error		
			4 Add'l. error		
		8 Prepare Carriage			7 Liq. Paper
					8 Eraser
	6 Misc.	1 Set margin		1 Small Distance	1 Manual
		2 Check right margin		2 Medium Distance	2 Electric
		3 Check left margin		3 Large Distance	
		4 Remove			
		5 Replace			
			6 Typewriter cover		
		7 Open paper release			
		8 Close paper release			
		9 Turn on or off			

Figure 8–1 B

1	2	3	4	5	6
CATEGORY	PURPOSE		OBJECT		AID/CONTROL
7 Handling	1 Fold	1 Short	1 Sheet(s)	1 Small	1 One
	2 Unfold	2 Medium	2 Coin	2 Med. Avg.	2 Approx.
	3 Digit Count	3 Long	3 Currency	3 Large	3 Exact
	4 Tear/cut	4 Add'l.			4 Scissors
	5 Moisten		5 Postage stamp		5 Paper cutter
	6 Adhere		6 Envelope flap		6 Dispenser
	7 Hand stamp/ press		7 Label back		7 Desk sponge
	8 Remove withdraw		8 Tape		8 Tongue
	9 Misc.		9 Impression		9 Machine
8 Walk/ Body motions	1 Walk				1 Stationary chair
	2 Sit				2 Swivel chair
	3 Stand				3 Stool
	4 Bend & arise				4 Obstructed walk
	5 Turn body or side step				
	6 Move chair/ stool				
	7 Move foot				

Figure 8–1 C

A description would be to GET (not PLACE or GET/PLACE) with a medium reach and a small stack:

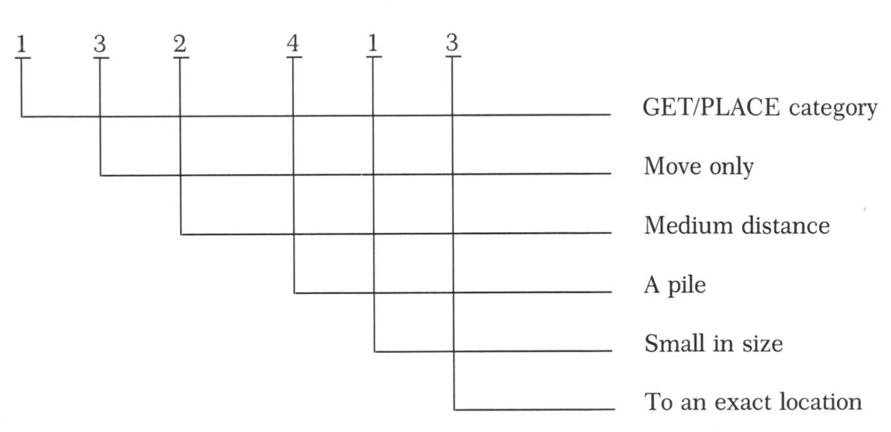

Perhaps it should be said that a "4" in the third position indicates that an *additional* action is taking place, like obtaining an additional card, an additional sheet, an improved grip (REGRASP), weight allowance, etc. Specific coding instructions for such occurrences are found in the data tables (see Fig. 8–2 for an example).

1	GET/PLACE			
CODE	TMU			Element Title
	1-Short	2-Medium	3-Long	
		GET		
11x 090	6	12	19	Any object with contact grasp
11x 100	10	17	24	Sheet - any single
11x 210	14	20	27	Stack - small size
11x 220	18	25	32	- medium size
11x 230	22	28	35	- large size
11x 310	11	17	25	Bundle - small size
11x 320	14	21	28	- medium size
11x 330	22	28	35	- large size
11x 410	16	22	30	Pile - small size
11x 420	26	29	36	- medium size
11x 430	33	39	46	- large size
11x 510	11	17	24	Single object - small size
11x 520	8	14	21	- medium size
11x 530	22	28	35	- large size
11x 610	21	26	33	Jumbled object - small size
11x 620	17	22	30	- medium size
11x 630	15	21	28	- large size
11x 700	11	17	25	Pad - any single (no size)
11x 800	11	17	24	Card - any single
11x 910	11	17	25	Folder - small size
11x 920	14	21	28	- medium size
11x 930	22	28	35	- large size
11x 000	6	11	17	Hand away
	4-Add'l.			
114 000	6	Get Improved Grip (Regrasp)		
114 100	10	Get additional sheet		
114 800	12	Get additional card		
		ASIDE		
	1-Short	2-Medium	3-Long	
12x 001	11	18	27	Aside only - to other hand
12x 002	8	15	22	- to approx. location
12x 003	23	31	41	- to exact place
	4-Add'l.			
104 010	3	Added for small weight (>5≤12) ENW		
104 020	6	Added for medium weight (>12≤19) ENW		
104 030	9	Added for large weight (>19≤25) ENW		
		MOVE		
	1-Short	2-Medium	3-Long	
13x 002	6	13	20	Move only - to approx. location
13x 003	21	29	39	- to exact place
	4-Add'l.			
104 010	3	Added for small weight (>5≤12) ENW		
104 020	6	Added for medium weight (>12≤19) ENW		
104 030	9	Added for large weight (>19≤25) ENW		

Figure 8–2. MTM-C Level 1 data elements for GET/PLACE. © MTM Association for Standards & Research, Fair Lawn, NJ.

It probably should be stated here that RELEASE is not included in GET/PLACE data as in many data systems. RELEASE is shown in an ASIDE element, and this makes it possible to maintain MTM-1 accuracy in Level 1 data. Also, an analysis of the frequency of occurrence of RL1 and RL2 shows that

MTM ASSOCIATION FOR STANDARDS AND RESEARCH

MTM-1 ANALYSIS OF MTM-C

CODE: 11x 21 (PURPOSE OBJECT)

ELEMENT TITLE: Get only - small stack

STARTS: Hands ready to reach

INCLUDES: Reach and grasp a small stack

STOPS: Stack under control of hand

DATE: 7/77

REV. NO.:

LEFT HAND DESCRIPTION	F	LH MOTION	TMU	RH MOTION	F	RIGHT HAND DESCRIPTION
Reach to small stack		RxB	V_1	RxB		Reach to small stack
Contact stack		G5	0.0	G5		Contact stack at edge with index finger
			2.9	M1B		Lift
			–	~~R1A~~		Free finger to underneath
			0.0	G5		Contact underneath
			0.0	RL2	2	Release one finger
			5.0	R1A	2	Reach with finger underneath
Release stack		RL2	0.0	G5	2	Contact with finger

CODE:	111 210	112 210	113 210
USE TMU:	14	20	27
VARIABLE MOTION-1:	R(0-6)B	R(6-16)B	R(>16)B
VARIABLE MOTION-2:			
CONSTANT VALUE:	7.9	7.9	7.9
VARIABLE VALUE-1:	5.7	12.0	19.4
VARIABLE VALUE-2:			
TOTAL TMU:	13.6	19.9	27.3

Figure 8–3. Illustrates the kind of backup material available. The actual illustration is of MTM-C1 GET/PLACE data for "get only—small stack."

1. Sheet

2. Stacks

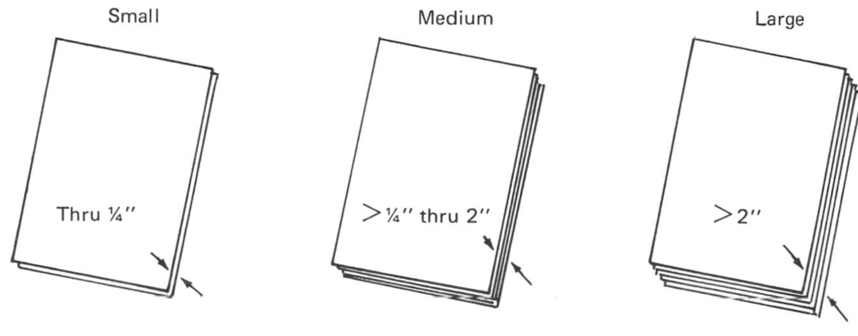

Symbols — < is "less than", > is "greater than"

3. Bundles

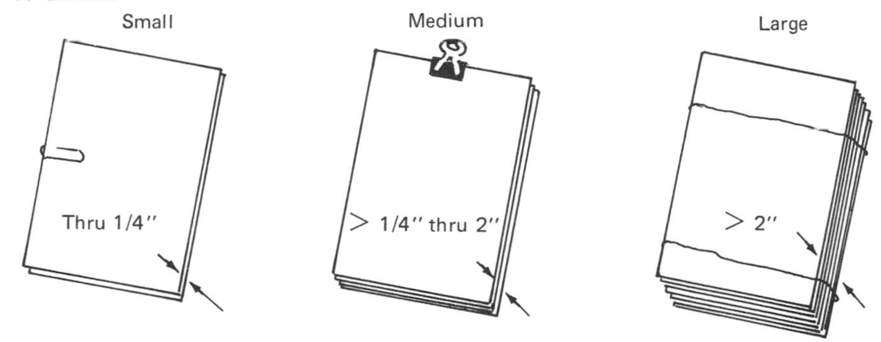

Figure 8–4 (A&B). GET and PLACE illustration. © MTM Association for Standards & Research, Fair Lawn, NJ.

there is only a negligible impact upon accuracy when using only RL1 in the ASIDE data. Therefore, all RELEASEs are RL1 in MTM-C.

The MTM-C manual carefully defines all objects, both in words and in illustrations where this is deemed desirable. An example of the illustrations used was presented in Fig. 8–4. Here are some word-type examples of definitions:

4. Piles

Small

Medium

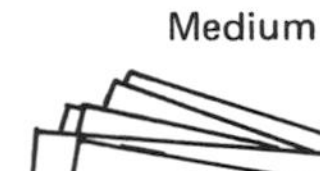

Large

5. Single Object

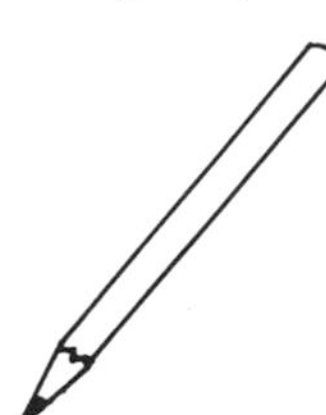

6. Jumbled Objects

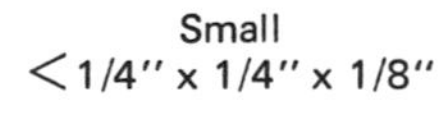
Small
<1/4" x 1/4" x 1/8"

Medium
>1/4" x 1/4" x 1/8"
<1" x 1" x 1"

Large
>1" x 1" x 1"

7. Pad

8. Card

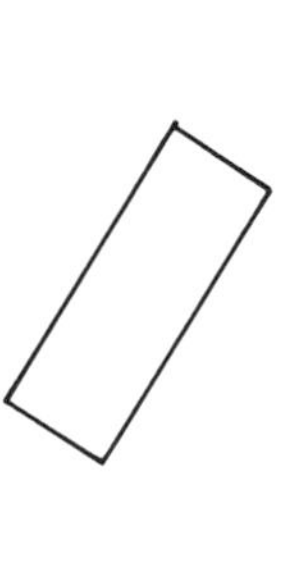

9. Folder

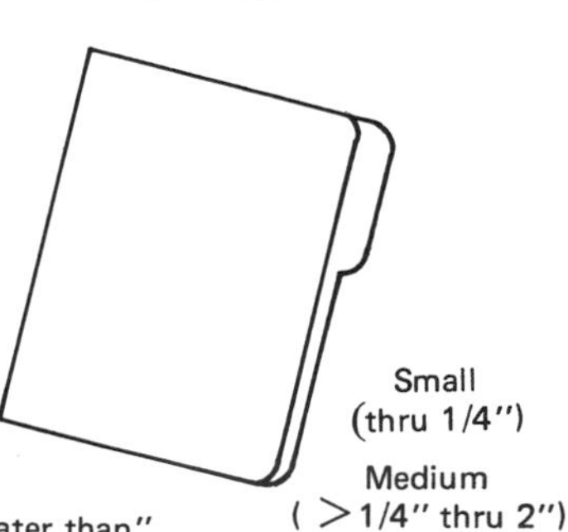

Symbols – < is "less than", > is "greater than"

Figure 8–4 B

1. A SHEET is a piece of paper lying flat entirely on a surface (not overhanging). Size is not a variable.
2. A STACK is an unbound organized group of items. Size designations are given below.
3. A BUNDLE is a group of items secured or bound at a point of easy grasp. Size designations are shown below.[5]

The above definitions apply when material is completely on a flat surface precluding simple grasping. If the material location allows for or facilitates simple grasping, then treat the material as a single object.

Other modifying statements illustrating the detail covered in the manual are the following:

Small and medium sizes may have assisting motions by other hand during GET, but are not limiting.

Large size specified using both hands with overlapping grasp in the GET.

The difficulty in handling paper objects relative to their vertical depth size is considered in the development of the element.[6]

Before proceeding to describe Level 2 data, it probably should be mentioned that every element and item of Level 1 has been as carefully defined and described as the examples given. Also, proper consideration was given to variability, equipment usage, kinds of equipment, etc. The equipment standards and definitions, for example, cover most all of the more ordinary machines found in a modern office.

MTM-C2 (Level 2)

The second level data, MTM-C2, was designed to increase the speed of application and to maintain a known accuracy of the data. It is three times as fast as MTM-1—actually the proportion for GET/PLACE is 24:7.9. This makes it 1.5 times as fast as MTM-C1. On a TMU basis it requires 120 TMU/TMU. You may recall that the same figure for MTM-C1 was 175 TMU/TMU and that MTM-1 requires about 350 TMU/TMU.

In summary, it is believed that when all categories have been validated, MTM-C2 will be as fast as MTM-3.

MTM-C2 was designed to meet many of the same objectives as MTM-C1. The coding system is alphanumeric, where the letters relate to the category and the numbers correspond to the objects. Naturally, object codes from Level 1 were used as much as possible. Level 2 uses only the medium distance range and the average size of the object. Again, these are taken from Level 1 wherever possible. MTM-C2 is directly traceable to the elements of MTM-C1 by reference on the Level 2 data development sheets.

In most of the categories, the GET/PLACE Level 1 elements have been combined with the action elements. Therefore, the major handling elements of the

[5] Courtesy of U.S./Canada MTM Association. See footnote 1, Chapter 6.
[6] *Ibid.*

object are included with the action required. Combined GET and ASIDE elements are applied as Level 2 GET and ASIDE, all others are Level 1 elements. Weight is not a separate consideration in Level 2. The GET and ASIDE elements include GET and RELEASE of the object. The GET TO USE element includes GET of the object and placing it in an approximate location with no release involved.

The GET TO USE and ASIDE element includes getting the object, placing it to use with no release, and asiding it to an approximate location or exact location as required. The release is in the ASIDE element.

A CONTACT GRASP is covered by specifying G0; it can be applied individually. G0 also is used to cover "hand away."

Some objects have been combined in MTM-C2.

Coding MTM-C2

Naturally the alphanumeric coding is similar to Level 1 (coding methods for standard data are discussed in Chapter 27 of *Engineered Work Measurement*). The first coding position is always a letter, and it indicates the category or subcategory. These are:

A—Aside (with release)
B—Body Motions (includes walk)
C—Close
F—Fasten
G—Get
H—Handling
I—Identify
K—Keypunch
L—Locate/File
M—Machines
O—Open
P—Place (no release)
R—Read
T—Typing
U—Unfasten
W—Write

The remaining coding positions can be either alpha or numeric and will actually depend upon the detail or variables to be included. The alpha coding as presented below is relatively easy to learn:

GET/PLACE—G, A, P
OPEN/CLOSE—O & C
FASTEN/UNFASTEN—F & U
FILING—L, I
READ/WRITE—R, W
TYPING—T
HANDLING—H
BODY MOTIONS—B
MACHINES—M, K

The numbers 1 through 7 of the objects of GET are the same as in Level 1. For example, G2 is "get a stack." G4 is "get a pile."

GET AND ASIDE can be combined in Level 2. In this situation G5A1 is "get an object and aside it to other hand." G2A3 is "get a stack of papers and aside them to a tight file box."

GET AND MOVE also are combined as in GET AND ASIDE; they are referred to as GET AND PLACE. Coding is the same as above, except that approximate location is coded P2 and exact position in PLACE is coded P3. G2A1 is "get and aside a stack to other hand." G2PA2 is "get a stack, aside to an approximate location." G6P3 is "get to use a jumbled object, exact location."

The entire MTM-C2 data card or table is shown in Fig. 8–5. In the Training and Data Manual each possible data element is shown and described.

Figure 8–6 illustrates the kind of application notes for each Level 2 data action element in the U.S./Canada Association *MTM-C Users Manual* (data and training manual). The actual illustration is for GET AND PLACE.

MTM-C APPLICATION DATA

LEVEL 2

Do not attempt to use this chart or apply Methods-Time Measurement in any way unless you understand the proper application of the data. This statement is included as a word of caution to prevent difficulties resulting from misapplication of the data.

1 TMU	= .00001	hour	1 hour	=	100,000.0 TMU
	= .0006	minute	1 minute	=	1,666.7 TMU
	= .036	seconds	1 second	=	27.8 TMU

GET-PLACE-ASIDE

G, P, A, PA		A1	A2	A3	P2	P3	PA2	PA3
GP, GPA		18	15	31	13	29	28	44
G0	12		27	43				
G1	17	35	32	48	30	46	45	61
G2	25	43	40	56	38	54	53	69
G3	21	39	36	52	34	50	49	65
G4	29	47	44	60	42	58	57	73
G5	14	32	29	45	27	43	42	58
G6	22	40	37	53	35	51	50	66

G7 10 Add'l Sheet	G8 12 Add'l Card
-Sheet, pad, card	5-Single Object
-Stack	6-Jumbled Object
-Bundle, folder	7-Additional Sheet
-Pile	8-Additional Card

BODY MOTIONS

B1	17	Walk, per pace
B2	188	Sit and stand to/from
B4	61	Bend and arise
B5	19	Horizontal body motion
B6	67	Move or turn in chair
B21	188+17N	B2 + B1
B41	61+17N	B4 + B1
B51	19+17N	B5 + B1
B214	249+17N	B2 + B1 + B4

OPEN-CLOSE

OPEN		CLOSE		OPEN-CLOSE	
O1	29	C1	27	OC1	56
O2	29	C2	26	OC2	55
O24	41	C24	39	OC24	80
O4	35	C4	35	OC4	70
O65	61	C65	75	OC65	136
O67	81				

OBJECTS:
1-Book, cover, folder
2-Door, drawer
4-Binder rings
6-Envelopes

AIDS: 4-Latch
5-String tie
7-Letter opener

IDENTIFY

With ET	I1	18	I2	25	I3	32
Without ET	I10	7	I20	15	I30	22

1-One word or 1, 2, or 3 digits
2-Two words or 4, 5, or 6 digits
3-Three words or 7, 8 or 9 digits

READ & COMPARE/TRANSCRIBE

(x)	RCW(x)	RCN(x)	RTW(x)	RTN(x)
1	35 per	35 per	101 per	46 per
2	25 word	35 num-	91 word	64 num-
3	21 "	35 ber	87 "	83 ber
4	16 "	49 "	84 "	108 "
5	16 "	49 "	83 "	127 "
6	16 "	49 "	83 "	145 "
7	13 "	64 "	81 "	171 "
8	13 "	64 "	81 "	189 "
9	13 "	64 "	81 "	208 "
10	12 "	79 "	80 "	233 "
RN	7 Read 1-3 numbers or characters Read 1-10 words, per word			
	5 Read av. prose per word (over10)			

104-80

Figure 8–5(A&B). MTM-C Level 2 data. © MTM Association for Standards & Research, Fair Lawn, NJ.

FASTEN-UNFASTEN

F1	56	**F54**	18	**U1**	28	**U5**	67
F3	82	**F6**	69	**U3**	32	**U54**	40
F4	78	**F64**	45	**U4**	33	**U6**	44
F5	54						
F52	122						

OBJECTS: 1-Clip; 3-Clamp; 4-Acco fastener; 5-Staple; 6-Rubber band

LOCATE

Number of Items, Folders, Files or Tabs	SHEET		CARD		FOLDER	
1-10	**LC11**	82	**LC81**	60	**LC91**	58
11-30	**LC12**	129	**LC82**	110	**LC92**	103
31-50	**LC14**	148	**LC84**	133	**LC94**	124
51-70	**LC16**	162	**LC86**	143	**LC96**	133
71-100	**LC18**	165	**LC88**	149	**LC98**	139
OBJECT	INSERT		REMOVE		TILT	U/D
1-Sheet	**LI1**	36	**LR1**	18	**LT8**	46
3-Bundle	**LI3**	42	**LR3**	42	**LT9**	73
8-Card	**LI8**	35	**LR8**	27		
9-Folder	**LI9**	41	**LR9**	42		

WRITE-READ & TRANSCRIBE

WD	18	Write digits, per digit		
WA	784	**RTA**	885	address
WD1	216	**RTD1**	237	November 18, 1980
WD2	173	**RTD2**	194	Nov. 18, 1980
WD3	138	**RTD3**	152	November 18
WD4	131	**RTD4**	152	18 Nov. 80
WD5	105	**RTD5**	126	11/18/80
WD6	95	**RTD6**	109	Nov. 18
WD7	69	**RTD7**	83	11/18
WI	67	**RTI**	74	initials
WN	245	**RTN**	266	name
WW	81	Write words, per word		

HANDLING

HA5	75	ADHERE stamp	**HJO**	*50	JOG only
HA6	102	env. flap	**HJ2**	88	stack
HA7	102	label	**HJ4**	55	ARR'GE pil
HA77	95	self-stick	**HP1**	65	PUNCH she
HA8	88	tape	**HP2**	98	stack
HC1	95+10N	COUNT	**HR16**	*38	REMOVE She
HC2	*18	coins, ea.			from env
HF11	66	FOLD one	**HR17**	35	from pad
HF12	130	two	**HR36**	67	Bndl fro
HH92	38	STAMP appr.			envelop
HH93	131	exact	**HS12**	80	INTERSORT
HI04	*64	INSERT shts.	**HT1**	42	TEAR shee
HI14	84	to rings	**HT14**	86	CUT, scis
HI16	91	to env.	**HT15**	70	p'p'r cutt
HI36	78	Bndl to env.	**HU11**	32	UNFOLD o
*GET not included			**HU12**	55	two fol

TYPING

THI1	221	**TCE5**	900	**TKE4E**	222	**TKM4E**	2
THI2	486	**TCL1**	324	**TKE5E**	275	**TKM5E**	3
THI3	556	**TCL2**	434	**TKE6E**	328	**TKM6E**	3
THI4	626	**TCL3**	544	**TKE7E**	382	**TKM7E**	4
THI5	696	**TCL4**	654	**TKE4P**	187	**TKM4P**	2
TCE1	233	**TCL5**	764	**TKE5P**	232	**TKM5P**	2
TCE2	444	**TCLA**	27	**TKE6P**	276	**TKM6P**	3
TCE3	596	**TCEA**	64	**TKE7P**	321	**TKM7P**	3
TCE4	748	**TCP**	46	**TKEW**	26	**TKMW**	

THI-Insert & Remove Sheets & Carbon
TCE-Erase single error
TCL-Liquid correction, single error
TCP-Correction paper, single error
K-Keystroke
E-Electric, Elite (12/pitch)
M-Manual
P-Pica (10/pitch)
4,5,6,7-Line length (per inch)
A-Additional W-Word

Figure 8–5B

Concluding Remarks

The foregoing represents a *practical* way to determine usable and predictably accurate time standards for most ordinary repetitive-type office work. However, the mere establishment of standards will acomplish little in the way of improved labor efficiency. The worker must be measured against the standard and informed regularly and consistently (as to time) about his/her performance. Details on incentive systems will be found in *Engineered Work Measurement*.[7]

[7]D. W. Karger and F. H. Bayha, *Engineered Work Measurement,* 3rd ed., Industrial Press, New York, 1977.

C. *Application Notes*

1. There is no weight consideration in Level 2.
2. To get and place any object with contact grasp and no release includes the medium distance for both the Get and Place with no time included for grasping. The control of this element is to an approximate location only (G0).
3. The Get and Aside elements include a grasp and release of the object and are the medium range distances for both the Reach and Move involved.
4. The Get to Use element includes Get of the object and place it to an approximate or exact location with no release involved. The medium distances are utilized.
5. The Get to Use and Aside element includes Get the object, place it to use with no release, and aside it to an approximate location or exact location as required. The release is included in the aside movement.
6. If a contact grasp is required by the applicator, the element G0 can be applied individually. The hand away is also applied with a G0. Asiding an object to a stop will use the approximate location (A2).
7. Some of the objects have been combined in Level 2. For example, a folder is the same as a bundle and classified as a G3. A pad or card is classified the same as a single sheet and is coded G1.
8. Place To Use in PA2 and PA3 is to an approximate location only.

Figure 8–6. Illustrates the kind of Application Notes available in the **User's Manual** ***(data and training manual). These specific notes are for GET/PLACE © MTM Association for Standards & Research, Fair Lawn, N.J.***

Perhaps it should also be said that developing and applying labor standards does *not* lead to worker alienation if it is done in the proper manner.

CHAPTER 9

MTM-V: A High-Level and Fast to Use MTM System for Setting Labor Standards in Machine Shop Work—A Specialized Standard Data System

MTM-V is a high-level MTM system for use and application in machine shops and related work. It was developed by the Swedish MTM Association (Svenska MTM Gruppen). The V designation was derived from the Swedish word *Verktygsmaskiner,* which means "machine tool."

A word of caution should be given. This discussion of MTM-V is limited in scope. If the reader wishes to use the system, he/she should take an MTM Association approved training course. If this is not done, the results will tend to be disastrous since not all of the "rules of thumb" and other details needed to make application a success are presented. This particular material is copyrighted by the Swedish Association and is not released for general publication, one of the primary reasons for the presentation of limited information. However, through the U.S./Canada Association permission was given to provide this material. Another reason that this system's presentation is limited is that it is a sort of cross between a simple high-level system and a standard data system (like MTM-C). This meant that the chapter would have been greater in length than available space had everything been included.

MTM-V was primarily aimed at firms with relatively short runs where a higher speed of manual application of the system and lower work measurement costs were a virtual necessity if engineered standards were to be used.

It was the product of a 14 company consortium which included the Swedish MTM Association, who provided guidance for the project. The group consisted of the following:

ASEA
Atlas Copco
AB Bofors
Hägglund & Söner
Lauritz Knudsen
Lidköpings Mekaniska Verkstad
NEBB
SKF
Svenska Flygmotor
Ekonomisk Företagsledning (EF)
Rationellt Näringsliv
Svenska MEC

MTSM-Selskapet A/L
Svenska MTM Gruppen

The effort began in 1967 and was completed late in 1969. Essentially, the consortium members provided MTM-1 analyses which were then synthesized and statistically combined and averaged to form the MTM-V elements. The MTM-V system was developed in the same general manner as MTM-2 and MTM-3.

This higher-level system can be used to set standards for most *manual work* associated with machines. It cannot be used to cover machine process times or allowances. Neither will it determine speeds and feeds.

Basically, it can be used in the same manner as the other higher-level MTM systems. Since it is traceable back to MTM-1 it is combinable with the other members of the MTM family of systems. Like MTM-3 it is fast to use—about 23 times as fast as manual MTM-1. For example, a 12–15 minute work cycle can be analyzed and a standard developed in about 2 hours. Its primary applications are in the analysis and setting of manual work time standards for:

Manual operation of machines
Setting up equipment
Handling material (workpieces)
Tool handling
Measuring or gaging

What is covered as to parameters is in general determined by whether MTM-1 basic motions can be *directly used* by the body members, especially the manual members. The use of tweezers, for example, to grasp and obtain a part is not coverable in MTM-V.

Batch or manufacturing lot size is discussed in this text with regard to the MTM family of systems, and this material should be reviewed by the reader, especially the chapters on learning and on system selection. It should also be recalled by the reader who contemplates applying MTM-V to small lot size manufacture that the underlying MTM-1 standards are low task standards based upon a fully trained operator working with average skill and effort.

There are other limitations to MTM-V. For example, checking is limited to those inspection operations at the machine or within the operator's normal working area that are coverable by the MTM system.

Another example, cleaning, is limited to the operator's workplace using normal devices such as a cloth, brush, air hose, chip hook, etc.

Deburring and polishing are not included as a covered activity in the system, neither are they covered in any of the MTM family of systems.

When MTM-V is to be used in conjunction with another MTM system, remember to be sure that "breakpoints," the ending and beginning points of the underlying MTM-1 motions involved at the point where another system's output is to be inserted, are correctly identified so that the other system elements used in conjunction with MTM-V will properly mesh—so that the work is covered, but only once. "Breakpoints" (starting and ending points) are fully discussed for MTM-1 in *Engineered Work Measurement* (EWM).[1] In fact, these writers be-

[1]See *EWM* (footnote 1, Chapter 1).

lieve that such knowledge is needed for MTM-V to be used with any of the other MTM family of systems since it is almost impossible to define MTM-V "breakpoints" in a simple manner because there are so many variations in each of the system elements in addition to the substantial number of underlying MTM-1 elements. Even the MTM-V Training Manual does not clearly identify the end points, and certainly does not cover the "breakpoints" in the underlying MTM-1 elements. Unless the reader knows MTM-1 very well, he/she would have to go back to the source documents (they are available) to secure first the data on the underlying MTM-1 elements, and then to another source like EWM for the breakpoint information on the involved elements.

Accuracy vs system selection is thoroughly covered in Chapter 12. However, it should be said that MTM-V, with 95 percent confidence, has an absolute total deviation of ±5 percent at approximately 60,000 TMU (36 minutes).

The relative (to MTM-1) balance time of MTM-V is 40,000 TMU (24 minutes), which means that at 40,000 TMU cycle time the standard produced by MTM-V will be within ±5 percent of that produced by MTM-1 at the 95 percent confidence level.

The reader who has studied the text up to this point will know how to interpret the accuracy graph shown in Fig. 12–2. If in doubt, consult Chapters 2, 3, and 12.

It is possible that the high system balance and high relative balance (in terms of TMU) times will trigger operator grievance problems if MTM-V is applied to short cycle jobs. With regard to this application it is suggested that the reader read Chapter 12 on system selection and also to read Chapter 2 to see how learning time adjustments could possibly be made.

The MTM-V System

First, term usage should be defined. An *activity* is a sequence of manual motions performed to achieve a specific task (purpose). An *object* is anything upon which an activity is performed (workpiece, equipment, surface area, etc.). A *tool* is an implement used in the hand to achieve the purpose of an activity. Examples are hammer, wrench, gage, cloth, brush, etc. In MTM-V a tool may be classified as an object when an activity is performed *on it* (like adjusting a "monkey wrench") or when it is not intended to be used to achieve the purpose of an activity.

There are two types of elements in MTM-V, Simple and Complex.

Simple Elements have

1. one purpose
2. one object

Some simple elements associated with the handling category are called construction elements because they are used in constructing other simple and all Complex Elements. An illustration of such an element is illustrated in Fig. 9–1.

The basic MTM-V data card is not essentially different from the other MTM data cards, except that there is a companion card containing decision diagrams. Each simple element has either a two- or three-bit alpha code followed by a

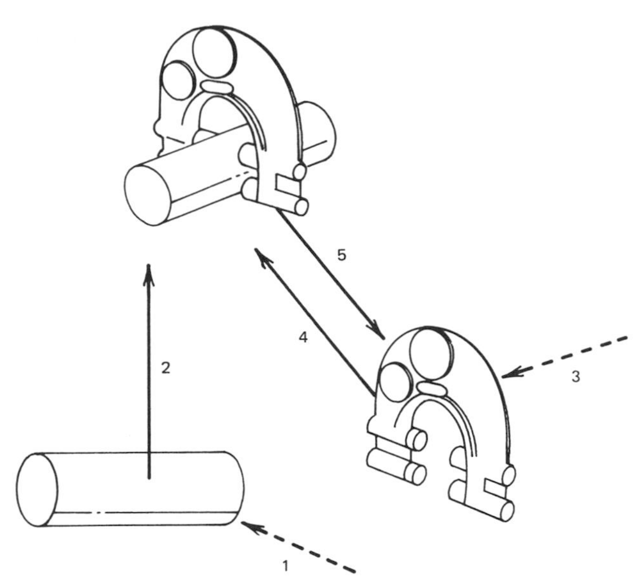

Figure 9–1. An illustration of a complex element. © MTM Association for Standards & Research, Inc., Fair Lawn, NJ.

single numeric code. All Complex Element codes (except Couple) are composed of three alpha and two alpha characters—a total of five positions. Tool handling is covered only in the Complex Elements utilizing the fourth (numeric) position. Although a Complex Element, Couple is coded similarly to the Simple Elements.

The difference between Simple Elements, Complex Elements, and work outside the scope of the system requiring Special Measures is illustrated by the decision diagram shown in Fig. 9–2.

The Special Measures designation occurs in each decision model. Such measures could include the use of another MTM system or they might require the use of time study procedures to determine the time for such work as process time.

Simple Elements

The first two Simple Elements to be discussed involve HANDLING. What it covers is generally in accord with what the average industrial engineering reader visualizes from the word "handle." While the work is necessary to the accomplishment of the operation, the reader normally does not "see" significant meaningful work being accomplished—and this largely is the correct view.

HANDLE HAND TOOL (HH) is used to cover the obtaining, placing, and returning of a hand tool, which at some point in the work cycle will be used in a meaningful manner. However, HH is also used to move a tool between places of use.

HANDLE OBJECT (HO) is used to cover the obtaining (getting) *and/or*

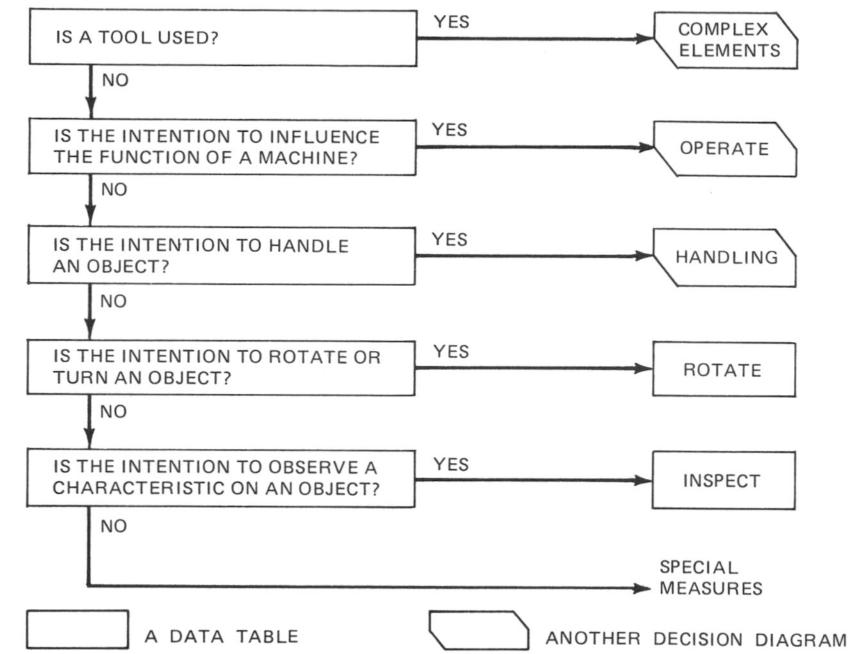

Figure 9–2. Decision Diagram used to determine the type of element. © MTM Association for Standards & Research, Inc., Fair Lawn, NJ.

placing a freely accessible object upon which some work or activity is to be performed, immediately or later. Careful thought about this definition will make clear that it may cover (1) the obtaining of an object or (2) the placing of the object or (3) the obtaining *and* placing of the object. As we shall see later, this will illustrate just one of the difficulties of identifying the so-called "breakpoints."

HANDLE LATCH (HL) covers the opening and closing activity required to make objects accessible, but, interestingly, LATCH as used here does not mean the commonly understood latching devices exclusively. Instead it defines (1) a drawer or (2) a cabinet door *and* a drawer in the cabinet or (3) a cabinet door (it also covers a double door cabinet with the doors hinged, respectively, on the right and the left sides) to make shelves accessible (and objects stored on the shelves, but not an object buried in a jumbled pile on the shelves) or (4) a hinged lid on a storage box. (All of these items are usually equipped with latching devices.) *Moreover,* it covers the activity to obtain *and/or* replace an object or to merely look for an object—numerical codes indicate the case.

The decision diagram is shown in Fig. 9–3, where all the decisions in the figure lead directly to data tables.

It is suggested that the reader now study Fig. 9–4, Simple Element Structure, and the related Simple Element Data Card shown in Fig. 9–5.

The actual codes used are mnemonic in Swedish. The listing below makes most of them somewhat mnemonic in English.

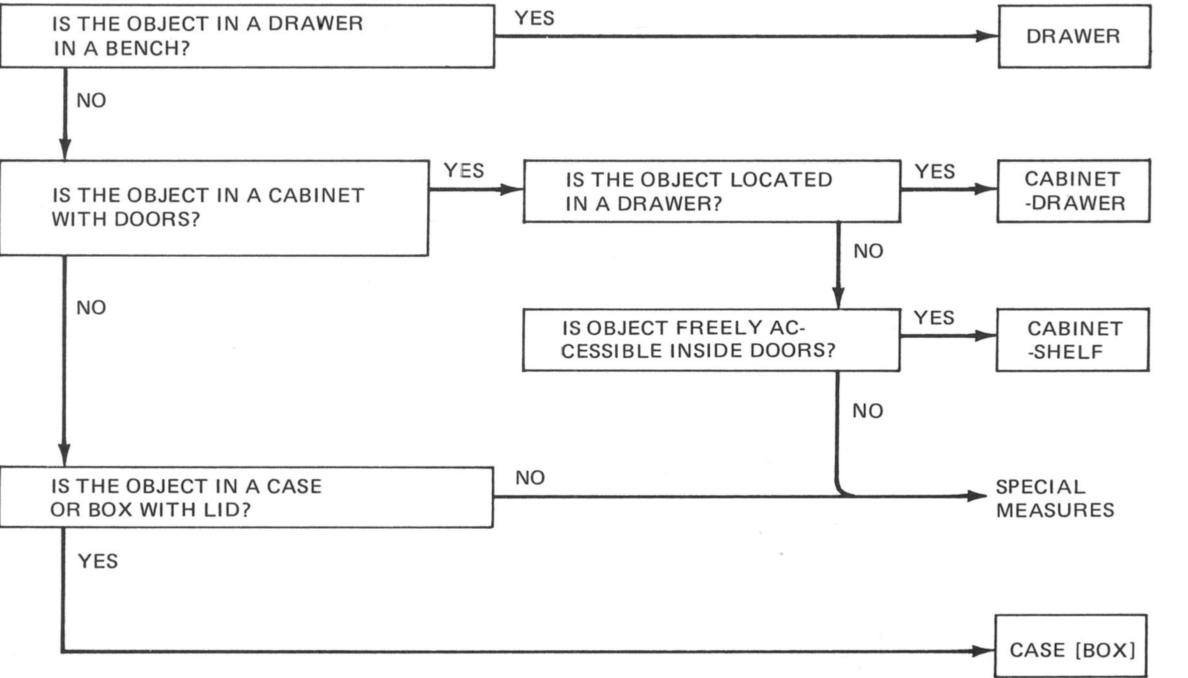

Figure 9–3. Decision Diagram for HANDLE LATCH. © MTM Association for Standards & Research, Inc., Fair Lawn, N.J.

ELEMENT DESCRIPTION	ACTIVITY DESCRIPTION	ELEMENT CODE	ACTIVITY CODE	CASE OF HANDLING 0: No Object Handling	1: Tool Already in Hand or in Place	2: Get and/or Place Object or Control W/O Step or Bend	3: Get and/or Place Object With Step But No Bend	4: Get and/or Place Object or Control w/Step and/or Bend
Handle Object	—	HO				40	90	160
Handle Hand Tool	—	HH			20	60	110	
Handle Latch / Open & Close Door, Drawer or Hinged Lid	Open & Close Drawer	HL	A	90		130	190	260
	Open & Close Door, Drawer Inside Cab.	HL	B	200		230	290	360
	Open & Close Cabinet Door	HL	C	130		170	220	300
	Open & Close Box (Case) Lid	HL	D	70		110	170	240
Rotate with Fingers (Screw)	1 or 2 turns	SK	A	30		70	130	200
	3 to 5 turns	SK	B	50		90	150	220
	6 to 12 turns	SK	C	100		140	190	260
Inspect (GRade)	—	GR		40		80	140	210
Operate MAchine Controls	Two-position Control	MA	A	10		20		50
	Lever ⩽180 Deg.	MA	B	10		30		50
	CRANK >1/2 ⩽ 2 Revs.	MA	C	40		50		80
	CRANK >2 ⩽5 Revs.	MA	D	70		90		110
	CRANK >5 ⩽12 Revs.	MA	E	130		140		170
	Set	MA	F	30		40		70

Figure 9–4. The structure of the Simple Elements. © MTM Association for Standards & Research, Inc., Fair Lawn, NJ.

MTM-V DATA CARD

Do not use MTM-V data without training

SIMPLE ELEMENTS

HANDLE OBJECT	CODE	TMU
WITHOUT STEP	HO2	40
WITH STEP, WITHOUT BEND	HO3	90
WITH STEP, WITH ONE BEND	HO4	160

HANDLE TOOL	CODE	TMU
MOVE	HH1	20
WITHOUT STEP	HH2	60
WITH STEP	HH3	110

GET or RETURN	CODE	OBJECT HANDLING 0	2	3	4
DRAWER	HLA–	90	130	190	260
CABINET-DRAWER	HLB–	200	230	290	360
CABINET-SHELF	HLC–	130	170	220	300
BOX (CASE)	HLD–	70	110	170	240

ROTATE	CODE	OBJECT HANDLING 0	2	3	4
≤ 2 TURNS	SKA–	30	70	130	200
≤ 5 TURNS	SKB–	50	90	150	220
≤ 12 TURNS	SKC–	100	140	190	260

INSPECT	CODE	OBJECT HANDLING 0	2	3	4
	GR–	40	80	140	210

OPERATE	GET CONTROL	CODE	TMU
TRIGGER	0	MAA0	10
	2	MAA2	20
	4	MAA4	50
LEVER (Movement ≤ 180°)	0	MAB0	10
	2	MAB2	30
	4	MAB4	50
CRANK 2 ≤ 2 Revolutions	0	MAC0	40
	2	MAC2	50
	4	MAC4	80
CRANK 5 ≤ 5 Revolutions	0	MAD0	70
	2	MAD2	90
	4	MAD4	110
CRANK 12 ≤ 12 Revolutions	0	MAE0	130
	2	MAE2	140
	4	MAE4	170
SET	0	MAF0	30
	2	MAF2	40
	4	MAF4	70

Coding Principles: *Simple Elements*

HO2
- Object Handling: Without Body Motions
- Purpose: Handle Object

HL B2
- Object Handling: Without Body Motions
- Object Location: Cabinet Drawer
- Purpose: Get/Return

Figure 9–5. The illustration presents the MTM-V data card's presentation of the Simple Elements. © MTM Association for Standards & Research, Inc., Fair Lawn, NJ.

HO HANDLE OBJECT
HH HANDLE HAND TOOL
HL HANDLE LATCH
SK ROTATE (Screw)
GR INSPECT (GRade)
MA OPERATE MACHINE CONTROLS

Examination of Figs. 9–4 and 9–5 will make it clear that numeric codes "3" and "4" contain the MTM-1 element STEP. The maximum distance coverable in MTM-V by the STEP designation is 80 inches—but this really means a coverable 80 inches to Obtain an Object or Tool and a further maximum of 80 inches to Place the Object or Tool at the intended location. If more STEPS are needed, cover these with another MTM system such as MTM-1 or MTM-2.

Code 4 not only covers the STEP described above, but it also covers one BEND and ARISE. More BENDs and/or ARISEs must be covered by another MTM system.

As stated before, motion probabilities (frequencies) were involved in determining the time allowed for each motion category—just as in MTM-2 and MTM-3.

Since STEP covering walking time has been mentioned, it should be stated that walking time should be considered only when other required motions or elements cannot be *completed* due to the walk requirement—meaning that it is a limiting requirement or action. Often some action or some motions can occur during walk, like a partial REACH or MOVE. Do *not* confuse body motions, including minor steps (forward, backward, and side steps), that are normal to an operator working in the standing position—so long as they are not "limiting" (explained in an earlier chapter; also, this aspect of working in a standing position is further discussed in the previously mentioned reference *Engineered Work Measurement*[2]).

STEP, especially side steps, that *move the torso* less than 12 inches are seldom limiting, they merely provide a form of body assistance. These are quite prevalent for a worker operating a machine tool or standing at a bench while working.

Good methods procedures dictate that the analyst should try to reduce the STEPs to minimum, especially any excess over the 80 inch limit.

HANDLE OBJECT—HO

The definition of HO was previously given. However, it should by now be understood that it also covers required body motions in accord with the previous discussion.

Again it should be emphasized that this text was written with the assumption that the reader is already familiar with MTM-1. This is one reason why "breakpoints" will not be discussed for MTM-V (previously explained as to reasons involved).

The decision diagram for HO is illustrated in Fig. 9–6.

[2]*Ibid.*

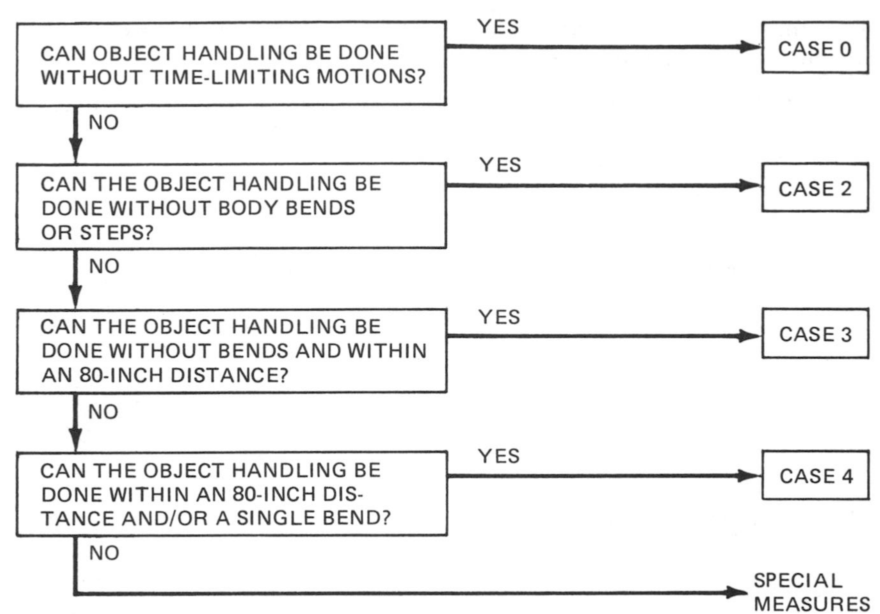

Figure 9–6. Decision Diagram for HANDLE OBJECT (HO). © MTM Association for Standards & Research, Inc., Fair Lawn, NJ.

The object or workpiece size must be such that it can be "handled" by the manual members. However, it can, for example, in some elements be hanging from a crane hook or sling so that one man can control or "handle" it.

If the reader will examine Figs. 9–4 and 9–5, he/she should note that HO has no associated code 0 and 1. HH does use code 1, but has no code 4. In the last case (HH) if a body motion is required, use MTM-1 or MTM-2. As to no object handling, HO naturally has no case "0" since the element has to do with handling objects. HH covers *tool* handling, not object handling, therefore it too has no case "0."

A further examination of Fig. 9–4 will lead to the conclusion that the remainder of the coding differences are equally logical. For example, HL has no case or code "1" since, as will be seen later, the operator either merely looks for the object (usually a tool) in which case code "0" is assigned or an object (usually a tool) is obtained—codes 2, 3, or 4.

A few examples may help make clear how work situations can be analyzed, especially when the situations are a little complex.

The operator steps 100 inches to a pallet of parts and bends to obtain one from the pallet, arises, steps 100 inches, and bends to place it on another pallet.

Extra distance:	100 − 80 = 20		
Extra steps:	20 ÷ 30 = 1 each way		
Analysis:	HO4	160 TMU	
	B	61	(covers bend and arise,
	2S	36	an MTM-2 element)
		257 TMU	

In the above example both extra steps and extra body motion were required. One more example, in this case involving several objects.

An operator obtains two T-slot screws from his work bench next to his radial drill. The walk distance to the bench is less than 80 inches. No bending is required.

Obtain two screws and place one	HO3	90 TMU
Place second screw	HO2	40
		130 TMU

The last example: here the operator bends over a pallet of parts and picks up with the right hand and puts into the left hand 10 parts, one at a time. He then arises and puts the 10 parts, together, in a jumbled pile, more or less on the bench.

Obtain and place first part in left hand	HO4	160 TMU
Obtain and place nine parts in left hand	9HO2	360
Place 10 parts on bench	HO4	160
		680 TMU

HANDLE HAND TOOL—HH

Code HH is seldom used as an independent element, usually it is connected with the Complex Element FASTEN/LOOSEN, or when the activity is accomplished during machine time (process time).

An example might help the reader understand the use of HH.

An operator, after having chamfered on edge of a work piece, moves the file to another edge for chamfering the second edge.

Move tool	HH1	20 TMU (tool already in possession of operator)
Chamfering		Not coverable with MTM-V, use another system or time study

HANDLE LATCH—HL

HL is used to gain access to a storage container in order to obtain or replace an object or to both obtain and replace an object or to just look to see if the object is, for example, lying on a cabinet shelf. It obviously covers opening a door(s) (as previously explained in the general discussion of MTM-V), a drawer, a hinged box or chest lid, and any other similar closure device. It even covers in one category the opening of both a cabinet door(s) and a drawer.

Latch obviously means a covering device (like a door or lid) and any associated latch.

Objects are presumed to be freely accessible once the latch(s) involved is opened. If this is not true, it is another situation in which the analyst must use Special Measures—meaning use another MTM system or time study.

Opening the door(s) or lid includes the time required to release simple "catches," but not such a device as a padlock or a key-locked door.

The decision diagram for this element was presented earlier in Fig. 9–3 where it was used to illustrate what an element decision diagram looks like.

Examples will help clarify any questions regarding the application of HL. While these examples concentrate on HLA, Drawer Handling, the other codes such as HLB, HLC, and HLD are applied in a similar manner.

1. Open a bench drawer to look for an object, nothing is removed. The drawer is then closed.
   Open drawer, look for object, and close drawer. HLA0 90 TMU
2. Open a bench drawer and take out a wrench and *use it*. Return the wrench to the drawer and close it. No body motions are involved.
   Open and close drawer HLA0 90 TMU
   Get, use, and return tool ? ?
   Obviously, another MTM-V element has to be used that has not yet been defined and/or described. HL does not cover USE of a tool.
3. Operator opens a bench drawer, takes out a wrench, and lays it on the bench. He/she closes the drawer; this may or may not be simo with the laying of the tool on the bench—but likely it will be.
   Open drawer, take out wrench, lay wrench on bench, and close drawer. HLA2 130 TMU
4. Operator steps 72 inches to a drawer, opens it, takes out a wrench to use it, and returns it after use and closes the drawer (it was left open during the use of the wrench).
   Open and close drawer HLA0 90
   Get, use, and return tool ? ?
   The use of a Complex Element is required to cover the get, use, and return of the tool.
5. Operator steps 72 inches to a drawer, opens it, takes out a wrench, closes the drawer, and lays it on the bench.
   Walk to drawer, open it, remove tool, close drawer, and lay wrench on bench HLA3 190 TMU
6. Operator uses wrench on a bolt in a machine on the work bench, removes wrench from bolt, opens a bench drawer, places wrench inside, and closes drawer.
   Tighten bolt (another GET/USE element, which also covers the return of wrench) ? ?
   Open and close drawer HLA0 90 TMU
7. Operation requires a part to be picked up with the left hand, hold it, then open a drawer with the right hand, place the part in the drawer, and close it.
   Obtain part HO2 40 TMU
   Open drawer, place part, and close drawer HLA2 130
   170 TMU

ROTATE—SK

SK covers screwing or unscrewing a threaded object with the hands or fingers.

The number of turns covered is delineated by the alphabetical letters A, B, and C in the third position.

SKA Not over 2 turns.
SKB Greater than 2 and not over 5 turns.
SKC Over 5 but not over 12 turns.

If more turns are needed, assign the smallest combination of covered turns that still cover the required number of turns. For example, 13–14 turns would call for an SKC + SKA.

None of the turning included in SK is to cover turning an object to an exact position (rotary or axial).

The numeric codes are used to cover the other major variables.

Code 0 covers the case where the two objects are already engaged and after "the turning" they are still engaged. It covers grasping the nut or bolt to be turned, but includes no disposal.

Codes 2, 3, or 4 involve bringing the objects together (like a bolt and a nut) or separating them (by screwing) and placing one or both in one approximate location (approximate as defined in MTM-1)—it does not cover the combination just described. Only one object handling is covered.

Codes 3 and 4 involve the above-defined activities and, in addition, the body motion variables previously described as being associated with these codes.

A few examples will help make the aforementioned material more understandable.

1. Grasp a nut already engaged on a bolt and turn it one turn.

Turn nut one turn	SKA0	30 TMU

Often the other hand, but not always, is already on the combination in some fashion.

2. Operator picks up a bolt from the lathe cross-slide without body motions, places it in a threaded hole, and turns it down two and one-half turns.

Get bolt, engage, and turn down two and one-half turns	SKB2	90 TMU

3. Operator takes hold of an engaged nut, unscrews it eight turns, and lays it on a bench 96 inches away.

Unscrew bolt, place on bench	SKC3	190 TMU
Extra step (96 − 80 = 16)	S	18
		208 TMU

4. Operator holds a bolt in his/her left hand, picks up a nut without body motions, and places it on the bolt with two turns. He/she then lays the assembled bolt and nut on a pallet on the floor close beside him (body motion required as well as a small step or two).

One Solution

Get nut, place on bolt, turn nut two turns	SKA2	70
Aside assembly to pallet	HO4	160
		230 TMU

A Second Solution

Get nut and place on bolt	HO2	40 TMU
Turn nut two turns and aside assembly to pallet	SKA4	200
		240 TMU

Note that when object handling can be analyzed in two ways, the total times may vary by ± 10 TMU. This is due to the data card values being rounded to the nearest 10 TMU of their backup values.

5. Operator picks up a nut without body motions, places it on a bolt, and turns it down 14 turns.

Place nut and turn 12 turns	SKC2	140
Accomplish two additional turns	SKA0	30
		170 TMU

In the above no disposal was required, hence the SKA0 element.

INSPECT—GR

It is suggested that readers review the discussion of Eye Time in *Engineered Work Measurement*[3] and also review the chapter on Decision Time before studying the material under this subtitle unless they are already familiar with the reference material or are participating in an MTM Association approved training course.

GR is used to cover the time required to recognize an easily recognizable characteristic of an object (one characteristic only). The reference material defines an easily recognized characteristic.

The only variable in GR is the case of the Object Handling (numeric code), which also includes a consideration of body motions as previously discussed.

GR covers only the obtaining of the object *or* its disposal.

Case 0 (GR0) covers the situation which involves no obtain or disposal. Either the object is in the control of the hand(s) and remains there after GR *or* it can be inspected without an obtain or a disposal.

Cases 2, 3, and 4 cover *either* an obtain or a disposal.

Cases 3 and 4 again cover our familiar combination of body motions.

The above seems so clear that no examples will be given.

OPERATE–MA

MA is used when actuating machine controls which prepare, start, set, stop, etc., the machine.

It starts when the body member makes contact with the control (lever, crank, or trigger).

MA stops when the control device is released.

The Decision Diagram for MA is illustrated in Fig. 9–7.

At this point it is again suggested that readers review Figs. 9–4 and 9–5 before proceeding further.

The GET CONTROL in the Decision Diagram is a special application of HO, but only uses three of the five cases or codes (0, 2, and 4). Hence, these are the only kinds of "handling" covered by MA. As a result, the three cases are depen-

[3]*Ibid.*

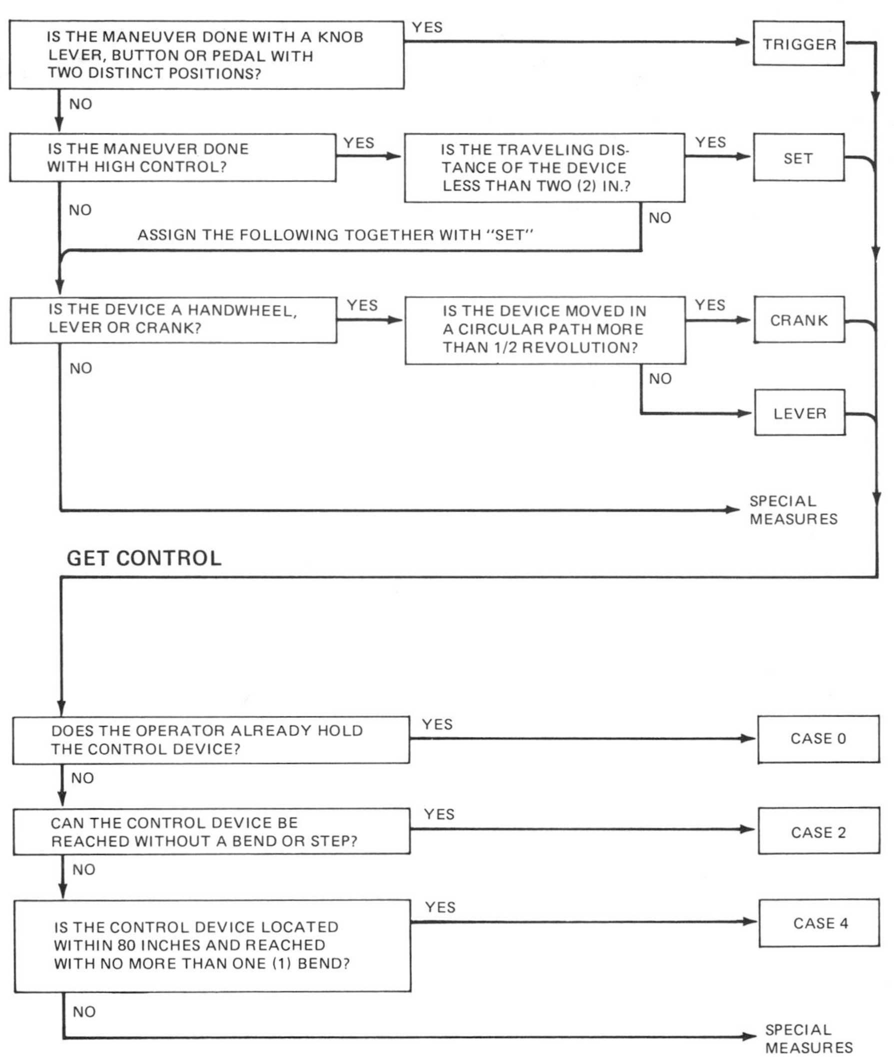

Figure 9–7. Decision Diagram for OPERATE (MA). © MTM Association for Standards & Research, Inc., Fair Lawn, NJ.

dent on the time-limiting hand or body motions required. Time-limiting steps are rarely required to handle most machine controls, so Case 3 is not needed. If time-limiting steps are required, they are handled separately.

GET CONTROL never occurs here as a pure Simple Element, and it is applicable *before* the actuate activity is performed. The object in this element is rather obviously the control device.

OPERATE (MA) Case 0 is assigned when the hand or foot is already in *contact* with the control device.

Case 2 is assigned when the control device is to be *contacted* with the hand or foot without time-limiting steps or bends.

Case 4 is assigned when time-limiting steps within the 80-inch rule and/or one bend are required to contact the control. The MTM-1 element ARISE FROM BEND (AB) is not included since it normally takes place during machining time (a special usage of Case 4) If AB is time limiting, it must be analyzed and shown separately. Its value is 31.9 TMU.

MA, Type A (MAA), TRIGGER

MAA *is not* what it sounds like. TRIGGER means any two-fixed-position control device (such as an on/off control) that can be actuated by the finger(s), hand, or foot movement where the movement required does not exceed 2 inches. Examples of such controls are:

Push Button
Toggle Switch
Knob
Foot Pedal

MA, Type B (MAB), LEVER

LEVER (MAB) is assigned to a control device where the finger or hand movements needed to actuate it *exceed* 2 inches, but it does not require a circular path of 180 degrees. Included *are* necessary hand motions to overcome resistance to movement *or* the release of locking devices.

The examples which are given below include elements not yet discussed, but which will be in following subtitles. If the reader wishes, he/she may turn ahead for the discussions of CRANK (MAC, MAD, MAE), SET (MAF), and Grade (GR).

1. Operate start button	MAA2	20 TMU
2. Stop spindle [operator's hand positioned on button during machine (process) time]	MAA0	10 TMU
3. Lift lockring on rapid chuck to change drill	MAA2	20 TMU
4. Operator bends to turn on pump with a two-position knob	MAA4	50 TMU
5. Operator lowers spindle to start cut on drilling machine	MAB2	30 TMU
Care must be taken	MAF0	30
		60 TMU
6. Operator moves top slide by turning crank 150 degrees	MAB2	30 TMU
Sets to a line	MAF0	30
		60 TMU
7. Operator loosens tailstock spindle lock	MAB2	30 TMU
Moves spindle about 5/8 inches, three revolutions	MAD2	90
Relocks spindle lock	MAB2	30
		150 TMU

8. After tool is about in position (by cranking cross-slide)	MADO	70 TMU
Set tool to axle periphery	MAF0	30
		100 TMU
9. Operator observes (reads) voltage setting	GR0	40 TMU
Get knob (requires steps) and adjust to new setting	MAF4	70
		110 TMU

If a high degree of control is required in actuating a lever, the additional element SET (MAF0) must be used. In fact it applies in many situations as illustrated in the examples.

CRANK (MAC, MAD, and MAE)

There are three types of CRANK:

Type C (MAC) covers >1/2 but not over 2 revolutions
Type D (MAD) covers >2 but not over 5 revolutions
Type E (MAE) covers > 5 but not over 12 revolutions

CRANK is assigned to a control device which moves in a circular path greater than 180 degrees. Included are the hand motions to gain control of the handwheel. This also includes, where it is necessary, to *engage* the handwheel prior to moving it.

If a LEVER movement is more than 180 degrees, MAC is used to cover the motion.

Type F (MAF), SET

SET is assigned to a Control Device where a high degree of control is required during actuate (operation of the Control Device). Examples include setting a handwheel to scale, or moving a cutting tool into proper position with respect to the workpiece, etc. It is assigned as an independent element where the finger or hand movements do not exceed 2 inches *or* in addition to LEVER or CRANK where appropriate.

MAF is commonly used with LEVER or CRANK which means it will usually be Case 0. The element can also be used separately.

Limits on SET (MAF) are as follows:

1. Set to scale through movement of a handwheel, crank, handle, etc. The "setting" position(s) (only one setting covered in MAF) is not supplied with locks, stoppers, or anything facilitating the "setting." Also, SET is based upon the scale line being set within ± 0.08 inches.
2. Set machine component or tool with the help of a device to a careful position that is determined by sight.
3. Engage cutting tool against workpiece when there is risk of damage to the tool if contact is not accomplished with care. Such engaging is accomplished without an automatic control.

The following is an example of the application of MAB, MAC, MAD, and MAF:

The operator goes through the following work sequence:
(a) Cranks carriage forward three revolutions
(b) Cranks in cross-slide four revolutions so that the tool reaches the axle's (part being turned) periphery
(c) Moves carriage away by turning carriage wheel 145 degrees
(d) Cranks in cross-slide three-quarters of a revolution, and
(e) Sets it to a scale line.
The following is the MTM-V motion or element pattern for the above:

Carriage forward three revolutions	MAD2	90 TMU
Cross-slide in four revolutions (cross-slide crank obtained while carriage was cranked forward)	MAD0	70
Set tool to axle	MAF0	30
Carriage away 145 degrees (hand remained on crank)	MAB0	10
Cross-slide in (hand remained on crank)	MAC0	40
Set to scale line	MAF0	30
		270 TMU

COMPLEX ELEMENTS

There are six:

FL	*FASTEN/LOOSEN*
BE	PROCESS (*Blow off Equipment; Brush off Equipment*)
MT	MEASURE (*Meter*)
KO	GAGE
MR	*MARK*
KP	COUPLE AND UNCOUPLE

Again the Complex Elements are mnemonic in Swedish, but only partially so in English.

All Complex Elements except KP utilize all five coding positions:

ELEMENT		ACTIVITY	TOOL HANDLING	OBJECT HANDLING
alpha	alpha	alpha	numeric	numeric

The Decision Diagram in Fig. 9–8 will guide the reader to a proper choice of the Complex Elements. The data card display for the Complex Elements are as shown in Fig. 9–9.

The three groups of variables associated with all the Complex Elements except KP are as follows:

1. *Type of Activity*. The type of hand tool used or its frequency of usage is alpha coded in the third position.
2. *Case of Tool Handling*. This is coded 1, 2, or 3, and the code is found in the fourth position; KP is the exception.
3. *Case of Object Handling*. Coded 0, 2, 3, or 4 in the fifth or last position.

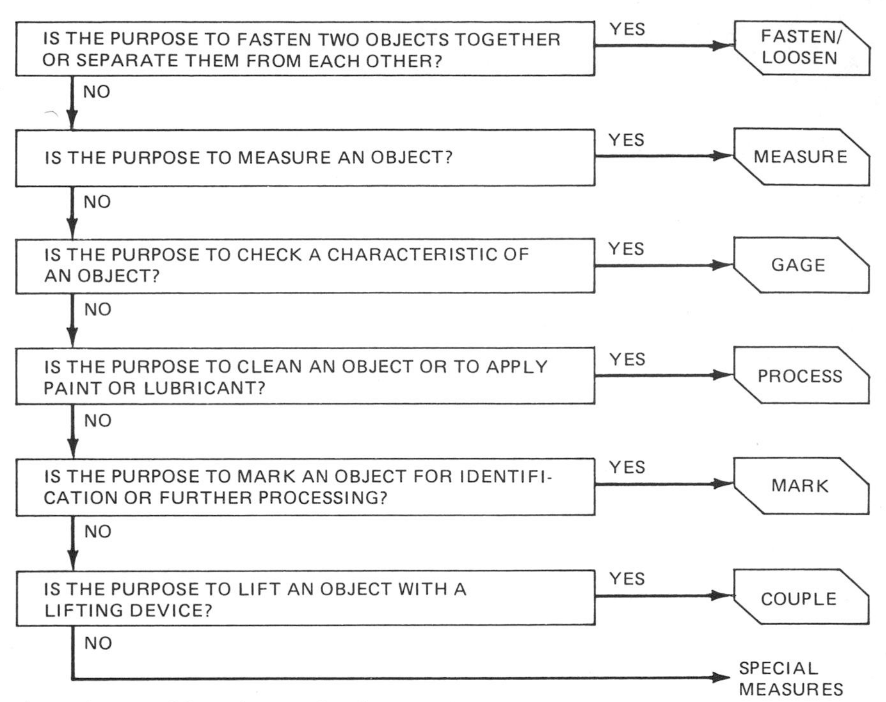

Figure 9–8. Decision Diagram for the Complex Elements. © MTM Association for Standards & Research, Inc., Fair Lawn, N.J.

KP does not include Tool Handling, but it is a part of all the other Complex Elements.

The *included* Object Handling may occur *either* before or after the element activity. If it occurs both before and after, it requires coverage by another element, usually HO.

The fourth coding position (first numeric) covers the Case of Tool Handing.

Case 1	The hand tool involved is already in control of the hand
Case 2	Tool is to be procured and/or returned after use *without* body motions
Case 3	Tool is to be procured and/or returned after use when steps are required (within the 80 inch rule)

Steps are covered for all three phases of Tool Handling (GET, PLACE, RETURN), although they may actually occur in only one or two of the phases.

Example 1: Step 72 inches to bench, pick up a wrench, and use it on the bench, aside the wrench without further stepping: FL-3-.

Example 2: Step 72 inches to bench, pick up a wrench, step 72 inches to a machine, use the wrench, and aside it at the machine without further stepping: FL-3-.

COMPLEX ELEMENTS

FASTEN/LOOSEN	TOOL H'ND-LING	CODE	OBJECT HANDLING			
			0	2	3	4
STRIKING TOOLS ≤ 3 strokes	1	FLA1–	70	110	160	230
	2	FLA2–	110	140	200	270
	3	FLA3–	160	200	250	320
PRYING TOOLS ≤ 3 pries	1	FLB1–	80	120	170	250
	2	FLB2–	120	150	210	280
	3	FLB3–	170	210	260	340
CHUCK KEY ≤ 2 turns	1	FLC1–	100	140	190	260
	2	FLC2–	140	170	230	300
	3	FLC3–	190	230	280	350
ROTATION TOOLS-2 ≤ 2 turns	1	FLD1–	80	110	170	240
	2	FLD2–	110	150	210	280
	3	FLD3–	170	200	260	330
ROTATION TOOLS-5 ≤ 5 turns	1	FLE1–	130	160	220	290
	2	FLE2–	160	200	260	330
	3	FLE3–	220	260	310	380
ROTATION TOOLS-12 ≤ 12 turns	1	FLF1–	290	330	390	460
	2	FLF2–	330	370	420	490
	3	FLF3–	380	420	480	550

MEASURE	TOOL H'ND-LING	CODE	OBJECT HANDLING			
			0	2	3	4
SCALES-1 ≤ 20" (500mm)	1	MTA1–	120	160	220	290
	2	MTA2–	160	200	250	330
	3	MTA3–	220	250	310	380
SCALES-2 ≤ 120" (3000mm)	1	MTB1–	360	400	460	530
	2	MTB2–	400	440	490	570
	3	MTB3–	460	490	550	620
OUTSIDE MEAS'G TOOLS-1 ≤ 4" (100mm)	1	MTC1–	130	170	230	300
	2	MTC2–	170	210	260	340
	3	MTC3–	220	260	320	390
OUTSIDE MEAS'G TOOLS-2 ≤ 20" (500mm)	1	MTD1–	210	240	300	370
	2	MTD2–	240	280	340	410
	3	MTD3–	300	340	390	460
OUTSIDE MEAS'G TOOLS-3 ≤ 40" (1000mm)	1	MTE1–	340	380	440	510
	2	MTE2–	380	420	470	550
	3	MTE3–	440	470	530	600
INSIDE MEASURING TOOLS ≤ 40" (1000mm)	1	MTF1–	220	260	310	380
	2	MTF2–	260	290	350	420
	3	MTF3–	310	350	400	480
DEPTH MEASURING TOOLS ≤ 8" (200mm)	1	MTG1–	160	190	250	320
	2	MTG2–	190	230	290	360
	3	MTG3–	250	290	340	410

Figure 9–9. Data Card tables for the Complex Elements. © MTM Association for Standards & Research, Inc., Fair Lawn, N.J. *(Continued on pp. 176–177.)*

MARK	TOOL H'ND-LING	CODE	OBJECT HANDLING			
			0	2	3	4
PUNCH AND HAMMER	1	MRA1–	90	130	190	260
	2	MRA2–	130	170	220	300
	3	MRA3–	190	220	280	350
COMPASSES ≤ 12" Dia. (300mm)	1	MRB1–	150	190	240	310
	2	MRB2–	180	220	280	350
	3	MRB3–	240	280	330	400
MARKER & TEMPLATE ≤ 40" (1000mm)	1	MRC1–	270	310	360	440
	2	MRC2–	310	350	400	470
	3	MRC3–	360	400	460	530
HEIGHT GAGE Scribing Length ≤ 20" (500mm)	1	MRD1–	70	110	160	240
	2	MRD2–	110	150	200	270
	3	MRD3–	160	200	260	330
WRITING TOOLS	1	MRE1–	70	110	170	240
	2	MRE2–	110	150	200	270
	3	MRE3–	160	200	260	330

PROCESS	TOOL H'ND-LING	CODE	OBJECT HANDLING			
			0	2	3	4
AIR-4 ≤ 64 In^2 ($4dm^2$)	1	BEA1–	50	90	150	220
	2	BEA2–	90	130	180	260
	3	BEA3–	140	180	240	310
AIR-20 ≤ 320 In^2 ($20dm^2$)	1	BEB1–	130	160	220	290
	2	BEB2–	160	200	260	330
	3	BEB3–	220	250	310	380
BRUSH-4 ≤ 64 In^2 ($4dm^2$)	1	BEC1–	30	70	130	200
	2	BEC2–	70	110	160	240
	3	BEC3–	120	160	220	290
BRUSH-20 ≤ 320 In^2 ($20dm^2$)	1	BED1–	190	230	280	350
	2	BED2–	220	260	320	390
	3	BED3–	280	320	370	440
CLOTH ≤ 64 In^2 ($4dm^2$)	1	BEE1–	100	130	190	260
	2	BEE2–	130	170	230	300
	3	BEE3–	190	220	280	350
CHIP HOOK ≤ 3 Motions	1	BEF1–	130	160	220	290
	2	BEF2–	160	200	260	330
	3	BEF3–	220	260	310	380
BROOM ≤ 11 Ft^2 (1 m^2)	1	BEG1–	270	310	360	430
	2	BEG2–	310	340	400	470
	3	BEG3–	360	400	450	520
PAINTING TOOLS ≤ 64 In^2 ($4dm^2$)	1	BEH1–	140	180	230	300
	2	BEH2–	170	210	270	340
	3	BEH3–	230	270	320	390
LUBRICATING TOOLS ≤ 64 In^2 ($4dm^2$)	1	BEI1–	50	90	150	220
	2	BEI2–	90	130	180	250
	3	BEI3–	140	180	240	310

Figure 9–9 (Continued)

COUPLE TO OBJECT	MAIN HOIST HANDLING	CODE	TMU
HOOK, Per Hook	0	KPA0	130
	2	KPA2	210
	3	KPA3	320
	4	KPA4	460
LOOP (Chain or Strap) Per Couple Point	0	KPB0	160
	2	KPB2	230
	3	KPB3	340
	4	KPB4	490
SNARE (Chain or Strap) Per Couple Point	0	KPC0	400
	2	KPC2	470
	3	KPC3	590
	4	KPC4	730

Figure 9–9 (Continued)

Example 3: Step 72 inches to bench, pick up a wrench, step 72 inches to a machine, use the wrench, step 72 inches to return the wrench to the bench: Fl-3-.

No provision is made in the tool handling coding to cover bending and arising on the theory that tools are generally not kept on the floor. If bending does occur (when the tool box is on the floor, for example), it should be analyzed separately with MTM-2.

The fifth coding position, except for COUPLE TO OBJECT (KP), contains the OBJECT HANDLING (HO) codes or cases 0, 2, 3, and 4. See the discussion for HO if these are not remembered.

In general, the same concepts, limitations, exceptions, etc., apply to the Complex Elements as to the Simple Elements. These elements will therefore not be discussed in great detail. A few elements, however, will be briefly treated so as to enable the reader to understand their usage.

FASTEN/LOOSEN—FL

Glancing down the left column of the FL Table indicates that there are six alpha cases, “A” through “F”.

The Decision Diagram for FL is illustrated in Fig. 9–10. Note that the other Complex Element Diagrams are similar and all of them can be obtained from the U.S./Canada MTM Association.

A few amplifying facts. STRIKING TOOL is assigned to *any* tool used to strike blows on an object, even something not thought of as a striking tool.

The “tapping” of an object to ensure contact or to align a workpiece is included in the concept of FL.

Do not strike a revolving workpiece, it is not safe and is not included in FL.

The following examples will make clear the use of FL:

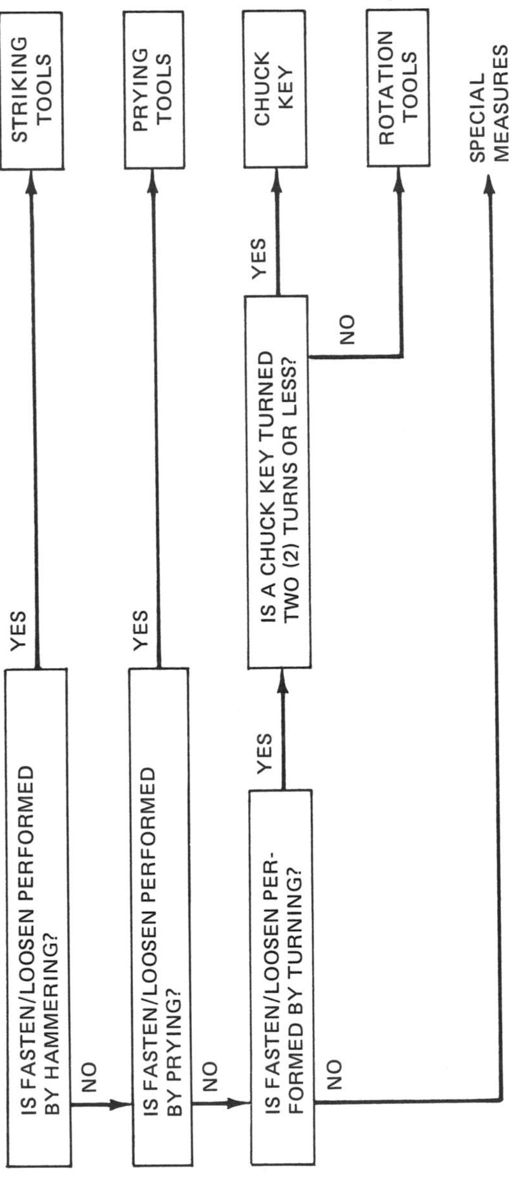

Figure 9–10. Decision Diagram for FASTEN/LOOSEN. © MTM Association for Standards & Research, Inc., Fair Lawn, N.J.

1. Operator picks up a rubber mallet within reach and taps a workpiece in a chuck twice to align it.

Tap, two strokes	FLA20	110 TMU

2. Operator steps 6 feet to get a hammer, then steps back to work place and strikes a wedge eight times to dislodge it. He/she then places the wedge on the bench and returns the hammer to its original location.

Get, place, and return hammer, strike wedge three times	FLA30	160 TMU
Strike three additional strokes	FLA10	70
Strike two final strokes and aside wedge	FLA12	110
		340 TMU

Note that hammer return was covered in FLA30. The wedge handling (object handling) was covered in FLA12.

3. Operator picks up a hammer within reach, strikes a wedge twice to remove it, and lays the wedge aside. Then he/she strikes a second wedge four times and removes it. Finally, he/she asides the wedge and steps five feet to lay the hammer aside.

Get, place, and return hammer, strike and aside wedge	FLA32	200 TMU
Strike second wedge three times	FLA10	70
Strike second wedge once more and aside it	FLA12	110
		380 TMU

PRYING TOOL

No amplifying remarks are considered necessary.

CHUCK KEY

FLC is assigned to a tool that usually requires two hands to FASTEN/LOOSEN. A maximum of two turns is included, since this usually is ample. If chuck key can be turned with one hand or when more turns are needed, the element is analyzed by using an appropriate ROTATION TOOL (FL0, FLE, or FLF).

If additional tightening or loosening is needed by moving the key to another place on the chuck, ROTATION TOOL (FLD) is used for the second activity.

If simply more turns are required at a given chuck key location, assign ROTATION TOOL FLE or FLF (or a combination of FLD, FLE, and FLF), whichever is correct.

ROTATION TOOL

This element is assigned to a hand tool which is used to FASTEN/LOOSEN a threaded object which joins, or is joined, to other objects by mated threads. It is also used when more turns are required on a chuck key than two turns.

PROCESS—BE

BE is used when an object is cleaned or treated with lubricant or a coloring agent is applied with the help of a hand tool.

The cleaning is limited to the work the operator does at his/her regular place of work and in connection with his/her machine. The cleaning might even be drying by using a cloth.

The use of the BROOM is *not* to clean the floor, but is assigned to a tool which is a brush that requires (normally) two hands to sweep loose deposits from an object.

Painting tool not only covers the conventional coloring process, but it is also assigned where the "agent" is transferred to the fingers for application or applied with a stick, brush, oil can, etc. It includes "dipping" the applicator. It is assigned *only* when no high degree of control is required. It allows process time for allowing oil to pour onto an object.

MEASURE—MT

This code is assigned when a tool is used to determine *a* dimension of an object and to *mentally* record it. It starts when the object and tool are in position against each other, ready to perform the measurement. It includes all motions to fit and adjust the tool, to determine the measurement, and to break contact between the tool and the object.

Scales 1 and 2

This measurement involves using a tool without a moving part. Basically, it is used to measure between two parallel surfaces, edges, or between two points.

It is likely that a case of HO will be included to handle the part being measured and/or a case of HH for handling to tool (scale).

Inside, Outside, and Depth Measurements

These cover the conventionally understood measurements associated with these terms. The measurements involve using vernier calipers and micrometers—outside, inside, and depth varieties for both.

Maximum ID and OD covered is 40 inches—but note the increments identified in the table. With Scale 2 the measurement is up to 10 feet.

Depth can be measured up to a maximum of 8 inches with MTG.

GAGE—KO

KO is used when checking if an object is within given tolerance(s) with the help of a tool.

It starts when the gage is in contact with the object, ready for positioning, and ends upon separation of the part and the gage.

This Decision Diagram for KO is interesting and is illustrated in Fig. 9–11. This set of elements is assigned when a tool is used to check the minimum and/or maximum dimension between internal faces of an object (for holes up to 4 inches in diameter). It includes the use of feeler gages, plug gages, taper gages, step gages, etc.

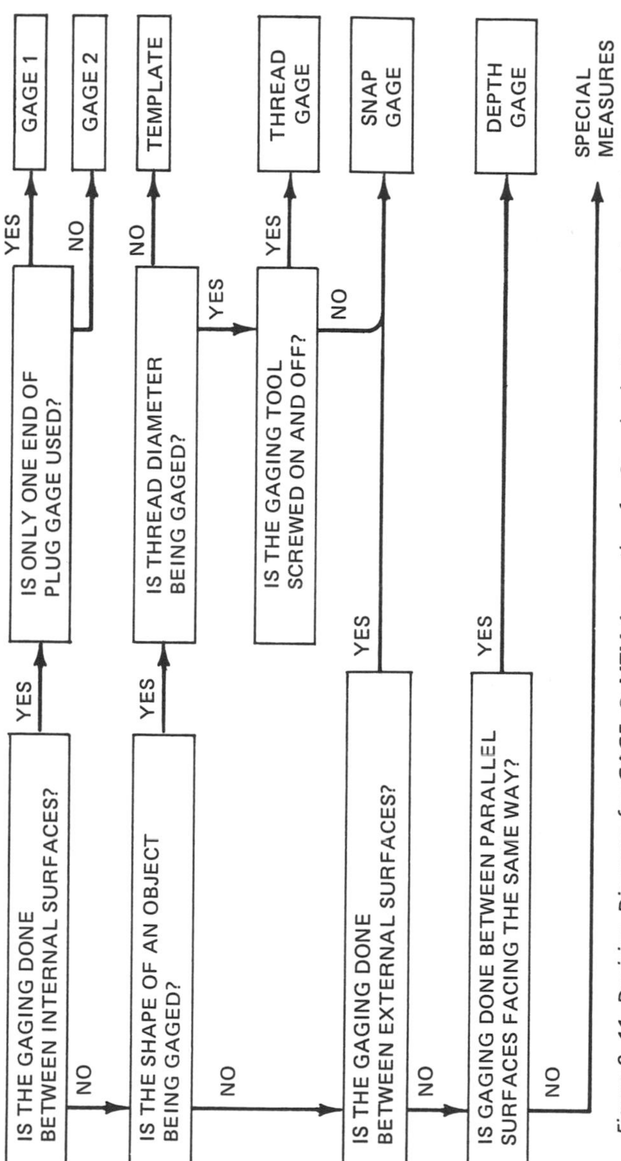

Figure 9–11. Decision Diagram for GAGE. © MTM Assocation for Standards & Research, Inc., Fair Lawn, N.J.

A typical application is "step from bench to get gage, get part from bench, apply gage, aside part (HO2)." If the aside had been "simo gage and part," the HO2 would not have been required, only KOB32, assuming, for example, that a plug gage was used.

TEMPLATE covers a tool used to check the profile of an object to a maximum length of 20 inches (there are actually two categories, up to 8 inches and above 8 inches to 20 inches). Templates include radius gages and contour gages, but not thread gages.

THREAD GAGE is assigned to the use of a tool which is screwed onto or into a threaded object in order to establish whether the fit (thread) is within specified tolerance—up to a diameter of 4 inches.

"Running" the control device (gage) on or off the object with the object rotating is not covered.

A typical example is to check the threads in a hole in a part (object) mounted on a threader. The operator has the gage in his/her hand and walks to the machine. He/she screws in the gage 10 turns and then unscrews it. The gage is asided on the machine.

Walk to machine	2S	36 TMU
Check thread	KOE10	340
		376 TMU

SNAP GAGE is used to check the dimension between external surfaces up to an 8 inch length.

DEPTH GAGE is assigned when a tool is used to check a dimension between parallel head and bottom seating of an object (usually one external surface and one internal surface)—to a depth of 4 inches.

Gaging with gage blocks is analyzed by using the DEPTH GAGE category. An example of this application would be as follows:

Open and close (after use of gage block) gage block case (within reach)	HLD0	70 TMU
Get block, place shaft in fixture, gage shaft, return block	KOH22	170
Aside shaft to machine table	HO3	90
8 feet away	S	18
		348 TMU

MARK—MR

MR is used when covering the marking of an object *with a tool* to identify an exact position.

It starts when the object and the tool are in position against each other ready for marking, the tool handling covers movements to select and get, for example, a punch and return it (say, to a case). This "tool handling" does not cover such things as taking out a case of punches from a cabinet and removing or opening the lid of the case, etc. Two examples will make this clear.

1. Get a hammer and letter punch "E" lying on top of a tool cabinet 60 inches away and mark an object in the machine.

Mark part	MRA30	190 TMU

2. An operator picks up two punches ("5" and "2") from a reachable open case and lays them on his/her bench. He/she then punches (in sequence) 5525 on a part which is *not* handled.

Get punches and punch first "5"	MRA22	170 TMU
Punch second "5"	MRA10	90
Punch "2"	MRA20	130
Punch third "5" and return punch and hammer	MRA20	130
		520 TMU

COMPASS is assigned to an instrument used to describe an arc or circle of a specified diameter. The element does not include adjusting the two points of the compass; use MTD__ for this.

MARKER AND TEMPLATE is assigned when a scribe is used to mark a maximum of 40 inches along a profile, template, or rule. Scribe *and* template are considered one tool.

HEIGHT GAGE is assigned to a tool or device used to mark a line around an object (maximum length 20 inches) at a specified height. The element does *not* include setting the scribing height; use MTG__ for this action.

WRITING TOOL is assigned to a tool used to form a maximum of seven characters or symbols on an object. Writing with a vibrating tool is not covered, and special measures must be used to cover such action.

COUPLE–UNCOUPLE—KP

This code is used when attaching and subsequently detaching an object from mechanical handling equipment using a specified coupling device (see table in Fig. 9–9). Handling devices include swinging crane, hoist, or overhead crane.

This code includes all time-limiting hand, arm, and body motions necessary to attach a coupling device to mechanical handling equipment or to one coupling point on an object before lifting it and subsquently detaching it.

It does not, for example, include putting a chain around a piece of heavy equipment, then bringing up the mechanical handling equipment by pulling it (or traversing it by operating a walkalong set of push buttons) to the correct location for attaching and lifting. What is covered is coupling a hoist hook, for example, to the chain and detaching it at the new location. Moving the loaded crane is not covered, neither are the actions to lift the load.

COUPLING DEVICE HANDLING uses the same case structure as Object handling; the values, however, are doubled to allow for both the coupling and uncoupling.

KP has only one handling category since the object itself is never handled manually; it is therefore not necessary to specify a case for it in the fifth code position. (If a code were shown, it would always be "0.")

Case 0 is assigned when the coupling device is permanently joined to or already located on the mechanical handling equipment and is not removed on uncoupling.

Case 2 is used when no body motions are required to obtain, attach, and/or replace the coupling device.

Case 3 is assigned, like in other such numbered cases, when time-limiting steps (within the 80 inch rule) are required to obtain, attach, and/or replace the coupling device.

Case 4 is assigned, as in other such numbered cases, when time-limiting steps and/or one bend are required in order to obtain, attach, and/or replace the coupling device.

KP is the only MTM-V element which may be assigned with a frequency of one-half. Whenever coupling alone or uncoupling alone is analyzed, it is assigned as only one-half KP. Also, when the coupling device handling case is different for uncoupling than for couple, half elements are used.

One other aspect of KP is that the times are based upon *one* coupling point. If there are *two places of attachment* (like two hooks or two places a "snare" or a loop is attached), each coupling point must be covered. The second coupling point code would carry a Case 0, since obtaining/returning the coupling device is included with the first coupling point.

One final example is now presented:

> Operator lifts an axle to a machining table with the help of a traveling crane and two endless straps. They are lying on the floor next to the operator. Travel time (for the crane) is 100 TMU. The operator walks three steps *between the coupling points.*

Get strap, couple and uncouple first point	KPB4		490 TMU
Couple and uncouple second point	KPBO		160
Time between the coupling points	3S	2	108
Travel time	MT		100
			858 TMU

Concluding Admonition

Do not attempt to apply MTM-V without proper Association-approved training. The two previous paragraphs and the example illustrate the importance of "rules of thumb" and special conventions. Moreover, the backup MTM-1 patterns for the MTM-V are not available for reference in this text.

CHAPTER 10

MTM-M: A High-Level Data System for Setting Labor Standards on Work Involving Magnification for Part or All of the Work[1]

Definition[2]:

MTM-M is a system of objective methods-time standard data, based on regression analysis of empirical data, for evaluation of operator work using a stereoscopic microscope.

The authors would like to offer a somewhat more direct and different definition:

MTM-M is a specialized methods-time functionally oriented standard data system that is not a higher-level MTM-1 based system, yet it is designed to produce time standards that are compatible with MTM-1 standards for work performed partially or totally under a binocular stereoscopic microscope having a magnification range not exceeding 30 power

That the above definitions are compatible will become clear upon a study of this chapter.

The need for this new system was brought about by the great increase in microminiaturization in the electronics industry which created a special set of problems for the work analyst trying to set standards of performance. The stereoscopic microscope soon was in common use by operators who were placing and positioning parts as small as twenty thousandths of an inch to tolerances as small as one thousandth of an inch.

Traditional work measurement techniques just do not work in this environment. Stopwatch time study is not applicable since the analyst cannot see what is happening. Even if a special microscope was employed so that a time study analyst could actually view the actions of the operator under the microscope, the movements are often so minute and delicate that accurate timing would be

[1]MTM-M is a very specialized functionally oriented system that is useful only to organizations performing work under magnification, primarily under binocular stereoscopic microscopes. We have included it in this text because it is an excellent example of a unique and important work measurement research project that discovered new research findings. The reader should understand that only a very few of the necessary "rules of thumb" required for the system's successful use are included, as well as an incomplete data card. Do NOT attempt to apply this system without proper training in an approved course. Information on such courses is available from the MTM Association for Standards and Research, Fair Lawn, N.J. Much of the material presented herein was taken from the Approved Training Manual for MTM-M (direct quotations are cited), from other cited references, and from the personal knowledge of the authors, especially that of Dr. Walton Hancock.

[2]Courtesy of U.S./Canada MTM Association. See footnote 5, Chapter 6.

impossible. Moreover, the operator has to move into and out of the "field of vision." Breakpoint identification normally cannot be seen, certainly cannot be heard or easily "sensed." Conventional predetermined time systems, including MTM-1, seem to underpredict microassembly performance, sometimes by as much as 50 percent.[3, 4]

The results are also very variable. Essentially, the results cannot be used. The size of the parts, the movement distance, and the performance time effects of magnification caused the need for a new and fresh approach.

Before discussing the new approach it should be outlined as to why MTM-M and MTM-1 and other MTM-1 based higher-level systems are compatible. First, both systems produce low task standards. Low task standards are defined by the American National Standards Institute as: "LOW TASK—a term used to indicate that performance rating or production standards are based on daywork levels as contrasted to high task or incentive work performance . . . ."[5] This is in contrast to their definition for "high task": "HIGH TASK—performance of an average experienced operator working at an efficient pace, over an eight-hour day under incentive conditions, without undue or cumulative fatigue . . . ."[6]

The other reason they are compatible is that the "end points" of the combinations of motions are in accord with the end points of MTM-1 motions.

System accuracy is essentially defined in terms of MTM-1 system accuracy, a final reason.

MTM-M was developed by an ad hoc committee of the U.S./Canada MTM Association.[7] The committee members were from organizations that had a need for more information in the area of concern. Also included in the membership was the Association's research activity sponsored at the University of Michigan.[8] The committee was known as the MTM Magnification Research Committee until the committee changed its name from Committee to Consortium, which more nearly described their general activities.

The following organizations had participating members:

Company	*Location*
Autonetics Division, Rockwell International	Anaheim, CA
Bausch & Lomb	Rochester, NY

[3] S. Konz, G. Dickey, C. McCutchan, and B Koe, "Manufacturing Assembly Instructions; Part III Abstraction, Complexity and Information Theory," Journal of Industrial Engineering, Nov., 1967.

[4] *MTM Magnification Research Consortium, Final Report,* MTM Association for Standards and Research, 1973. This document is the source of much of the information presented in this chapter and is available through the MTM Association for Standards and Research, Fair Lawn, N. J.

[5] *American National Standard, Industrial Engineering Terminology, Work Measurement and Methods,* ANSI Z94.12, American National Standard, Secretariat American Institute of Industrial Engineers, Inc. and The American Society of Mechanical Engineers (secure copies from these societies).

[6] *Ibid.*

[7] MTM Association for Standards and Research, Fair Lawn, NJ.

[8] Coauthor Dr. Walton Hancock headed the sponsored research activity at the University of Michigan at the time of MTM-M's development. Much of the work was done by Gary D. Langolf and was the basis for his PhD thesis, *Human Motor Performance in Precise Microscopic Assembly Work,* PhD dissertation, University of Michigan, 1971.

Hughes Aircraft Company	El Segundo, CA
IBM Corporation	Poughkeepsie, NY
Microsystems International, Ltd.	Ottawa, Ontario, Canada
MTM Association for Standards & Research	Fair Lawn, NJ
Sylvania Electric Products, Inc.	Warren, PA
Systems Group of TRW, Inc.	Redondo Beach, CA
University of Michigan	Ann Arbor, MI

This consortium represented an outstanding example of intercompany cooperation and collective action. The participating organizations contributed time, talent, and funds—in fact, virtually all of the funds required were provided by the industrial members.

During the research, approximately 1000 technical, engineering, and management people participated. Moreover, 114 operational personnel cooperated in the case studies and experiments. In total, 20 case studies were conducted that involved 48 operations cycles and represented about 80,000 TMU.

The developed system was named MTM-M since it was developed by the MTM Association and involves magnification. Training is available to the users or the system through the Association.

While MTM-1 basic times were not used, the behavioral motion elements were used in the analysis of the experiments, case studies, and operations. In fact, the beginning and end point definitions of MTM-M motion combinations, as stated before, are compatible with the beginning and end point definitions of MTM-1. This is especially important for MTM-1 covered work performance outside the field of the microscope prior to and/or subsequent to MTM-M type activity.

This data system is the first standard data system based on original data developed and acquired under the auspices of the U.S./Canada MTM Association.

Some Important Facts: Design Criteria, Advantages, and Limitations

Like other MTM systems developed after MTM-1, a set of design criteria were established early in the project. These are

- Promote consistency in application
- Be descriptive of the method
- Be combinable with other MTM systems
- Provide a base for higher-level data

MTM-M has the following advantages when dealing with the kind of work covered:

- Forces in-depth methods analysis
- Provides assistance in product and tool design

Provides methods description which can be used in operator training
Produces consistent time standards
Utilizes motion combinations to increase speed of application without sacrificing accuracy
User's element codes are mnemonic and easy to understand

As is true for almost any system, MTM-M does have limitations:

Requires skillful application
Designed for stereoscopic microscopes with a magnification range not over 30 power
Applicable with common types of tools only
Does not measure process time or limiting eye time

The Research Project Organization and a General Description of its Activities

Figure 10–1 presents the essential facts of the development and validation system model. A flow chart illustrating the primary sequence of events is shown in Figure 10–2. This illustration does not indicate the presence of very important, but ancillary, work. Including such events would have made the figure extremely complex and confusing. The ancillary work included library research (an integral part of any good research project), laboratory studies, employee studies (vision, tremor, etc.), production equipment studies, research equipment selection and/or construction, methods and training applications, etc. These activities involved both the industrial organization members of the consortium and the University of Michigan research team.

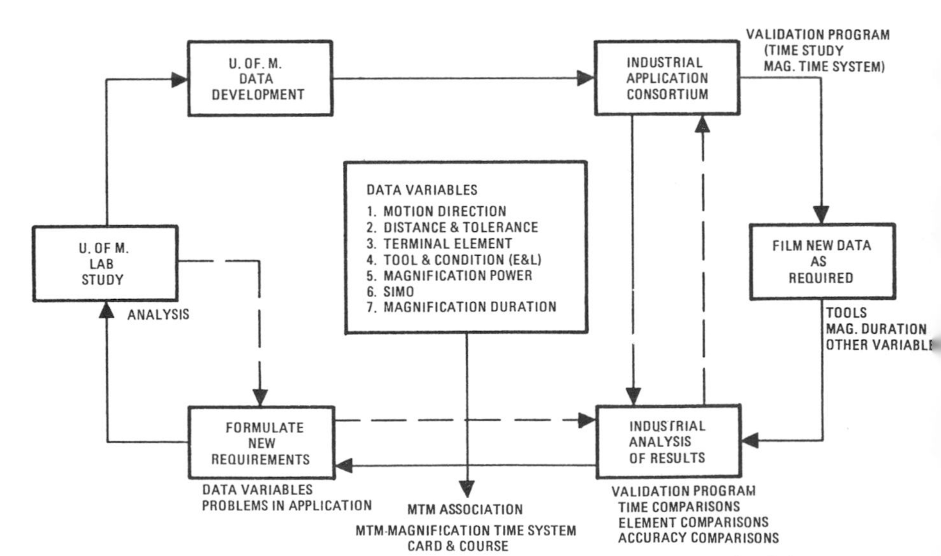

Figure 10–1. MTM - Magnification Time System. Development and validation system model. © MTM Association for Standards & Research, Fair Lawn, NJ.

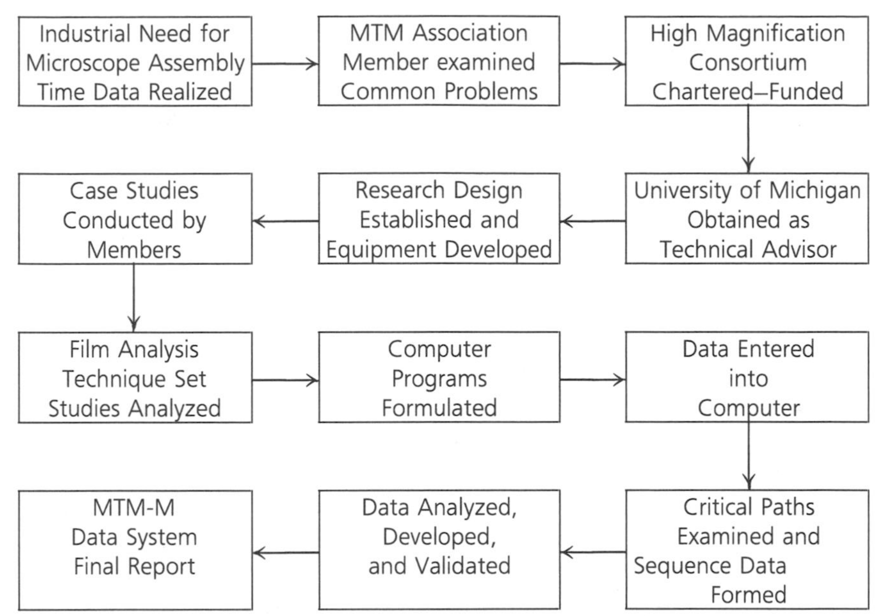

Figure 10–2. The primary sequence of events in the MTM-M Research and Development project. © MTM Association for Standards and Research, NJ.

Consortium members surveyed their organization's usage of microscopes to determine the types of work being performed and the importance of such work as reflected by the number of employees assigned to each task. An assignment matrix reflecting these data vs the research needs is presented in Fig. 10–3.

Employees selected for participation in the project were restricted to those who were satisfactory performers and skilled in the particular operation to be studied. Eye examinations were given to each subject to establish the subject's degree of depth perception, astigmatism, and far-and-near sightedness. The fact that glasses were often normally required by an operator did not exclude that person from being included in the study.

Preliminary eye–hand coordination tests and tremor tests were also recorded for each individual through the use of a Dexterity Maze (developed by the University of Michigan research team) and a Tremor Transducer connected to a Data Storage Oscilloscope. In fact, these two tests were periodically given during the case study procedures.

As to the tremor test data, since all humans have a certain degree or amount of tremor in their hands and fingers, it was therefore decided that information should be collected documenting the amplitude and frequency of such movements in each person. The device used to detect this was developed at the University of Michigan and was used during the filming to document the operator tremor most clearly approximating the operations as they were filmed.

Another new and somewhat different data set was obtained on all case study data acquisitions, by means of a general technique previously used in MTM research. This technique, developed by Dr. Hancock and his associates at the

Power Range	Classes of Work			
	Direct Manual	Tool Use	Micro-manipulators	Inspection
7–10	Autonetics-5 Autonetics-10	TRW-7 Hughes-10 Hughes-10 Micro-10[a] Autonetics-10		TRW-10 Sylvania-7
11–20	Autonetics-15	TRW-20 Hughes-15 Hughes-20 TRW-20	Hughes-20 Micro-15[a] Micro-15[a]	Autonetics-15 Autonetics-20
21–40		TRW-30 TRW-30 TRW-30	Micro-40[a]	TRW-30 Sylvania-30
41–60			Micro-50[a] Autonetics-45	Autonetics-50
61–100				IBM-80
100–400				Micro-100+[a]

[a]*Microsystems.*

Figure 10–3. Assignment matrix in terms of company, magnification power, and classes of work. © MTM Association for Standards & Research, Fair Lawn, NJ.

university, made it possible to photographically capture eye movements. This was accomplished by photographing the "tracings" on an oscilloscope at the same instant the other data were being captured on film by means of a special camera described later. The "tracings" were produced by placing electrodes on the operator's facial eye muscles which signal muscular action. The technique is known as electrooculargraphy (EOG).

Data on *all* operations and case studies were gathered by filming. Two high-speed electric Milliken cameras were synchronized, and one camera was used to record the actions as seen through the microscope and the other filmed the actions taking place outside the microscope's field of vision. Naturally a special microscope was used that permitted both the camera and the operator to "see" the same thing.

When filming a case study, the cameras were driven from a single control box to assure electrical unification. The outside of the field camera also recorded the tracings on the oscilloscope which indicated the eye movement of the operator. Following the filming, frame by frame analyses were made of each operation to permit documenting all motions. This was accomplished by utilizing a pair of L & W Photo Optical Data Analyzer Projectors which provided flickerless projection and accurate frame count registry. So as to make the analyses in a consistent manner, a training program was developed which incorporated the analysis of film clips from the first two case studies. Each team analyzed the training films by applying established analytical rules. This was followed by audits of these results which enabled the committee to qualify analysts and resulted in consistent and precise data from the film analysis. All data were stored in the computer.

In order to assure the proper collection and recording of data a set of standardized data sheets had to be completed for each case study. Three of these data sheets concentrated on the general and personal operator data. The other two dealt with the operation data.

The Microscope

Since this MTM system deals primarily with work under a microscope, some related general and specific information is presented here. It will enable the reader to be better able to perceive the problems encountered and understand how some of these were overcome. More information is given in the training course.

Industry has used various means to improve the visibility and quality of assembly operations. One of the first instruments used was the common magnifying glass, and it is still being used. Generally these magnifying glasses provide no higher than double power (2×).

As electronic assemblies became smaller, industry began to be forced into the use of microscopes. These were originally designed for use in the medical science or biological fields.

The type of instrument used today is known as a binocular microscope, so called because both eyes are used for viewing. Some of the microscope trademarks are Stereozoom and Stereo-Star. There are others, including Japanese makes and styles. The essential external features are as in Fig. 10–4.

Most binocular microscopes contain a focusing knob which is used for raising and lowering the main body or pod of the microscope. A magnification knob is used to increase or decrease the magnification power. One of the eyepieces has a focusing ring built into the eyepiece tube so that the operator can adjust the microscope to accommodate differing degrees of vision between the two eyes—just like one can with a good pair of binoculars. Also, the distance between the eyepiece tubes can be adjusted to the viewers physical dimensions.

When training operators who normally wear glasses, they should be encouraged to continue to wear them when using the microscope. The main reason for this is that much of their work is performed off the microscope and the operators will need their glasses for this activity when it occurs. In order to accommodate the wearing of glasses it is only necessary to remove the eye guards and substitute felt eyeglass protectors on the eyepieces.

The magnification knob works with the eye pieces to obtain the desired magnifying power. The total magnification of a microscope equals the power of the eyepieces (these come in various sizes, assume, for example, 10×) times the power setting of the magnification knob. If the magnification knob in our example is set at 0.8, the total power is 8× (10 × 0.8). Most microscopes used in industry have a power range of 7–30. As the magnification is increased, the viewing area becomes smaller and smaller, but the components *appear* to get bigger and bigger.

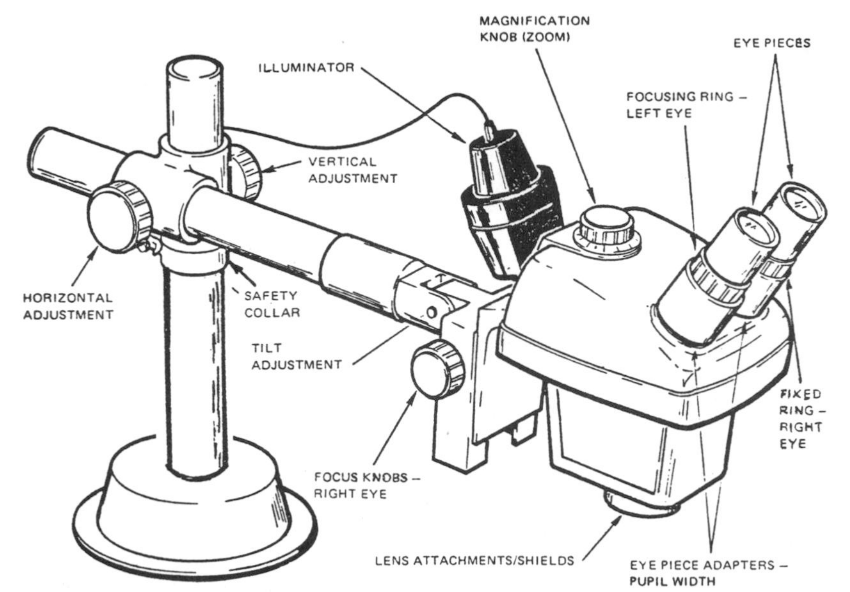

Figure 10–4. The essential features of a Stereozoom microscope. © MTM Association for Standards & Research, Fair Lawn, NJ. Manufactured by Bausch & Lomb Co. and American Optical Co.

The focusing knob of the microscope is used to bring the object into focus. With the Stereozoom microscopes the operator needs to focus on the working area only once. Thereafter, the magnification knob is used to get whatever power is required without refocusing as long as the plane of the desired focus has not changed.

The following are important guidelines for the type of work being discussed:

1. The microscope must be positioned over the work and adjusted to a comfortable working height.
2. The operator's chair must be adjustable to the fixed height of the workbench so that the work surface is even with the elbow.
3. The chair back rest should be adjusted to support the small of the back.
4. The foot rest should be adjustable to a comfortable position for the operator; the knees should be slightly higher than the hips.
5. Next determine the proper microscope eyepiece level and angle and make such adjustment. This adjustment should be such that when the operator swivels away from the microscope and back again, the eyes will remain at the same level.
6. Angle the pod so there is a minimum of head bending involved in looking through the eyepieces.
7. The final adjustment is to fit the working surface of the work to the microscope working distance which is dependent upon the focal length of the microscope. The plane (surface) where the work is in focus is called the working distance, and it varies from a few inches to about 7 inches. The greater this distance, generally, the more the operator will like the instrument (neglecting magnification).

8. Fit the working surface to the microscope working distance, which is done by raising and lowering the work stage surface directly under the microscope lens and its field of view.
9. Adjust the microscope for proper viewing.

The exact methods used to accomplish these adjustments will be found in the approved *Training Course Manual.*

Operator Tremor

Every individual has some tremor which can be measured in amplitude and in its frequency per second. Tremor varies with age and other physical characteristics and becomes more noticeable visually as magnification power is increased.

It was noted during this research that the higher the power and the closer the tolerance, the longer it took an operator to accomplish a position—a fact related to tremor.

Illumination

While this subject is not a microscope characteristic, it is very closely associated with its use. It was found that the operators seemed to accept higher and higher light levels—all that would be provided.

The majority of microscopes have a built-in spotlight whose illumination can be varied by the operator. However, many operators like to use supplementary lighting in the form of high intensity lamps.

Room lighting should be similar to the lighting through the microscope in order to minimize pupillary contraction or enlargement when working "off" the microscope, or more specifically when moving into and out of the microscope field. By similar it is meant that it should have roughly the same intensity (foot candle power at the viewing level).

In conclusion, it is important to remember that as magnification increases

the field of view decreases,
the depth of field decreases (small depth of field is a hindrance to the operator), and
the working distance decreases (a small working distance, distance from lens to working plane, is a handicap in many operations).

Fitts' Index of Difficulty (ID)[9]

This index was developed in a human performance laboratory and was reported on in 1954. ID is a function of the movement distance (or amplitude as

[9]P. M. Fitts, "The Information Capacity of the Human Motor System in Controlling the Amplitude of Movement," *Journal of Experimental Psychology* **47** (1954).

Fitts stated the discovery) symbolized as A and the tolerance (the way it is used and defined in MTM-M means target diameter or width vs that of the object being placed) at the terminal point. The tolerance is symbolized as W. The ID and other similar measures have often been used in experimental psychology studies of motor performance.[10, 11, 12] It has also been used to indicate task difficulty in industrial engineering studies.[13, 14] Movement times were found to be proportional to Fitts' ID.[15, 16] In fact, it is a well recognized and much reported on relationship that many have found useful.[17, 18, 19]

Footnote 17 is an early report on the beginning of the MTM-M study. It is important to understand some of the essential facts of Fitts' Index of Difficulty since it is a principal parameter of the MTM-M data tables. Why this is so will be developed later. Moreover, we believe it important since it states some fundamental and underlying facts surrounding human performance. It could well be used in other data systems and even in some forms of standard data.

One of the facts discovered in MTM-M is that the movement times were *linearly* related to ID in accordance with Fitts' law[20]:

$$\mathrm{MT} = a + b \log_2 (2A/W)$$

where MT is movement time. In the above a and b are empirically determined constant parameters resulting from regression analysis, but they do vary with the type of movement studied. The amplitude A and the target width W have already been identified. Since this so-called "law" was developed from information theory, the ID has units of "bits." The original expression of his work was

$$\mathrm{ID} = \log_2 \frac{2A}{W}$$

Once Fitts' ID was singled out for attention and studied extensively by the consortium, it became obvious that Fitts' ID might be used as the key element

[10]E. Crossman, "The Information Capacity of the Human Motor System in Pursuit Tracking," *Quarterly Journal of Experimental Psychology* **12**, 1–16 (1960).

[11]P. Fitts and J. Peterson, "Information Capacity of Discrete Motor Responses," *Journal of Experimental Psychology* **67**, 103 (1964).

[12]A. Welford, "Measurement of Sensory Motor Performance," *Ergonomics* **3**, 182–230 (1960).

[13]S. Konz, C. Jeans, and R. Rathore, "Arm Motions in the Horizontal Plane," *AIIE Transactions* (Dec., 1970).

[14]S. Konz, G. Dickey, C. McCutchan, and B. Koe, "Manufacturing Assembly Instructions: Part II. Abstraction, complexity and Information Theory," *Journal of Industrial Engineering* (Nov. 1967) (previously referenced).

[15]See footnotes 3, 4, and 5.

[16]Jannet, C. W. Golby, and H. Kay, "The Measurement of Elements in an Assembly Task, the Information Output of the Human Motor System," *Quarterly Journal of Experimental Psychology,* **10** (1958).

[17]D. O. Clark, "Industry Launching MTM Microscope Research," *Journal of Methods Time Measurement* **XIV**, 4 (1969).

[18]P. M. Fitts and J. R. Peterson, "Information Capacity of Discrete Motor Responses Under Different Cognitive Sets," *Journal of Experimental Psychology,* **69**, 71 (1966).

[19]S. W. Keele, "Movement Control in Skilled Motor Performance," *Psychological Bulletin* **70** (1968).

[20]W. M. Hancock, G. Langolf, and D. O. Clark, "The Development of Standard Data for Stereoscopic Microscope Work," *AIIE Transactions* **5**, No. 2 (1973). Note: This paper was another principal source used for the writing of this chapter.

(predictive unit) in a predetermined time system—especially MTM-M. Fitts' law states that the time required for a movement is a function of the ratio of distance to tolerance (*A/W* is the ratio described). Therefore, it could be expected that for a movement (*A*) of 5 inches where the associated destination tolerance (*W*) is 1 inch the time required would be the same as for a movement (*A*) of 0.5 inch if the tolerance (*W*) was 0.1 inch. The law implicitly gives equal weight to distance and tolerance. This means that doubling the distance has the same effect as reducing the tolerance by 50 percent.

It was mentioned that the ID seemed to be a good predictor of the time required to perform certain tasks. For example, in an early laboratory experiment at the University of Michigan it was found that the ID was a very good predictor of the time required to perform a miniature peg transfer task within the field of a microscope. Correlation coefficients between the ID and the performance time for three subjects ranged from 0.88 to 0.97.[21] More specific data on ID as a predictor will follow.

No existing predetermined time system at the time MTM-M was being developed had been designed to use ID as a key element. When ID was calculated for all distance–tolerance combinations of MTM-1 C Moves and the supplementary Position Data of MTM-1 and compared to data card times, it was found that the correlation between the ID and time was approximately 0.80. The slope constant *b* in the equation was 3.0 TMU/bit or 108 ms/bit. This compares favorably to values in the 100-ms/bit range observed in Fitts' laboratory studies of similar motions. Keele[22] in a similar analysis of another data system found that the relationship between ID and movement time in the data of Bailey and Presgrave[23] showed an even higher correlation, 0.97.

Since the ID "explained" about 80 percent of the time required to perform the kind of work covered by MTM-M, it was logical to use subdivisions of *A/W* in setting up the data tables. This ratio is designated as "Range," but *W* or tolerance is defined as in MTM-1 POSITION. More will be said on this later.

Instrumentation and Equipment Utilized

The industrial data were recorded on two high speed cameras, one recording the activity inside the microscope field using a split beam adaptor attached to the right eyepiece of the microscope. The other camera recorded data in the workplace surrounding the microscope. Electrooculographic signals from electrodes attached near the eyes of the operator permitted the recording of eye movements on an oscilloscope, which were in turn recorded by the "outside the field" camera. This set of equipment assured the recording of all essential data associated with the operator's performance.

Laboratory experimental data at the University of Michigan were collected on three separate input channels via electronic instrumentation. These data

---

[21] *Ibid.*

[22] *Ibid.*

[23] G. B. Bailey and R. Presgrave, *Basic Motion Time Study*, McGraw-Hill, New York, 1958.

were run on-line using a Hewlett-Packard 2115A computer which almost instantaneously determined all movement times. The input channels were:

1. Electrical contact via the operator, the operator's tweezers, and the targets on the experimental circuit board (workpiece), which is later described.
2. Size of tweezer opening was signaled by strain gauges.
3. A special University of Michigan computerized television "Eye" system looking through a split image device relayed the location of the part to the computer every 1/30 second. The "Eye" system also determined the accuracy of part placement positioning.

The Key Experimental Study at the University of Michigan

One of the first actions of the consortium was to further test and explore the ID via direct experimentation at the University of Michigan via the instrumentation previously described.

An experimental task layout was used that involved working inside the microscope field, performing a variety of very specific tasks.[24]

Although part size does have an effect on "reach plus grasp" and "move plus position" time as discovered here and in later studies, the effects are considered minor and are not taken into account with MTM-M. The work analyst, however, should be aware that there are such effects. This also probably means such effects are present in all of the other MTM family of systems.

Other factors possibly having minor effects are part position uncertainty and the difficulty of subsequent motions.

In addition to ID, the researchers developed and used a somewhat related index, called the Index of Positional Certainty (IPC):

$$\text{IPC} = \log_2 (\text{part diameter/radial tolerance})$$

where radial tolerance = $\frac{1}{2}$ (target diameter − part diameter).

In MTM-M the tolerance is calculated the same way. It is equal to one-half of the amount obtained by subtracting from the target diameter the diameter of the part to be placed—or the best approximation of this kind of determination for nonround objects.

It was found, for example, that IPC improved the quality of the prediction of reach plus grasp rather considerably. R + G time decreases as the positional certainty of the part, as measured by the IPC, increases.

When both the ID and IPC were used in regression, these two factors accounted for 92 percent of the mean reach plus grasp time variance.

When ID was used in the regression relationship to predict Move plus Position time (using MTM-1 terminology), the relationship was expressed as

[24] G. D. Langolf, *Human Motor Performance in Precise Microscopic Work* (PhD dissertation, Industrial Engineering), University of Michigan, Ann Arbor, Michigan, 1971.

$$M + P \text{ time (ms)} = -136 + 2381D$$
$$r^2 = 0.87$$
$$SE = 198 \text{ ms (where SE is the standard error)}$$

ID seemed to explain most of the variance.

Since part size does have a *minor* effect, the analyst can take into account the over or under prediction of time if the parts are extremely small or extremely large.

Data Construction

By now it should be evident that MTM-M was developed systematically and reflected advanced thinking as well as excellent research procedures. Analysis of variance and regression analysis procedures were used to secure the best fit of the empirical data to the data categories and/or subdivisions. The process consumed several years and was concluded in 1973.

The film analysis was carried out by subdividing each operation into MTM-1 type motions such as REACH, GRASP, MOVE, and POSITION. However, as the reader by now realizes, it would be impossible to apply a microscopic time system expressed in such terms. Hence, the data tables combine many motion elements and tool-material characteristics. In this sense it is a higher-level data system than MTM-1. Also, it was developed after MTM-1.

Data Variables

Motion Direction in Relation to the Microscope Field of View

There are four major and one supplementary directional movements involved. These classifications are of critical importance to proper MTM-M application.

The directional movements are:

1. *Inside-to-Inside (of the field of view) Motion (II).* This kind of motion covers any motion with the predominant purpose of moving an object or a tool (tool category includes the fingers if they are to be used as a tool) from one point to another point within the microscope field of view. It starts with the first movement (within the field of view that is not associated with an entering movement) and includes all actions between this point and the completion of the movement at the target point.

   This kind of motion is *very* difficult to observe and must sometimes be visualized. Also, frequencies of II motions must be carefully determined.
2. *Inside-to-Outside Motions (IO).* This category includes any motion with the predominant purpose of moving an object or tool from within the microscope field of view to a location outside the field of view. It starts with the first movement within the field of view, includes all intermediate actions such as acceleration, orientation, deceleration, grasp, or release, as required to perform *one distinct motion,* and it is completed with the final movement of completion outside the field of view.

3. *Outside-to-Outside Motion (OO).* A motion with the predominant purpose of moving a tool or object to another position outside of the microscope field of view—the entire sequence being one distinct motion. The motion starts with the first movement outside of the field of view, includes all intermediate actions such as acceleration, orientation, deceleration, grasp, or release, and ends with the movement completed outside of the field of view.
4. *Outside-to-Inside Motions (OI).* A motion with the predominant purpose of moving an object or tool from a location outside the field of view to *within* the field. It begins with the first movement outside to within the field and ends when the tool or object is clearly visible. As before, it includes all acceleration, alignment, and deceleration to get the tool or object in the microscope field of view. It is a very difficult motion to perform and such a move is always required prior to an In-Field-to-Final Target Move (IF).

4A. *In-Field-to-Final Target Motion (IF).* This motion is required to continue (complete) an OI motion to the final target position. As is logical, it starts when the tool or object is in the view of the operator and ends with the movement completed at the target in the field of view. All mental adjustments, acceleration, deceleration, grasp, or release are included again.

Tool Group and Condition

During the consortium's study and research it was discovered that many different kinds of tools were in use and that they often were not used as originally intended. Therefore, in order to better accommodate new tools or new tool usage, it was decided to establish a *functional tool grouping*. For example, X-acto knives were commonly used as probes as were closed tweezers. As to grasping, tweezers, the fingers, or any tool capable of having opposed tool jaws close upon an object can be used. The current functional usage classifications are:

1. Grasping Tools naturally are a general functional class, which would include tweezers, fingers, diagonal cutters, etc. The "cut through" move is included in this category.
2. Probing Tools include any device that logically can be used to perform pushing, touching, digging, or spreading motions with no jaw manipulation required. Common probes are toothpicks, metal and plastic probes, pencils, solder irons, strand of solder, X-acto knives, closed tweezers, etc.

   The process time for soldering is not included. Process time begins when the soldering iron contacts the object.

   Also, if a grasping tool is used to probe, the motion could be analyzed as a contact grasp, but this is also true for the other probing tools.
3. Cutting Tools (razor blades, knives, etc.) are used to cut or trim with a slicing motion; such motions are normally made with considerable care. No jaw manipulation occurs, such as with diagonal cutters—but these are considered as grasping tools. Remember, X-acto knives are often used for probing as well as for cutting.
4. Stripping Tools are devices designed or used to remove insulation from wires by *thermal* means. Such devices have opposing jaws, but this does not permit using

a prior tool classification since they have special handling characteristics requiring special analysis and data values. Process time (when the jaws are closed) is not included in the data.

Tool Condition essentially concerns itself with the empty vs loaded condition. However, when the hand is transporting a tool the predominant purpose of the motion should be the determining factor for judging the tool empty or loaded. However, also involved in the judging is deciding which motion is most critical.

An example might help in understanding these statements. The hand, which is considered a tool in MTM-M (if it is used as a tool), can be transporting a tweezers to pick up a part. The question that must be answered is "are we seeing a loaded finger move or an empty tweezer being transported?" Since the tweezers is essential to performing the defined task, it is an empty tweezers move.

In fact, the tool is generally predominant or of most importance to the motion pattern except for the effect of the "terminating point."

Terminating Points

The following are MTM-M termination possibilities:

1. Grasp
2. Release
3. Abrupt change of direction (greater than 90°, usually about 180°.
4. Motion stop

The predominant purpose as well as the terminating point affect the time required. For example, if an operator is to pick up a part with a tweezers, the tweezers must be so oriented that this action is possible and the eye–hand coordination prior to closing the jaws is time consuming. A complicating factor for the analyst is that he/she cannot always see the termination in microassembly and must visualize the action.

The following types of grasp and release are in MTM-M and are therefore found in the data tables:

Position 5 code (explained later)	GRASP		
G	GT	Grasp *T*ool	Close tool on object, a pick-up grasp
G	GM	Grasp *M*anual	Close fingers on object, a pick-up grasp
C	GTC	Grasp *T*ool, Contact	Gain control by touching object with tool
C	GMC	Grasp *M*anual, Contact	Gain control by having one or more fingers touch object

RELEASE

O	RLT	*R*elease, *T*ool	Release object
O	RLM	*R*elease, *M*anual	Release object by opening fingers
S	RLTC-2	*R*elease, *T*ool, *C*ontact	Surface movement of object across a two-dimensional surface followed by removing tool with a contact release from object
V	RLTC-3	*R*elease, *T*ool, *C*ontact	Same as RLTC-2, but on a three-dimensional space followed by a contact release
M	RLMC	*R*elease, *M*anual, *C*ontact	Release object by removing finger contact

Other Types of Termination

	Motion Abbreviation	Mnemonic Word	
A	—	All	All conditions, type of grasp or release has no effect on data values
N	NONE	None	No grasp or release motion at end of element
M	—	Manual	Aside and RLM of tool (IO and OO tables only)
N	SCRUB	None	A rapid back and forth motion of high frequency which contains no grasp. Generally, it requires little alignment or visual control

Distance Moved

The motion distance to use is the "real" distance traveled, not the apparent distance due to magnification. There are special rules to follow in measuring it since some of these distances are as small as 0.100 inch. They are to be measured or scaled according to the following instructions:

1. Macro motion distances are measured by the standard arc-of-travel techniques of MTM-1.
2. Short finger and tool motion distances are measured from the arc of the tool travel or finger tip travel.
3. Sliding two-dimensional motion distances are measured or scaled on a straight line basis.

4. Drawings, blueprints, and/or scaled photographs should be referred to for those motions too small to be seen clearly or measured easily.

Target Tolerance

This is synonymous with the radial clearance of MTM-1—one-half of the total clearance. It enters into Fitts' ID as used in the data tables, hence, the need to determine this dimension. It is the denominator in the distance/tolerance dimension.

Some examples may help in understanding this measurement for the very small dimensions. Refer to *Engineered Work Measurement* for the larger cases. If a probe is 0.100 inch in diameter and the part to be touched is 0.300 inch in diameter, the target tolerance is (0.300 − 0.100)/2 = 0.100 inch. If a tweezers prior to closing has a jaw gap of 0.050 inch and a part 0.001 inch is to be grasped with it, the target tolerance is (0.050 − 0.001)/2 = 0.025 inch (rounding off the fourth place).

Target tolerance measurement must deal with "real" dimensions, not apparent dimensions.

Additional Dimensional Relationships

The OI motion is *really* made up of two motions that must be combined, an OI plus and IF to the "actual target."

The first target (for the OI motion) is the field-of-view diameter, which varies with the amount of magnification.

To determine the OI table code range (line 1 of table) the user must divide the distance moved by the diameter of the microscope field of view. Enter the table (on the second line) with the quotient obtained in the process just described and read the code range above the group category containing the quotient.

The microscope's field of view will essentially be the IF move distance. This will be in error by some amount owing to actual target location. However, there may also be some vertical or Z-axis travel which causes a larger distance. As a result of this and other factors the table is based upon the field diameter and the move distance.

Therefore,

$$\text{OI TMU} + \text{IF TMU} = \text{Total Move Time}$$

$$\frac{\text{Distance}}{\text{Field Size}}\ \text{TMU} + \frac{\text{Field Size}}{\text{Target Tolerance}}\ \text{TMU} = \text{Total TMU}$$

Simo Motions

Virtually all MTM-M simo motions require an adjustment for the effects of their occurrence. In part, this is due to the higher perceptual loads when working under a microscope. The performance of simo motions generally requires switching attention from one hand to the other. Therefore, in MTM-M a motion is considered simo if any motion is occurring in the opposite hand while within the field of view. Outside of the field, use MTM-1 simo rules only for work not associated with the microscope.

The simo effect is an "adder" (added value in TMU) to the basic motion value. The basic motion TMU value to use is that resulting from the application of the simo rule to the predominant motion—that motion most critical (of most importance) to the work pattern (not the most difficult to perform). Incidently, this predominance or importance will vary from hand to hand, so do *not* choose a hand and stick to it.

Some examples are

1. Hand moving aside and out of the way while the other hand performs a more important task.
2. Manipulation of a tool or part while the other hand is performing a critical motion.

Magnification Power

MTM-M tables are based upon the exact power used by the operator, and this power is a product of eyepiece power, power pod setting, *and the power of any adaptor lens used.*

Microscope power has little time effect for II motions in the 5× – 20× range. At high magnification much of the increased time requirement is probably due to the very small field (0.26 inch in diameter at 30×).

However, in spite of the relatively small variation for II motions in the 5× – 20× range, the overall cycle time of all varieties of MTM-M motions involving magnification increases with the amount of magnification. This is especially evident where the eye must adjust to the hand–object entering or leaving the microscope field. The greater the magnification, the greater the amount of adjustment required. Magnification power, therefore, has its greatest effect on "cross field" motions.

Lab studies show that the motion-slowing effect when moving outside the field lasts about 130 TMU (4.7 seconds)—hence, magnification's effect is finite and eventually disappears. Therefore, the POWER FACTORs contained in the OO and IO tables should *not* be used when the *continuous* off-scope activity equals or exceeds 130 TMU. If the off-scope time is less, use the data table value.

Determining if the continuous outside motions total 130 or more TMU can be a little "tricky." However, if you observe an IO motion, followed by an OO (to the start of the OI motion—there could be another OO motion here before the OI motion), then the rule to follow is to subtotal the IO and OO motion times and compare the sum to the 130 TMU. Do not use the power factors for the OO motions if the sum is over 130 TMU.

Notation and Data Card

An illustration of the data card will now be given. Figure 10–5 presents the II and IO data tables. The other data tables (OO, OI, and IF) are constructed in a similar manner.

MTM-M DATA CARD

II - INSIDE FIELD to INSIDE FIELD

CODE RANGE (D/T)			1	2	3	4	5	6	7	8	9	10	11	SIMO ADDITIVE
RANGE DISTANCE ÷ TOLERANCE			TO 0.75	0.75 TO 1.5	1.5 TO 3.0	3.0 TO 6.0	6.0 TO 12	12 TO 25	25 TO 50	50 TO 100	100 TO 200	200 TO 350	350 TO 725	
TOOL AND COND	CODE	CHARACTERISTIC												
GRASPING	ET	C* GTC (TOOL CONTACT GR)	4.3	5.7	7.5	9.2	11.0	12.8	14.6	16.3	18.1	19.6	21.3	2.1
		G GT (TOOL GRASP)	3.7	7.6	12.2	16.7	21.3	26.0	30.7	35.2	39.8	43.8	48.2	12.8
	LT	N* NO RELEASE – SCRUB	3.7	4.2	4.7	5.2	5.7	6.2	6.8	7.3	7.8	8.2	8.7	+
		S RLTC – 2 DIM MOVE	3.0	3.5	4.2	4.8	5.4	6.1	6.7	7.3	8.0	8.5	9.1	3.9
		V RLTC – 3 DIM MOVE	5.6	6.1	6.8	7.4	8,0	8.7	9.3	9.9	10.6	11.1	11.7	3.9
		O RLT (TOOL RELEASE)	4.5	6.4	8.6	10.7	12.9	15.2	17.4	19.5	21.7	23.6	25.7	9.6
	EF	A ALL CONDITIONS	10.6	12.6	15.0	17.5	19.9	22.4	24.9	27.3	29.7	31.8	34.2	0.0
	LF	N NO RELEASE	3.3	5.5	8.0	10.5	13.0	15.7	18.3	20.8	23.3	25.5	28.0	15.9
		M* RLM OR RLMC	4.1	7.9	12.5	17.0	21.5	26.3	30.9	35.4	40.0	43.9	48.3	5.6
	ED	A ALL CONDITIONS	5.0	8.6	12.7	16.9	21.1	25.4	29.7	33.9	38.1	41.7	45.8	6.0
PROBING	ES	A ALL CONDITIONS	8.9	11.8	15.3	18.9	22.4	26.0	29.6	33.1	36.6	39.7	43.1	0.0
	LS	A ALL CONDITIONS	2.0	3.1	6.7	10.3	13.9	17.7	21.4	25.0	28.6	31.8	35.3	21.5
	EP	A ALL CONDITIONS	4.6	6.4	8.6	10.8	12.9	15.2	17.4	19.6	21.7	23.6	25.7	0.0
	LP	A ALL CONDITIONS	2.8	6.5	10.8	15.2	19.5	24.0	28.4	32.8	37.1	40.9	45.1	2.2
CUTTING	EC	A ALL CONDITIONS	–	–	–	30.0	30.0	30.0	36.8	53.4	69.9	84.3	100.3	+
	LC	A ALL CONDITIONS	67.3	69.9	73.0	76.1	79.2	82.4	85.5	88.6	91.7	94.4	97.4	+
STRIPPING	EZ	A THERMAL – ALL CONDITIONS	2.0	4.7	10.0	15.3	20.5	26.0	31.3	36.6	41.9	46.4	51.5	+

+NO SIMULTANEOUS MOTIONS INCLUDED IN DATA BASE

*POWER FACTOR – MAGNIFICATION POWER HAS AN EFFECT ON TMU VALUES ON THOSE DATA LINES LISTED BELOW. EACH CHART VALUE IS ADJUSTED AS FOLLOWS:

ETC – WHEN MAG POWER EXCEEDS 20X ADD 0.68 TMU FOR EACH POWER GREATER THAN 20X, e.g., 30X, ADD 0.68 (30 - 20) = 6.8 TMU

LTN – WHEN MAG POWER EXCEEDS 5X, SUBTRACT 0.06 TMU FOR EACH POWER GREATER THAN 5X, e.g., 20X, SUBTRACT 0.06 [20 - 5] = 0.9 TMU

LFM – WHEN MAG POWER EXCEEDS 5X, ADD 0.94 TMU FOR EACH POWER GREATER THAN 5X, e.g., 20X, ADD 0.94 (20 - 5) = 14.1 TMU

IO - INSIDE FIELD to OUTSIDE FIELD

CODE RANGE (D/T)			1	2	3	4	5	6	7	8	9	10	11	12	SIMO ADDITIVE
RANGE DISTANCE ÷ TOLERANCE			TO 0.75	0.75 TO 1.5	1.5 TO 3.0	3.0 TO 6.0	6.0 TO 12.0	12.0 TO 25.0	25.0 TO 50.0	50.0 TO 100	100 TO 200	200 TO 350	350 TO 725	725 TO 3000	
TOOL AND COND	CODE	CHARACTERISTIC													
ALL	RL	M* ASIDE AND RLM OF TOOL OR OBJECT	10.5	14.0	18.1	22.2	26.3	30.6	34.7	38.8	42.9	46.5	50.5	57.8	5.3
GRASPING	ET	C GTC AND NONE	10.9	13.0	15.4	17.9	20.4	23.0	25.5	28.0	30.5	32.6	35.0	39.6	0
		G GT (TOOL GRASP)	43.2	44.9	46.9	49.0	51.0	53.2	55.3	57.3	59.4	61.2	63.1	66.9	+
	LT	A ALL EXCEPT RLM	15.3	17.4	19.8	22.3	24.7	27.3	29.8	32.2	34.7	36.8	39.2	43.6	0
	EF	A ALL EXCEPT RLM	14.8	17.4	20.5	23.7	26.7	30.0	33.1	36.2	39.4	42.1	45.1	50.7	0
	LF	A ALL EXCEPT RLM	12.5	14.7	17.4	20.0	22.6	25.3	28.0	30.6	33.2	35.5	38.1	42.9	0
	ED	A ALL EXCEPT RLM	7.4	12.5	18.6	24.7	30.7	37.0	43.2	49.3	55.4	60.6	66.5	77.4	0
PROBING	ES	A ALL EXCEPT RLM	14.5	18.1	22.3	26.5	30.8	35.2	39.5	43.7	48.0	51.7	55.8	63.5	0
	EP/LP	A ALL EXCEPT RLM	14.2	17.7	21.8	25.9	30.0	34.3	38.5	42.6	46.7	50.3	54.3	61.8	0
CUTTING	EC	A ALL EXCEPT RLM	7.1	10.6	14.8	18.9	23.0	27.3	31.5	35.7	39.8	43.4	47.4	54.9	0
STRIPPING	EZ	A ALL EXCEPT RLM	7.0	15.8	26.1	36.5	46.8	57.6	68.1	78.4	88.8	97.8	107.8	126.3	+

+NO SIMULTANEOUS MOTIONS INCLUDED IN DATA BASE

*POWER FACTOR: WHEN MAG POWER EXCEEDS 5X ADD 0.61 TMU FOR EACH POWER GREATER THAN 5X

Figure 10–5. MTM-M Data Card. © MTM Association for Standards & Research, Fair Lawn, NJ. *(Continued on p. 204.)*

OO - OUTSIDE FIELD to OUTSIDE FIELD

CODE RANGE (D/T)				1	2	3	4	5	6	7	8	9	10	11	12	13	
RANGE DISTANCE ÷ TOLERANCE				TO 0.75	0.75 TO 1.5	1.5 TO 3.0	3.0 TO 6.0	6.0 TO 12.0	12.0 TO 25.0	25.0 TO 50.0	50 TO 100	100 TO 200	200 TO 350	350 TO 725	725 TO 1500	1500 TO 6000	SIMO ADDITIVE
TOOL AND COND		CODE	CHARACTERISTIC														
ALL	RL	M*	ASIDE AND RLM OF TOOL OR OBJECT	6.2	9.8	14.0	18.2	22.5	26.9	31.2	35.4	39.6	43.3	47.4	51.9	59.3	1.1
GRASPING	ET	C	GTC OR NONE	6.8	8.2	9.8	11.5	13.1	14.8	16.5	18.1	19.7	21.1	22.7	24.4	27.3	4.0
		G	GT (TOOL GRASP)	21.6	22.5	23.6	24.7	25.8	27.0	28.1	29.2	30.3	31.3	32.3	33.5	35.4	2.2
	LT	A	ALL EXCEPT RLM	6.2	7.7	9.5	11.3	13.1	15.0	16.8	18.6	20.4	21.9	23.7	25.6	28.7	7.3
	EF	C	GMC OR NONE	5.0	7.1	9.5	11.8	14.2	16.7	19.2	21.5	23.9	26.0	28.3	30.8	35.0	5.6
		G	GM (PICK UP GRASP)	9.4	11.4	13.6	15.9	18.1	20.4	22.7	25.0	27.2	29.2	31.4	33.7	37.7	4.0
	LF	A	ALL EXCEPT RLM	7.5	8.8	10.2	11.7	13.2	14.7	16.2	17.7	19.2	20.4	21.9	23.4	26.0	4.3
	ED	A	ALL EXCEPT RLM	2.0	2.0	5.1	8.8	12.5	16.4	20.2	23.9	27.7	30.9	34.5	38.4	45.0	–
PROBING	ES	A	ALL EXCEPT RLM	3.3	5.6	8.3	11.1	13.8	16.6	19.4	22.2	24.9	27.3	30.0	32.8	37.6	6.9
	EP	A	ALL EXCEPT RLM	16.3	17.1	18.0	19.0	19.9	20.8	21.8	22.7	23.6	24.4	25.3	26.3	27.9	0
	LP	A	ALL EXCEPT RLM	7.6	8.9	10.6	12.2	13.8	15.5	17.1	18.8	20.4	21.8	23.4	25.1	27.9	16.1
CUTTING	EC	–	---	NO DATA – MUST SYNTHESIZE FROM MTM–1													–
STRIPPING	EZ	A	ALL EXCEPT RLM	2.0	3.9	10.7	17.5	24.4	31.5	38.5	45.3	52.1	58.1	64.7	71.9	83.9	–

*POWER FACTOR: WHEN MAG POWER EXCEEDS 5X ADD 0.33 TMU FOR EACH POWER GREATER THAN 5X, e.g., 10X, 0.33(10–5)=1.7 TMU

OI - OUTSIDE FIELD to INSIDE FIELD

CODE RANGE (D/FS)				1	2	3	4	5	6	7	8	9	10	11	
RANGE DISTANCE ÷ FIELD SIZE				TO 0.75	0.75 TO 1.50	1.5 TO 3.0	3.0 TO 6.0	6.0 TO 12.0	12.0 TO 25.0	25.0 TO 50.0	50 TO 100	100 TO 200	200 TO 350	350 TO 725	SIMO ADDITIVE
TOOL AND COND		CODE	CHARACTERISTIC												
GRASPING	ET	A	ALL CONDITIONS	7.1	14.1	22.4	30.7	39.0	47.7	56.2	64.5	72.8	80.0	88.1	9.3
	LT	A	ALL CONDITIONS	6.2	11.4	21.4	31.4	41.4	51.8	62.0	72.0	82.0	90.7	100.4	3.0
	EF	A	ALL CONDITIONS	12.0	14.3	17.1	20.0	22.7	25.6	28.5	31.3	34.1	36.5	39.3	+
	LF	A	ALL CONDITIONS	5.1	5.1	6.0	10.5	15.0	19.7	24.3	28.8	33.3	37.2	41.6	30.7
	ED	A	ALL CONDITIONS	2.0	2.0	2.0	2.0	2.0	2.1	4.6	7.0	9.3	11.4	13.7	7.1
PROBING	ES & LS	A	ALL CONDITIONS	6.1	13.9	23.2	32.5	41.8	51.5	61.0	70.3	79.6	87.7	96.7	22.1
	EP & LP	A	ALL CONDITIONS	23.6	29.3	36.0	42.8	49.5	56.5	63.4	70.1	76.8	82.7	89.2	+
CUTTING	EC	A	ALL CONDITIONS	3.0	6.7	11.0	15.3	19.6	24.1	28.5	32.9	37.2	40.9	45.1	15.8
STRIPPING	EZ	A	ALL CONDITIONS	14.4	14.6	14.7	14.9	15.1	15.2	15.4	15.6	15.8	15.9	16.1	+

IF - INSIDE FIELD to FINAL TARGET

CODE RANGE (FS/T)				1	2	3	4	5	6	7	8	9	10	11
RANGE FIELD SIZE ÷ TARGET TOLERANCE				TO 0.75	0.75 – 1.50	1.5 – 3.0	3.0 – 6.0	6.0 – 12.0	12.0 – 25.0	25.0 – 50.0	50 – 100	100 – 200	200 – 350	350 – 725
GRASPING	ET	A	ALL CONDITIONS	4.5	4.7	5.0	5.3	5.6	5.9	6.1	6.4	6.7	7.0	7.2
	LT	A	ALL CONDITIONS	4.5	4.5	4.5	5.4	7.2	9.1	10.9	12.7	14.5	16.1	17.9
	EF	A	ALL CONDITIONS	11.1	11.1	11.1	11.1	11.1	11.1	11.1	11.1	11.1	11.1	11.1
	LF	A	ALL CONDITIONS	4.1	4.1	7.9	13.3	18.7	24.3	29.8	35.2	40.6	45.3	50.6
	ED	A	ALL CONDITIONS	5.0	5.0	5.0	10.7	17.4	24.2	31.0	37.6	44.2	50.0	56.4
PROBING	ES+LS	A	ALL CONDITIONS	4.2	7.3	10.9	14.5	18.1	21.9	25.5	29.2	32.8	35.9	39.4
	EP+LP	A	ALL CONDITIONS	21.9	23.1	24.5	25.9	27.3	28.7	30.2	31.6	33.0	34.2	35.5
CUTTING	EC	A	ALL CONDITIONS	1.7	1.8	2.0	2.1	2.3	2.5	2.6	2.9	3.0	3.1	3.2
STRIPPING	EZ	A	ALL CONDITIONS	18.2	23.4	40.4	52.4	64.4	76.9	89.2	101.2	113.2	123.7	135.3

+ – INSUFFICIENT SAMPLE SIZE TO DETERMINE SIMO ADDITIVE
BOTH TABLES (OI AND IF) MUST BE USED TO DETERMINE TOTAL TIME VALUE, i.e., OI + IF
ADD 57.5 TMU TO TIME VALUE WHEN TRANSPORTING CHIPS WITH FLUID TENSION

Figure 10–5 (Continued)

Note that various symbols and codes appear in the tables. These should be easy to understand after reviewing some of the material just presented and after reviewing some of the coding examples which follow. For example, IIET - C5S presents the following information.

II = Inside to inside motion
E = Empty tool condition
T = Tweezers (the tool being used)
C = Contact grasp at terminal point
5 = Distance/tolerance range based upon dimensions and translated into the range number by entering line 1 of data table.
S = Simultaneous motion involved
TMU = 11.0 + 2.1

(which then must be adjusted for the magnification power used as indicated in the table).

From the above, the tables and the manual the following additional information can be stated:

Position 3
Indicates Tool
Load Condition
E = Empty
L = Loaded

Position 4
Indicates Name
of Tool
T = Tweezers
F = Fingers
D = Diagonal cutter
S = Soldering iron
P = Probe
C = Razor blade
(cutter or cutting tool)
Z = Thermal stripper

Position 5 covers the special motion termination points. These were previously given under "Terminating Points."

Some additional examples now follow:

IORL – M6 + PF for 10×
IO = Inside to Outside
RL = Aside and Release
M = Manual
6 = D/T code
30.6 TMU + 5 (0.61) TMU = 30.6 + 3.05 = 33.65 TMU

OIES – A7S = 83.1 TMU (includes simo addition of 22.1 TMU) plus
IFES – A6 = 21.9 TMU
Total 105.0 TMU
OI = Outside to inside
ES = Empty soldering iron
A = All
7 = D/FS, code 7 category

S = Simo motion involved
IF = Infield to target
ES = Still an empty soldering iron
A = All
6 = FS/T, code 6 category

Accuracy

The accuracy of MTM-M data values can be determined from existing regression equations, but this is difficult and time consuming. Data tables can also be used and are available, but the graph in Fig. 10–6 easily provides answers of usually sufficient quality at a 0.95 percent confidence level (95 out of 100 times). As one might expect, the accuracy increases with cycle time. At 1000 TMU of nonrepetitive work, the accuracy is ± 6 percent.

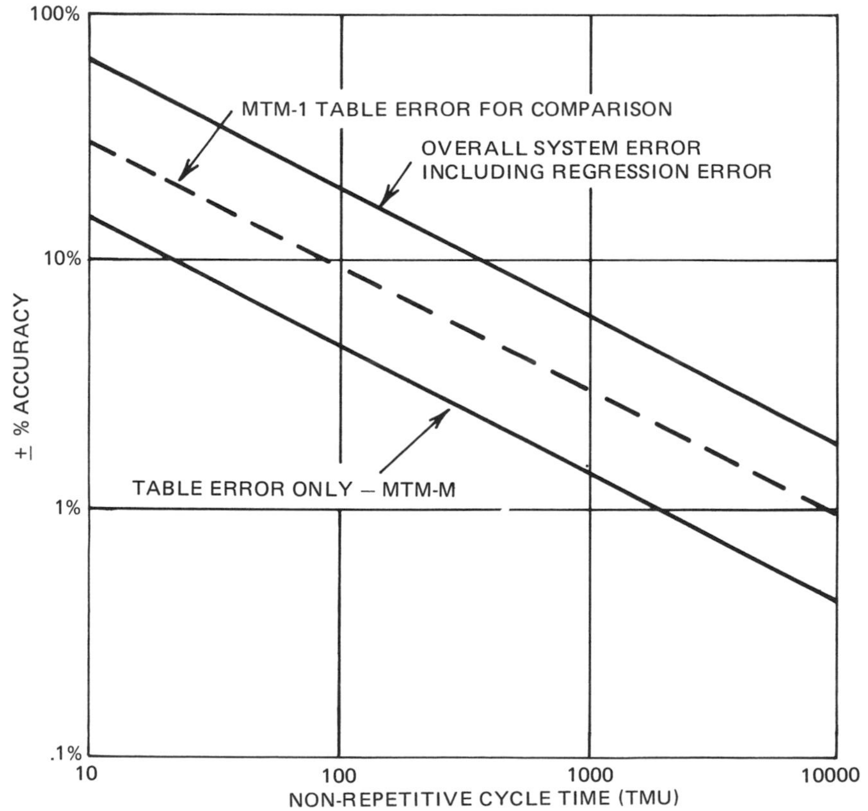

Figure 10–6. MTM-M accuracy at 95 percent confidence level. © MTM Association for Standards & Research, Fair Lawn, NJ.

The formula expressing the internal relationship is

$$\text{Percent A} = \frac{\text{Predicted Time} - \text{Actual Time}}{\text{Actual Time}} \times 100$$

In actuality, the system error calculation involves a restrictive set of statistical assumptions that include a normal distribution of performance times, and their variance is assumed to be homogeneous across the range of ID. Therefore, the error calculation must be considered a rough approximation. The real accuracy must be established by industry.

Application Rules

The vast majority of these will be reserved for coverage in the training course. However, two examples are given:

1. When a tool or object is asided to a table surface ending in an RLM, the target tolerance to use is 0.5 inch. However, if an aside to a fixture is involved, use regular target tolerance calculation procedures.
2. Allow eye time only when it is limiting. Eye time must be covered by MTM-1 or other MTM data.

Concluding Remarks

This discussion of MTM-M should provide the reader a good understanding of its development, the conceptual basis of the system, how it is structured, and how it can be utilized in a practical manner to set work standards for work performed under a microscope.

If the reader decides to really use the system, we strongly recommend the he/she first secure adequate and proper training.

CHAPTER 11

4M DATA MOD II: A Computer-Aided System Based Upon MTM-1 that is Faster to Use and More Accurate than MTM-1

Coauthor: Karl Eady, Regional Director, MTM Association for Standards and Research

There has been a long existing effort to computerize MTM-1 in such a manner as to provide overall advantages in its application. In general, the early efforts did computerize much of the system, but there was little resultant short-term advantage in reducing application time. Ultimately, the interactive system approach might have produced superior results.

The very early efforts at computerizing MTM-1 were made by private firms. One was International Business Machines (IBM), and they eventually offered the public their system through their then existing IBM Service Corporation—a subsidiary firm that is no longer a part of IBM. It was never a popular nor successful service.

An early U.S./Canada MTM Association effort was made at the University of Michigan under the guidance of Dr. Walton Hancock. The approach was creative and involved an interactive man–machine computer program that merely required the work analyst to enter the parameters of the workplace and the operation. However, the development costs were high, especially so since this approach required a continuing effort. As a result the sponsorship of the program ceased.

The expanding computer technology and its use by business and industry naturally provided a spur to various firms and individuals to again try to utilize the technology in applying both MTM-1 and portions of the companion family of systems. Also, the development of the companion MTM systems pointed the way to aggregate the MTM-1 elements in a successful manner from a time determination viewpoint.

One of these firms, a member of the U.S./Canada MTM Association, was successful in achieving significant overall benefits—Westinghouse. They turned over the copyright of the system to the Association, who not only made it publicly available, but also are continuing its further development and improvement. It is known as 4M DATA, an acronym adopted in 1973.

The original system has now been significantly revised and is titled 4M MOD II. MOD II is a "structured program" of *modular* construction. This makes it relatively easy to add additional future program functions. Currently, Advanced Machining Logic, a module which calculates speeds, feeds, and process times, is being prepared. Other major modules being considered are Advanced Assembly Logic, On-Line Application Screen, Performance Reporting System, and Work Center Routing. It is expected that MTM-2 and MTM-3 modules may ultimately be made available. However, the timing of these will depend upon user demand.

The 4M DATA system is so structured that its files may be "loaded" with the elements of functional MTM systems such as MTM-V and MTM-C. Users of these functional systems may utilize the 4M DATA System for standards application and operator instruction generation. Computerized MTM-V and MTM-C are not considered to be "modules" since no additional programming is necessary to use them.

This computer system from a feature or software viewpoint is rather completely explained as to how it is used and what it can accomplish for the user. If the reader is considering the use of this system, it is suggested that the chapter on system selection be carefully reviewed. This discussion, however, does not provide program lists, flow diagrams, program card decks, instructions on putting the system "on line," or some of the very fine details of usage. These are a part of the *User's Package of Information*. All of the above, and more, are available from the MTM Association for Standards and Research directly or through certain of its professional members.

A fairly complete knowledge of MTM-1 is desirable for proper and effective application and usage of 4M. In fact, one person, certified as an MTM-1 practitioner, at each location where it is used must be available to 4M practitioners for consultation. This discussion of the 4M system will be rather meaningless to the reader who does not have a good knowledge of MTM-1. MTM-1 training is available through the Association and its professional members. MTM-1 detailed knowledge is also available in the 3rd edition of *Engineered Work Measurement*.[1]

The 4M program deck is written in ANSI (American National Standards Institute) COBOL in either DOS or OS environment. A 50K storage partition is required, and auxiliary storage on a disk pack or a similar medium is also needed.

So far over 70 firms have purchased the 4M DATA System and are using it. It not only speeds the determination of standard times, but it even enhances the methods analysis power of MTM-1. How this occurs will be explained later in this chapter and in the chapter on system selection. 4M DATA is approximately four times faster than MTM-1 to apply and two to three times faster than MTM-2. The accuracy is greater than MTM-1.[2]

Users of 4M MOD I usually used punched cards as the input medium. However, a CRT data terminal with video can also be used, in conjunction with an interim storage medium such as a "floppy disc" when the program is run in the batch mode. This is and was the usual case, since it reduces computer requirements and costs in most cases. However, if desired, the 4M program can be run in the time-sharing mode.

[1]D. W. Karger and F. H. Bayha, *Engineered Work Measurement*, 3rd ed., Industrial Press, Inc., New York, 1978.

[2]*MTM System Accuracy*, MTM Association for Standards and Research, Inc., Document No. 2023-79, Fair Lawn, NJ.

Naturally, an output printer is needed. Moreover, a magnetic tape drive for back-up storage is very desirable.

The MOD-IIB program is similar to the original version with respect to the above computer requirements.

MOD IIA contains interactive programming which makes it suitable for "on-line" as well as "batch" operation. MOD IIB is the "batch only" version of 4M. The same *User's Manual* describes both systems, the principal difference between the two being the input procedures.

In the interactive mode inputting is by remote terminal (keyboard and CRT display). The analyst is "prompted" for the necessary input data. As soon as the data are entered the files are updated, which means the inputting can be done in one or more input sessions. Output is available immediately or later.

The batch mode is possible with both versions. Inputting may also be accomplished by Remote Job Entry (RJE) or by keypunching.

Conversion of any original system installation to MOD II is a minor problem and can be accomplished rather easily.

There are other computerization developments which will be discussed in Chapter 13. However, this chapter only focuses on 4M which *is* a successful computerization of MTM-1 and is compatible with the other members of the MTM family of systems.

Through proper user programming a computerized set of standard data can be built as the program is used. This approach not only retains all of the advantages of MTM-1, but it also can make its application more efficient and effective. Last, but not least, as previously mentioned, the 4M program enhances the methods analysis aspect of MTM-1. The economic considerations are advantageous both for the firm with a computer of the proper size and characteristics or through the use of leased or rental services (time-sharing) if there is a sufficient work measurement work load to justify the expense of acquisition, installation, training, and operation. For the larger firm the advantages are overwhelming.

It should be obvious that a one-chapter presentation cannot include logic diagrams, very detailed operating instructions, etc. It will, however, describe all of 4M's important features, and a considerable amount of the more significant detail from an overall view. It is assumed that the reader will ultimately utilize a *User's Manual* if he/she intends to use the system personally.

4M, The Derivation or Source of Its Name

The acronym 4M (or MMMM) was derived from the following four major attributes of the system:

Micro. The system retains all of the elements of MTM-1 in their complete detail and thereby is able to provide significant advantages and features.

Matic. Really automatic, since the analyst no longer needs to refer to the data card or to tables. This is possible because of the simple and easily remembered coding used by the analyst to describe the work. The coding is automatically translated by the computer, which determines a near-optimum MTM-1 method. Any compromises in optimization are analyst imposed through the degree of completeness of descriptions.

Methods. The 4M system develops and defines sound manual methods to the same degree as analyst-used MTM-1. In fact, it not only retains this advantage, but it also extends it by automatically providing improvement indexes (ratios) that indicate the approximate degree of optimization—or, conversely, additional methods opportunity.

Measurement. While GET and PLACE codes are used, the computer system interprets these and their special numerical descriptors so that it recognizes and applies the proper MTM-1 components for both hands. In fact, when necessary, it even automatically reduces C MOVE into two increments—a Class B MOVE and a short Class C MOVE. Furthermore, MTM-1 simo rules and the case of motions are all properly applied automatically by the computer.

Analysis of frequency of motion occurrence indicates that about 90 percent of the motions performed in industrial motions are those of REACH, GRASP, MOVE, POSITION, and RELEASE—essentially the design basis of MTM-2. The 4M GET classification is used to cover REACH, GRASP, and RELEASE elements. PLACE is used to cover the MOVE and POSITION elements. The remainder of the MTM-1 elements are applied using virtually unchanged MTM-1 notations.

GET—G

GET is the 4M element that includes MTM-1's REACH, GRASP, and RELEASE. Anyone who understands MTM-1 will readily see upon examination of Fig. 11–1 that the basic or underlying Case of REACH, Distance of REACH, and Class of GRASP are all specified by the coding. This leaves little to chance or aggregation.

The rules for distance measurement or estimation are those of the MTM-1 system. Distances of 3/4 inch or less are designated by "F." It perhaps should be said here the 4M DATA System will handle distance and weight notations in the metric as well as English systems.

Type II MTM-1 motions are handled by the 4M System in a manner that introduces almost no error. In such cases subtract either 4 inches (10 cm) or one-half the distance—whichever produces a longer remaining distance—and use this remainder in the GET notation. For clarity, the actual distance may be shown in parentheses along with the literal data, preceded or followed by the letter M as illustrated below:

RH NOTATION (46–49)	DIST. (50–52)	WT. (53–54)	X (55)	RH DESCRIPTION (56–70)		FREQUENCY (72–78)	P (79)	B/C (80)
G12	10			(M14) IN MOTION	A	T START		
G03	2			(4M) IN MOTION A	T	END		

4M analysts (applicators of the system) record pertinent operational facts and the needed 4M data notation on an Analyst Data Sheet which identifies

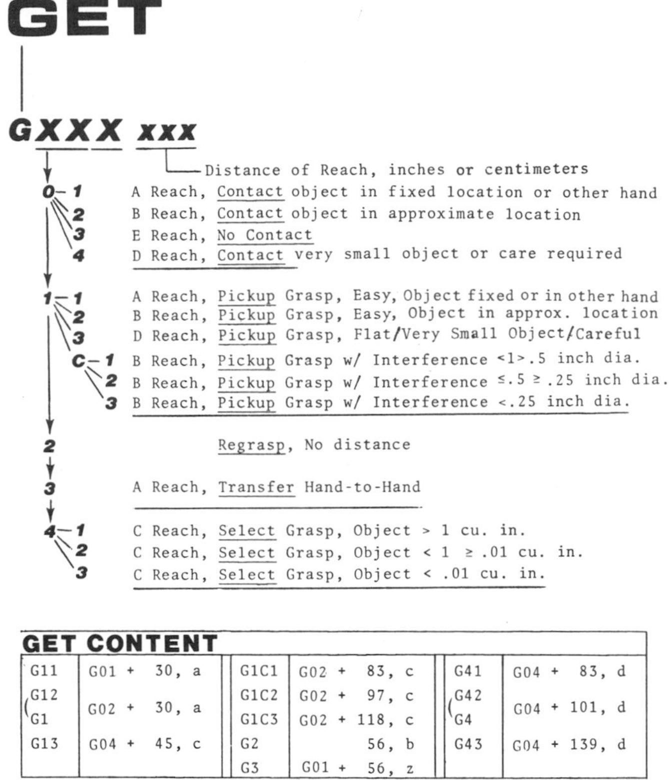

GET CONTENT

G11	G01 + 30, a	G1C1	G02 + 83, c	G41	G04 + 83, d
(G12 / G1	G02 + 30, a	G1C2	G02 + 97, c	(G42 / G4	G04 + 101, d
		G1C3	G02 + 118, c		
G13	G04 + 45, c	G2	56, b	G43	G04 + 139, d
		G3	G01 + 56, z		

Figure 11–1. The essentials of GET coding. © MTM Association for Standards & Research, Inc., Fair Lawn, NJ.

"fields" (columns) on the sheet by *number*. In a few cases where unusual field assignments occur we will specify the field numbers to use. However, the more nondistinctive field assignments are not specified in this text, but are readily available from the *User's Manual*.

PLACE—P

PLACE (P) is the 4M element that covers MTM-1's motion elements of MOVE, WEIGHT FACTOR (static and dynamic), POSITION, and SECONDARY DISENGAGE. Like GET, element P provides for identification of MOVE distance, class, and type (Type II). The coding also provides for specifying alignment factors and the fit of POSITION. Coding details are found in Fig. 11–2.

RELEASE—R: An MTM-1 Motion Element

At this point a number of items need to be discussed. One is that neither GET nor PLACE provide for the specification of an MTM-1 RELEASE. Contact

PLACE

PXXX xxx w

- w → ENW if over 2.5 lb or 1kg
- xxx → Distance of C Move, inches or cm

0 – 1	Move object against stop or to other hand, No Position
0 – 2	Move object to approx. location, No Position
0 – 2 T	Toss, No Position
0 – 3	Move object w/ care required, No Position
1 _ _	Clearance ≤ .700 ≥ .300
2 _ _	Clearance < .300 ≥ .050
3 _ _	Clearance < .050 ≥ .010
_ 1 _	Symmetrical(> 10 ways to engage)
_ 2 _	Semi-Symmetrical (2-10 ways to engage)
_ 3 _	Non-Symmetrical (1 way to engage)
_ _ A	Surface Alignment
_ _ 0	Easy Insertion to .125 inches inclusive
_ _ 1	Easy Insertion to .75 inches inclusive
_ _ 2	Easy Insertion to 1.25 inches inclusive
_ _ 3	Easy Insertion to 1.75 inches inclusive
_ _ 5	Difficult Insertion to .125 inches inclusive
_ _ 6	Difficult Insertion to .75 inches inclusive
_ _ 7	Difficult Insertion to 1.25 inches inclusive
_ _ 8	Difficult Insertion to 1.75 inches inclusive

PLACE CONTENT

P110	P03 + 34, s	P210	P03 + 72, u	P310	P03 + 95, w
P111	P03 + 66, s	P211 (P-1)	P03 + 119, u	P311	P03 + 163, ẇ
P112	P03 + 77, s	P212	P03 + 130, u	P312	P03 + 187, w
P113	P03 + 88, s	P213	P03 + 142, u	P313	P03 + 210, w
P115	P03 + 90, t	P215	P03 + 128, v	P315	P03 + 151, x
P116	P03 + 122, t	P216 (P-1D)	P03 + 175, v	P316	P03 + 219, x
P117	P03 + 133, t	P217	P03 + 186, v	P317	P03 + 243, x
P118	P03 + 144, t	P218	P03 + 198, v	P318	P03 + 266, x
P120	P03 + 103, u	P220 (P-2)	P03 + 149, w	P320	P03 + 173, y
P121 (P1)	P03 + 135, u	P221 (P2)	P03 + 196, w	P321 (P3)	P03 + 241, y
P122	P03 + 146, u	P222	P03 + 207, w	P322	P03 + 265, y
P123	P03 + 157, u	P223	P03 + 219, w	P323	P03 + 288, y
P125	P03 + 159, v	P225 (P-2D)	P03 + 205, x	P325	P03 + 229, z
P126 (P1D)	P03 + 191, v	P226 (P2D)	P03 + 252, x	P326 (P3D)	P03 + 297, z
P127	P03 + 202, v	P227	P03 + 263, x	P327	P03 + 321, z
P128	P03 + 213, v	P228	P03 + 275, x	P328	P03 + 344, z
P130	P03 + 155, u	P230	P03 + 202, w	P330 (P-3)	P03 + 229, y
P131	P03 + 187, u	P231	P03 + 249, w	P331	P03 + 297, y
P132	P03 + 198, u	P232	P03 + 260, w	P332	P03 + 321, y
P133	P03 + 209, u	P233	P03 + 272, w	P333	P03 + 344, y
P135	P03 + 211, v	P235	P03 + 258, x	P335 (P-3D)	P03 + 285, z
P136	P03 + 243, v	P236	P03 + 305, x	P336	P03 + 353, z
P137	P03 + 254, v	P237	P03 + 316, x	P337	P03 + 377, z
P138	P03 + 265, v	P238	P03 + 328, x	P338	P03 + 400, z

Figure 11–2. The essentials of PLACE coding. © MTM Association for Standards & Research, Inc., Fair Lawn, NJ.

RELEASE can be ignored from a time value view since it is 0.00 and its presence is usually obvious to a trained MTM-1 analyst. A "standard" RL1-type RELEASE is automatically included in the GET category even though it occurs after the GET and other intervening motions are completed. The time standard included, however, is modified, since a research study disclosed that many RELEASE motions are overlapped by longer simo motions and, therefore, do not add to the net motion time; it was found that only about 50 percent of the RELEASEs were limiting as there were limiting GRASP values. Therefore, the time allotment is 50 percent of an RL1 in the GET aggregate. The error introduced is negligible.

MOVE with Weight

We are here dealing with the Effective Net Weight of MTM-1. Its treatment in 4M is in accord with MTM-1 research findings and its application rules.

The computer does an even more accurate job of handling weight factors than the MTM-1 Data Card, which aggregates categories of weight. The computer applies the formula for both the Static and the Dynamic Factor.

TMU = 0.011 × Effective Net Weight where the Static factor is not needed (the weight is already "in control"); it is necessary to specify "Dynamic Only" to the computer as indicated below:

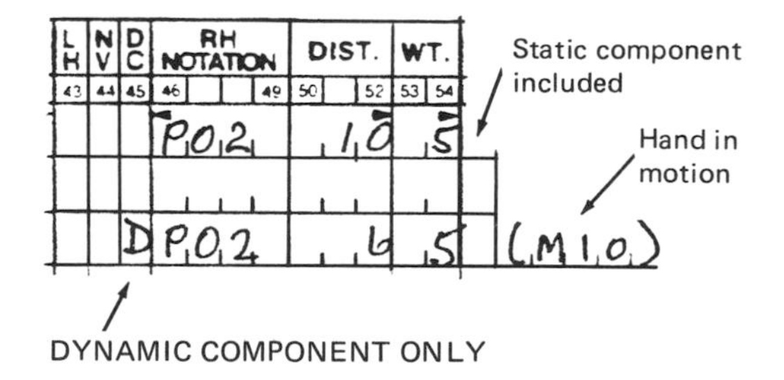

POSITION—P

This is an element of PLACE that may or may not exist, and the analyst must indicate whether it exists through the 4M coding. The coding also specifies the kind of POSITION that becomes a part of the PLACE.

The 4M DATA System was designed to use the newer Supplementary POSITION data rather than the original International POSITION data. However, if a user desires to use the International data, it is possible to do so when applying 4M. There is a category or element of PUT on the 4M data card, which is shown in Fig. 11–3, that is designed to allow the usage of the International POSITION data. A numerical entry in the distance field of a PLACE or PUT notation automatically provides for the inclusion of a prior C MOVE, otherwise only the POSITION value is provided.

Hand Interaction (the possibility of it) in POSITION must be discussed. Sometimes one hand holds a part while the other hand assembles a second part to the one being held—generally a poor assembly practice that should be eliminated through the use of proper fixtures. When it is necessary, the POSITION

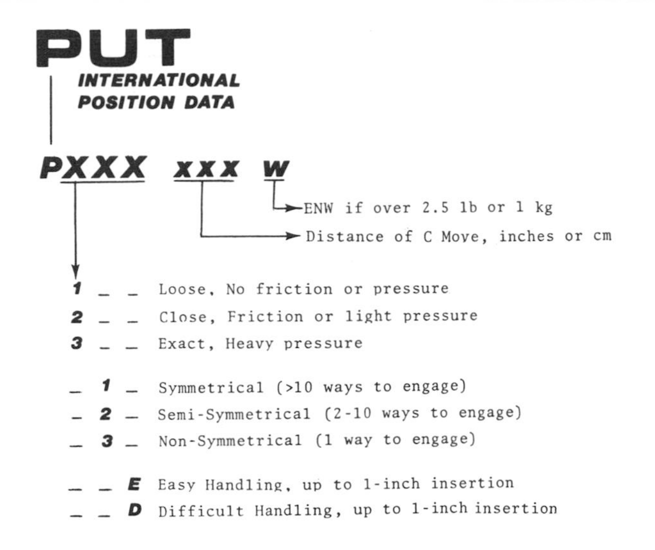

PUT CONTENT *(International Data)*

P11E	P03 + 56, s	P21E	P03 + 162, u	P31E	P03 + 430, z
P12E	P03 + 91, u	P22E	P03 + 197, x	P32E	P03 + 465, z
P13E	P03 + 104, x	P23E	P03 + 210, x	P33E	P03 + 478, z
P11D	P03 + 112, t	P21D	P03 + 218, v	P31D	P03 + 486, z
P12D	P03 + 147, v	P22D	P03 + 253, y	P32D	P03 + 521, z
P13D	P03 + 160, y	P23D	P03 + 266, y	P33D	P03 + 534, z

Figure 11–3. PUT, a place aggregate using International POSITION. The E and D in the fourth place signal to the computer that the old data are to be used. © MTM Association for Standards & Research, Inc., Fair Lawn, NJ.

(where such a motion element is required) must be indicated for *only one part*. The example below not only shows this being done, but it also shows how to handle such a "somewhat" dual secondary engage:

LH DESCRIPTION	LH NOTATION	DIST.	WT.	LH	NV	DC	RH NOTATION	DIST.	WT.	X	RH DESCRIPTION
EAR'G TO SHAFT	P01	12					P110	12			CHAMF. SHAFT
	SP						E202			X	INSERT SHAFT

The indication of a POSITION must only be done for one hand since the other hand is merely waiting or holding (could be doing something else). This will be reflected in the Improvement Index, which is described later. Such a procedure also prevents the otherwise sequential charging of two POSITIONs.

When both hands are used on the part being positioned, the SP shown for the left hand (could be the right) signifies "Same Part" and instructs the computer to give the same time credit to the other hand. More will be said about this later in the chapter.

Other MTM-1 Motion Categories

More than 90 percent of most MTM-1 analyses will usually be comprised of the MTM-1 elements included in the GET and PLACE (PUT) aggregates. This should make clear, with a little reflection by the reader, why there is a great simplification in MTM-1 analysis provided by 4M.

The remainder of the MTM-1 motion elements are used directly from MTM-1 with only a few notation changes required because of the 4M format.

TURN (Strictly Defined as in MTM-1)

Turn conforms to MTM-1 as to values. It is accounted for here in 30° increments (usually estimated). Under 30° use the GET and PLACE categories. WEIGHT (or resistance to motion) must be accounted for here as in MTM-1.

In 4M the time values supplied are in accord with the notations (to be entered on the analysis form) indicated below:

	30°	60°	90°	120°	150°	180°
No object	T1	T2	T3	T4	T5	T6
Small, 0–2 pounds	T1S	T2S	T3S	T4S	T5S	T6S
Medium, 2.1–10 pounds	T1M	T2M	T3M	T4M	T5M	T6M
Large, 10.1–35 pounds	T1L	T2L	T3L	T4L	T5L	T6L

A "sort of" 4M rule of thumb is that if a TURN (MTM-1 variety) is combined with a REACH or MOVE by the same hand (a so-called Combined Motion), only enter the longer motion (the limiting motion) on the analyst analysis sheet. If desired, the shorter motion may also be listed (always done with standard MTM-1, but it is crossed out) *along with* a symbol that shows the notation "for reference only." The following is an example:

L H	N V	D C	RH NOTATION	DIST.	WT.	X
43	44	45	46 49	50 52	53 54	55
			P02	1.0		
/			G2			

CRANK—C: Strictly as Defined in MTM-1

In 4M, the MTM-1 Master Crank Formula is used for all cases of CRANK. The formula is as follows:

$$MU = FNT + N(52F + C)$$

where

1 TMU = 10 MU or MU = 1/10 TMU

F = Dynamic factor of WEIGHT (in this case resistance) from the MOVE table, which equals 1 + 0.011W (range 1–1.5) where W is weight in pounds.

N = Number of revolutions, including any fractional revolutions (like 4.3 revolutions).

T = MU per revolution as a function of cranking diameter.

C = Static component of MU weight allowance from MOVE table or 4.75 + 3.45W (range 0–160 MU).

52 = MU for one start and stop.

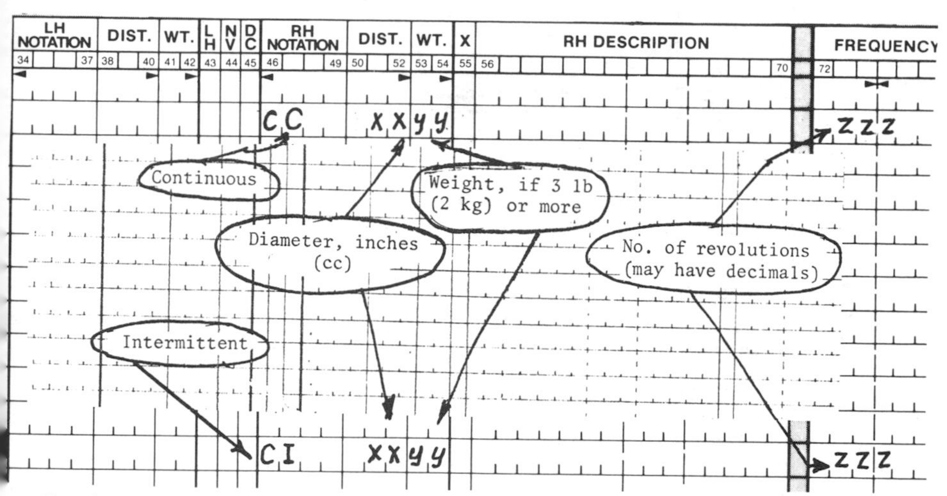

Figure 11–4. CRANK notations for 4M. © MTM Association for Standards & Research, Inc., Fair Lawn, NJ.

N′ depends upon the nature of the cranking as indicated below:

CASE A, continuous cranking where N′ = 1, always
CASE B, intermittent cranking where N involves a whole number—then N′ = N
CASE C, intermittent cranking where N involves a whole number plus a fraction (like 1.6 revolutions). In this case, N′ equals the next higher *whole* number (for example, 5.4 should be shown as N′ = 6).

If two CRANKs are performed simo, or if CRANK is performed Simo with other motions, the analyst in these situations (they will rarely occur) must determine the limiting motion because the frequency field is used to indicate N (actual revolutions) and the computer therefore is unable to calculate the limiting condition. This sounds like a problem, but think how seldom such a condition is encountered.

Use fields 72–74 to indicate the number of complete turns and any fractional portion of a turn (if any) in 75–78.

Two example recordings of CRANK motions are shown in Fig. 11–4.

APPLY PRESSURE, As Defined in MTM-1

The computer program will accept only the new data, APA and APB, and components of the new data—AF, DM, and RLF.

EYE ACTION, As Defined in MTM-1

The two involved actions are EYE FOCUS (EF) and EYE TRAVEL (ET). If one enters the Travel Distance (T) and the distance from the eye to the line of travel (D) on the analyst data recording form, the computer will properly compute the eye travel time. If these dimensions are not entered, the computer will automatically enter the ratio of 0.625 (essentially T = 10 inches and D = 16 inches). The recording of T and D for ET is shown on p. 218.

DC	RH NOTATION	DIST.	WT.	X	RH DESCRIPTION
45	46 49	50 52	53 54	55	56
	ET	18	14	←	D
		↑			T

If eye motions are controlling, they should be recorded on a line not used for another purpose, on the right side of the form.

The analyst should carefully adhere to the MTM-1 rules regarding EYE MOTIONS as concerned with the limiting question. Since eye usage is included in MTM-1 manual motion time (when needed), they are never noted as being performed simo with manual motions.

BODY MOTIONS, As Defined in MTM-1

Basically body motions are tiring and they require considerable time to perform. Hence, the presence of such motions is a possible opportunity for methods improvement.

The major body motions will be limiting in the 4M analysis. Where body motion assist to an arm motion occurs, the body motion assist is subtracted out before recording the arm–hand motion as later described. Standard MTM-1 Body Motion symbols are used and entered only on the RH side of the input. The left-hand side of the line involved must be left blank.

FOOT MOTION—FM

In 4M, the distance used is based on an average of 2–4 inches in one direction as measured at the toe. For a complete down and up motion of the foot it requires the indication or assignment of two FMs. If the end of the down motion must involve an APPLY PRESSURE, assign or record an APA. The APA may be combined with the second FOOT MOTION and recorded as FMP.

LEG MOTION—LM

Use the symbol LM followed by the motion distance, recording it in columns 51 and 52.

BEND—B; STOOP—S; KNEEL, ONE KNEE—KOK; KNEEL, BOTH KNEES—KBK

An easy and clear situation. Use the symbols indicated. For the Arise Motion, add an A. For example, ARISE from BOTH KNEES is recorded AKBK.

SIT—SIT; STAND—STD

Defined as in MTM-1, merely record the symbols when these motions are needed.

SIDE STEP—SS

Side step is recorded as SSC1 or SSC2. The side step distance is assumed to be 12 inches in either case and is not included in the notation as it is in MTM-1. No provision is made to record larger side steps.

TURN BODY—TB

This motion element has two MTM-1 cases, Case 1 and Case 2. Record as TBC1 or TBC2.

WALK

4M uses the MTM-1 "per pace" method. Average unloaded pace length is considered as 34 inches. Average loaded pace length is 30 inches for 5–35-pound loads and 24 inches for 35–50-pound loads. However, the easy way is merely to count the number of times the foot hits the floor.

Unobstructed walking is recorded as WP and obstructed walking is recorded as WPO. The number of paces is entered in the distance field on the input.

Arm Motion Times with Simo Body Motions

Body motions often assist arm motions; it is seldom that the reverse action occurs. In 4M, by a decision prompted by both practical and programming needs, recorded body motions are always to be limiting (chargeable) and are entered on the right-hand side of the motion pattern—the LH side of the line used must be left blank as stated earlier. This means that where an arm motion is involved, any arm motion before, prior to the body motion, should be measured and recorded separately. Secondly, where such body motion assists an arm motion (it could be a continuation of the motion described in the previous sentence), merely measure and record separately any arm motion that occurs *after* completion of the body motion.

However, unless the process described is fully understood, the action can lead the analyst into recording incorrect motion elements. First it is suggested that the unqualified MTM-1 analyst reader study pages 316–319 of EWM. One key paragraph from these pages is now quoted.

> One of the best and most commonly used ways of determining the amount of any kind of body assistance is to make the arm and body movements separately, so that the amount of each can be measured. To do this, first move the body to its final position while holding the arm stiff in its original position, then move the arm along to its final position and note the effective arm movement. This technique works especially well for radial assistance. In using this method one is not required to use the formula for calculating allowed distance, because the measurement is directly the allowed distance.

Actually, in most such cases of Body Assist, the body motion overlaps and is "internal" to the arm motion. Therefore it is not limiting and need not be recorded (the internal portion).

If the body motion is severe (meaning it *must* be accomplished if the arm motion is to be completed), there will likely be a chargeable Body Motion.

Only somewhat related is the fact that Foot Motions and Leg Motions may occur that will be overlapped by GET and/or PLACE motions (meaning a Simo Motion has occurred). It is possible to record the overlapped (not limiting) motions on the analysis form in such a manner that they appear in the computer printout, but without an associated time charge. A symbol (called a *slash code*) is used to signal "for reference only" in cases where such a Combined or Simo Motion would be shown for methods clarity. It is done by entering:

WT.	L H	N V	D C	RH NOTATION	DIST.	WT.	X	RH DESCRIPTION
42	43	44	45	46 – 49	50 – 52	53 54	55	56 – 70
				PØ3	10			PEN TO PAPER
	/			G2				WHILE MOVING

Denotes limited motion

Body Motions with Weight

MTM-1 does not provide a formal and fully researched adjustment factor for the above occurrence. Just remember that the heavier the weight being handled, the shorter the steps, and this can be properly "taken into account" as described by either counting or using the shorter lengths indicated. Weight could even affect TB.

SAME MOTION Code—SM

If each hand is to perform identical motions, provides for merely entering SM (on the same horizontal line) for one of the hands. This instructs the computer to perform simo comparisons and print out the proper motion pattern.

SAME PART Code—SP

This code goes one step further than SM, in effect it tells the computer that both hands *are to* perform a motion or motion aggregate together (as a unit) and the Simo Chart restrictions are *not* to be applied. The computer program will cause the phrase "assist (other) hand" to be printed in the motion pattern. Follow the same recording pattern as in the previous case.

SIMULTANEOUS MOTIONS

This section will discuss what needs to be done to get the computer to "put it all together." Only a small amount of mechanistically oriented action need be taken (many have already been identified). 4M saves much time and thought over manually applying MTM-1. For example, the 4M program will cause the computer to reduce automatically the classification of one of two simo R_Cs, so that one can be partially accomplished as an R_E while the other R_C is completed. This action reduces the chargeable remaining distance for the uncompleted move, an R_C. It also does this for simo M_C moves—and, in fact, wherever the MTM-1 data system rules permit actions that partially complete a motion in a lower class. In other words, the computer applies all MTM-1 simo rules to automatically present the lowest time cost motion pattern.

This action greatly speeds up the process over manual application of MTM-1 and eliminates many of the errors that often occur in the analyst application of simo rules. This latter subject is discussed in other chapters, especially in Chapter 12.

The above can perhaps be best made clear by some illustrations. A commonly encounted work requirement is illustrated below:

LH	RH
Get Jumbled Part	Get Jumbled Part
Place on Fixture	Place on Fixture

A breakdown of the above to components might logically appear as in the next illustration:

LH	RH
C-Reach	C-Reach
Grasp	Grasp
C-Move	C-Move
Position	Position

13 14 15	LH NOTATION (34–37)	DIST. (38–40)	WT. (41–42)	LH (43)	NV (44)	DC (45)	RH NOTATION (46–49)	DIST. (50–52)	W	B/C (80)
010	P03	20					G12	22		B
020	G42	15					P132	22		B
030	P131	7					G42	4		B
040	G42	7					P131	7		B
050	P131	8					G42	8		B
060	G11	4					P131	8		B

Figure 11–5. Input notations for a six-line set. © MTM Association for Standards & Research, Inc., Fair Lawn, NJ.

Unfortunately this is far from a correct arrangement because the MTM-1 rules do not permit considering any of these motions as simo, assuming the right- and left-hand motions to be outside of the range of normal vision.

A six-line set of input notation that essentially calls for the above situation is shown in Fig. 11–5.

Next shown is a correct analysis of the four motion aggregates, as would be developed automatically by the 4M DATA computer program:

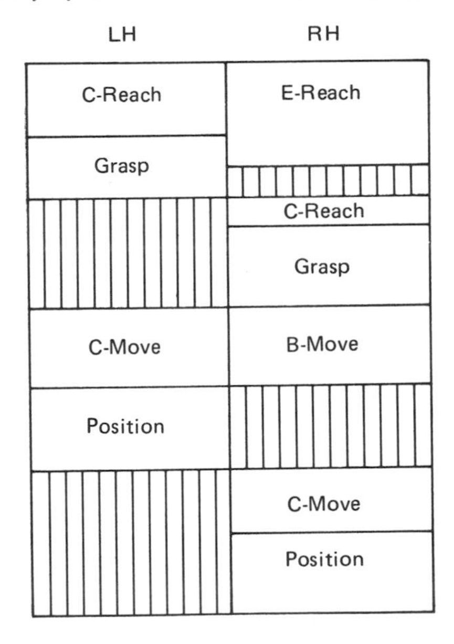

While this rearrangement of the motions into a practical and optimum motion pattern is accomplished automatically by 4M (Fig. 11–6), the analyst should know and understand enough MTM-1 to produce it on his/her own manually.

A 4M analyst would actually use the 4M analysis sheet to manually produce

```
ATA SYSTEM             4M OPERATION ANALYSIS                       REQUESTED BY TRNG  TIME 15.33.10  DATE 02/10/80  PAGE   1

 20F1521               METER FRAME                                                    ORIGINATED  06/02/72 BY TRNG
ATION  01       ASSEMBLE COMPENSATORS AND MAGNETS                                     REVISION  A 02/10/80 BY TRNG
. 12345   MACHINE          TOOLING              ASSEMBLY FIXTURE
                                           ALLOWANCE-MANUAL   .080            STANDARD HOURS-SETUP          RUN    .00233
                                                   -PROCESS                                      UNITS PER HOUR  429.20
                                                                                         PROCESS                  NET
          LH MOTIONS                                          RH OR BODY MOTIONS     FREQ.      TIME     LH    RH    MANUAL
                                                      RUN

                                                                         PO                                         2164

) MOVE     ASSEMBLY TO TABLE       P03 -20     G12 -22    GET     FRAME FROM CONVEYOR                                  B
) GET      LARGE COMPENSATOR       G42 -15     P132-22    PLACE   FRAME IN FIXTURE                                     B
) PLACE    COMPENSATOR IN CAVITY   P131- 7     G42 - 4    GET     SMALL COMPENSATOR                                    B
) GET      LARGE MAGNET            G42 - 7     P131- 7    PLACE   COMPENSATOR IN CAVITY                                B
) PLACE    MAGNET IN CAVITY        P131- 8     G42 - 8    GET     SMALL MAGNET                                         B
  GET      FRAME                   G11 - 4     P131- 8    PLACE   MAGNET IN CAVITY                 1400  1689       2164B

                                               TOTAL RUN MU-MANUAL       2164   WITH ALLOWANCE       2337
                                                           -PROCESS
                                                                  TOTAL STANDARD RUN TIME      .00233 HOURS
                                                                                                .140 MINUTES
                                                                           UNITS PER HOUR      429.20
71%         RMB  55%    GRA  15%    POS  30%         PROC   0%
```

Figure 11–6. 4M operation analysis report using Fig. 11–5 data. © MTM Association for Standards & Research, Inc., Fair Lawn, NJ.

the motion pattern shown in Fig. 11–7A. Figure 11–7B was produced by using the 4M Simo Chart, and it is in accord with MTM-1 simo rules. The only real difference is that it presents the Supplementary Position Data on the MTM-1 Data Card.

Combined Motions

These were previously discussed and the reader should consult EWM if the subject heading is not understood. It can be repeated that, in most cases when using 4M, the analyst only records the principal motion and adjusts the distance factor as required. This partial recording is based upon the premise that most of these situations would be almost self-evident and are therefore not usually needed for the methods description.

Not recording motions that occur and which have influenced associated motions is not in accord with MTM-1 rules. But, the manner of handling these in 4M does correctly take their effect into account.

If such internal motions need to be recorded for the sake of methods clarity, it is merely necessary to show the motion and record the slash code. It will then appear on the printout for methods description, but it will not enter into the time calculation.

Lead Hand

In a few cases certain jobs require one hand to *lead* the other during REACHs or MOVEs that could be simo if lowered in case. Where this is the situation, record L or H in column 43.

METHODS ANALYSIS CHART
WESTINGHOUSE FORM 29590 A

SHEET 1 OF 1

DEPT. / GROUP	OPER. NO.	OPERATION	DATE
Corresponds to #105	01	Assemble compensators & magnets	4/8/74
PART/APPARATUS: Meter frame		DWG. & ITEM	ANALYST

DESCRIPTION - LEFT HAND	NO.	LH	TIME UNITS	RH	NO.	DESCRIPTION - RIGHT HAND
Assembly to table		M20C	22.1	R22B		To frame in container
Release		RL1	2.0	G1A		Frame
To large compensator		R15C	16.3	M16B		Frame toward fixture
Large compensator		G4B	9.1			
			10.3	M6C		Frame to fixture
Comp. toward cavity		M7B	19.8	P2INS4		Frame in fixture
Comp. to cavity		MfC	2.0	RL1		Release
Compensator in cavity		P2INS2	18.7	R4E		Toward small comp.
Release		RL1	2.0	RfC		To small compensator
Toward large magnet		R7E	9.1	G4B		Small compensator
To large magnet		RfC	2.0			
Large magnet		G4B	9.1			
			11.1	M7C		Compensator to cavity
Magnet toward cavity		M8B	18.7	P2INS2		Compensator in cavity
Magnet to cavity		MfC	2.0	RL1		Release
Magnet in cavity		P2INS2	18.7	R8E		Toward small magnet
Release		RL1	2.0	RfC		To small magnet
To frame		R4A	9.1	G4B		Small magnet
Frame		G1A	11.8	M8C		Magnet to cavity
			18.7	P2INS2		Magnet in cavity
			2.0	RL1		Release
			216.6	TMU		

Figure 11–7A. Manual 4M analysis for the operation shown in Fig. 11–6. © MTM Association for Standards & Research, Inc., Fair Lawn, NJ.

Macro Notations

The 4M DATA for GET and PLACE may be shortened for maximum simplicity. This may be done without sacrificing significant accuracy.

For example, a G42 GET notation may be shown as G4 without any change in time assignment. The G4 may also be used in place of G41 and G43 with a slight decrease in accuracy. The Macro PLACE notations are based on symmetry alone. For example, a PLACE containing a semisymmetrical position would be noted by the Macro P-2. The time value for this Macro is the same as for the detailed notation P220. The 100 GET and PLACE notations of 4M are covered by 11 Macro notations, thereby reducing decision-making time considerably.

Output Information

Operation Analysis Reports (R820 and R830)

There are two operation analysis reports. Figure 11–8 illustrates R820 and Fig. 11–9 presents R830—both show the same operation. R830 is reproduced on 8 1/2 × 11 format and omits the first word (action) of the left and right hand descriptions. Otherwise, the reports are identical.

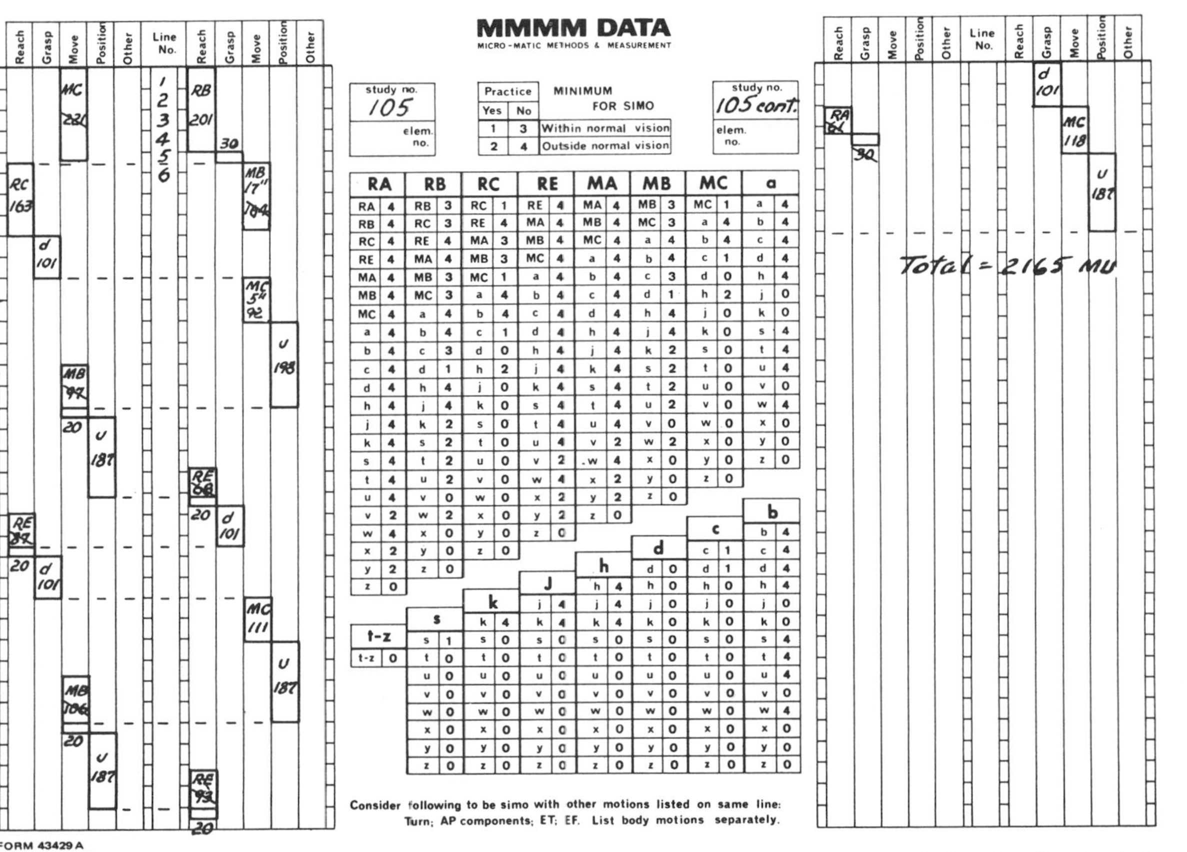

MMMM DATA
MICRO-MATIC METHODS & MEASUREMENT

study no. 105 — elem. no.

Practice		MINIMUM FOR SIMO
Yes	No	
1	3	Within normal vision
2	4	Outside normal vision

study no. 105 cont. — elem. no.

RA		RB		RC		RE		MA		MB		MC		a	
RA	4	RB	3	RC	1	RE	4	MA	4	MB	3	MC	1	a	4
RB	4	RC	3	RE	4	MA	4	MB	4	MC	3	a	4	b	4
RC	4	RE	4	MA	3	MB	4	MC	4	a	4	b	4	c	4
RE	4	MA	4	MB	3	MC	4	a	4	b	4	c	1	d	4
MA	4	MB	3	MC	1	a	4	b	4	c	3	d	0	h	4
MB	4	MC	3	a	4	b	4	c	4	d	1	h	2	j	0
MC	4	a	4	b	4	c	4	d	4	h	4	j	0	k	0
a	4	b	4	c	1	d	4	h	4	j	4	k	0	s	4
b	4	c	3	d	0	h	4	j	4	k	2	s	0	t	4
c	4	d	1	h	2	j	4	k	4	s	2	t	0	u	4
d	4	h	4	j	0	k	4	s	4	t	2	u	0	v	0
h	4	j	4	k	0	s	4	t	4	u	2	v	0	w	4
j	4	k	2	s	0	t	4	u	4	v	0	w	0	x	0
k	4	s	2	t	0	u	4	v	2	w	2	x	0	y	0
s	4	t	2	u	0	v	2	w	4	x	0	y	0	z	0
t	4	u	2	v	0	w	4	x	2	y	0	z	0		
u	4	v	0	w	0	x	2	y	2	z	0				
v	2	w	2	x	0	y	2	z	0						
w	4	x	0	y	0	z	0								
x	2	y	0	z	0										
y	2	z	0												
z	0														

t-z		s		k		J		h		d		c		b	
														b	4
												c	1	c	4
										d	0	d	1	d	4
								h	4	h	0	h	0	h	4
						j	4	j	4	j	0	j	0	j	0
				k	4	k	4	k	4	k	0	k	0	k	0
t-z	0	s	1	s	0	s	0	s	0	s	0	s	0	s	4
		t	0	t	0	t	0	t	0	t	0	t	0	t	4
		u	0	u	0	u	0	u	0	u	0	u	0	u	4
		v	0	v	0	v	0	v	0	v	0	v	0	v	0
		w	0	w	0	w	0	w	0	w	0	w	0	w	4
		x	0	x	0	x	0	x	0	x	0	x	0	x	0
		y	0	y	0	y	0	y	0	y	0	y	0	y	0
		z	0	z	0	z	0	z	0	z	0	z	0	z	0

Consider following to be simo with other motions listed on same line: Turn; AP components; ET; EF. List body motions separately.

FORM 43429 A

Figure 11–7B. Manual 4M analysis for the operation shown in Fig. 11–6. © MTM Association for Standards & Research, Inc., Fair Lawn, NJ.

4M DATA SYSTEM R0820 4M OPERATION ANALYSIS REQUESTED BY JPO TIME 11.01.03 DATE 02/21/80 PAGE 1

PART 25P1017 VOLT COIL ASSEMBLY ORIGINATED 06/16/79 BY DOC
OPERATION 03 ASSEMBLE AND STAKE 4 RIVETS IN VOLT COIL LAMINATI REVISION A 02/10/80 BY KEE
DEPT. 30 MACHINE 0300 TOOLING A-9675 ASSEMBLY FIXTURE
ALLOWANCE - MANUAL .080 STANDARD HOURS - SETUP .00000 RUN .00307
- PROCESS .200 UNITS PER HOUR 325.84

		LH MOTIONS				RH OR BODY MOTIONS	FREQ.	PROCESS TIME	LH	RH	NET MANUAL
				--- RUN ---							
001 103-01		ALIGN IRON AND PLACE 4 RIVETS				PO					
001	REACH	ALIGNING HANDLE	G01 -6						70		70
002	MOVE	ALIGNING HANDLE	P01 -2						36		36
003	GET	RIVET FROM TRAY	G43 -7	G43 -26	GET	RIVET FROM TRAY			247	363	4066
004	PLACE	RIVET IN HOLE	P311-6	N P311-6	PLACE	RIVET IN HOLE					8
005	GET	RIVET FROM TRAY	G43 -6	G43 -6	GET	RIVET FROM TRAY			505	505	7478
006	PLACE	RIVET IN HOLE	P311-6	N P311-6	PLACE	RIVET IN HOLE			266	266	4298
						ELEMENT SUBTOTAL					1688
002 103-02		STAKE RIVETS, PLACE NEXT COIL AND IRON IN FIXTURE				PO					
001	REACH	CONTROL BUTTON	G01 -9	G01 -9	REACH	CONTROL BUTTON			83	83	83
002	MOVE	BUTTON	P01 -1	P01 -1	MOVE	BUTTON			25	25	25
003					PROCESS	HOLD BUTTONS		190			
004					PROCESS	COMPLETE PRESS ACTION		250*			
005				G12 -14	GET	COIL, IRON IN TRAY				173P	173
006	RECEIVE	FROM RH	G3 -12	P01 -26	MOVE	PART TO LH			152	239P	295
007				G11 -8	GET	COIL, IRON IN FIXTURE				109P	109
008	PLACE	COIL,	E P121-6	P03 -24	MOVE	TO TRAY POCKET			238	227	258
						ELEMENT SUBTOTAL		190			943

TOTAL RUN MU - MANUAL 2631 WITH ALLOWANCE 2841
- PROCESS 190 WITH ALLOWANCE 228
TOTAL STANDARD RUN TIME 3069 MU
CYCLE QUANTITY 1.00
UNIT STANDARD RUN TIME 3069 MU
.00307 HOURS
.184 MINUTES
UNITS PER HOUR 325.84

MMMM MOD II

MAI 69 PERCENT RMB 60 PERCENT GRA 19 PERCENT POS 22 PERCENT PROC 7 PERCENT

Figure 11–8. Operation Analysis Report (R0820). © MTM Association for Standards & Research, Inc., Fair Lawn, NJ.

```
4M DATA SYSTEM    R 830       4M OPERATION ANALYSIS

PART  TF-25P1017              CURRENT COIL
OPERATION  03       ASSEMBLE & STAKE 4 RIVETS IN VOLT COIL LAMINATION
DEPT. 30       MACHINE           TOOLING
ALLOW.- MAN.  .080  STD HRS - SETUP   .00000                     ORIG.   07/01/81
      - PROC. .000           - RUN     .00303  UNITS/HR  329.92  REV.      /  /


      LH MOTIONS                     RH OR BODY MOTIONS   PROC    LH    RH       NET
                                                FREQ.     TIME                   MANUAL

                                       --- RUN ---

010 103-01     ALIGN IRON AND PLACE 4 RIVETS
                                                                 PO

 010 ALIGNING HANDLE   G01 -6                                     70              70
 020 ALIGNING HANDLE   P01 -2                                     36              36
 030 RIVET FROM TRAY   G43 -7      G43 -26  RIVET FROM TRAY      247   363       406B
 040 RIVET IN HOLE     P311-6     NP311-6   RIVET IN HOLE                           B
 050 RIVET FROM TRAY   G43 -6      G43 -6   RIVET FROM TRAY      505   505       747B
 060 RIVET IN HOLE     P311-6     NP311-6   RIVET IN HOLE        266   266       429B
                       ELEMENT SUBTOTAL                                         1688

020 103-02     STAKE RIVETS. PLACE NEXT COIL AND IRON IN FIXTURE
                                                                 PO

 010 CONTROL BUTTON    G01 -9      G01 -9   CONTROL BUTTON        83    83        83
 020 BUTTON            P01 -1      P01 -1   BUTTON                25    25        25
 030                                        HOLD BUT      190
 040                                        COMPLETE      250*
 050                               G12 -14  COIL, IRON IN FIX          173P      173
 060 TO LH             G3  -12     P01 -26  PART TO LH           152   239P      295
 070                               G11 -8   COIL, IRON IN FIX          109P      109
 080 COIL B IRON IN FI P121-6      P03 -24  TO TRAY POCKET       238   227       258

                       ELEMENT SUBTOTAL                 190                      943


          TOTAL RUN MU - MANUAL       2631   WITH ALLOWANCE       2841
                       - PROCESS       190   WITH ALLOWANCE        190
                                  TOTAL STANDARD RUN TIME         3031 MU
                                           CYCLE QUANTITY         1.00
                                   UNIT STANDARD RUN TIME        .3031 MU
                                                                .00303 HOURS
                                                                  .182 MINUTES


MAI  69 PERCENT        RMB  60   GRA  19   POS  22        PROC   6 PERCENT
```

Figure 11–9. 4M Operation Analysis—Abbreviated. © MTM Association for Standards & Research, Inc., Fair Lawn, NJ.

Operator Instruction Report (R5840)

This report only includes that information needed for operator instruction and which can be left with the operator for reference—a practice which these authors endorse. In fact, these writers suggest "hanging" them in the view of the operator. The information supplied does include the standard time and the run rate in units per hour. Figure 11–10 illustrates R5840.

Summary—Operators' Instruction (R0850)[3]

This report (Fig. 11–10) prints only the element codes, descriptions, and frequency.

Operation Standard Report (R0815)

This report again is a duplicate, this time of R0820, but it deletes all element detail lines. It only illustrates the central body of the report in order to save space where this information will suffice. It does indicate the forecast practice opportunity.

4M Element Analysis (R0715)

This report provides a detailed analysis of a given or desired portion of an operation. One such element analysis for an operation covered in the previous reports is shown in Fig. 11–11.

Process Completion Time

When process time is entered on the input, its location in the output will depend on whether the manual motions *which follow* can be performed during the process time.

If the manual motions can be performed internal to the process time, the analyst indicates this event by a "P" in the input, and the total process time will be shown in the output on the description with an asterisk (*). However, this value will not be included in the total cycle time, only the process time *left over* after the internal manual operations. This situation is shown in Fig. 11–12.

When process time can not be overlapped with manual motions, the process time is immediately shown in the total column and is included in the total cycle time. This situation is illustrated in Fig. 11–13.

If any process time is included in the total time required, the Waiting-on-Process Index (PROC) indicates the percentage of the unabsorbed time to the total cycle time.

PROC is an Improvement Index that clues the industrial engineer as to how much such wait time exists, and that he/she should possibly try to use this "wasted" (in a sense) operator time.

The PROC, as well as the other Method Improvement Indexes (next described), are calculated on both the element and on the operation level.

[3]Some of the reports are not illustrated in order to save space. Only the most significant and/or representative are shown.

4M DATA SYSTEM R0840 OPERATOR INSTRUCTION REPORT REQUESTED BY JPO TIME 11.01.03 DATE 02/21/80 PAGE 1

PART 25P1017 VOLT COIL ASSEMBLY ORIGINATED 06/16/79 BY DOC
OPERATION 03 ASSEMBLE AND STAKE 4 RIVETS IN VOLT COIL LAMINATI REVISION A 02/10/80 BY KEE
DEPT. 30 MACHINE 030D TOOLING A-9675 ASSEMBLY FIXTURE
STANDARD HOURS - SETUP .00000 RUN .00307
CYCLE QUANTITY 1.00 UNITS PER HOUR 325.84

		LH MOTIONS			RH OR BODY MOTIONS		FREQ.
					--- RUN ---		
001	ALIGN IRON AND PLACE 4 RIVETS						
	001	REACH	ALIGNING HANDLE				
	002	MOVE	ALIGNING HANDLE				
	003	GET	RIVET FROM TRAY		GET	RIVET FROM TRAY	
	004	PLACE	RIVET IN HOLE	N	PLACE	RIVET IN HOLE	
	005	GET	RIVET FROM TRAY		GET	RIVET FROM TRAY	
	006	PLACE	RIVET IN HOLE	N	PLACE	RIVET IN HOLE	
002	STAKE RIVETS, PLACE NEXT COIL AND IRON IN FIXTURE						
	001	REACH	CONTROL BUTTON		REACH	CONTROL BUTTON	
	002	MOVE	BUTTON		MOVE	BUTTON	
	003				PROCESS	HOLD BUTTONS	
	004				PROCESS	COMPLETE PRESS ACTION	
	005 *				GET	COIL, IRON IN TRAY	
	006 *	RECEIVE	FROM RH		MOVE	PART TO LH	
	007 *				GET	COIL, IRON IN FIXTURE	
	008	PLACE	COIL, IRON IN FIXTURE		MOVE	TO TRAY POCKET	

APPROVED BY:

* TO BE DONE DURING PROCESS TIME

Figure 11–10. Operators Instruction Report. © MTM Association for Standards & Research, Inc., Fair Lawn, NJ.

```
4M DATA SYSTEM    R5715B      4M ELEMENT ANALYSIS                    REQUESTED BY KEE   TIME 12.05.03  DATE 09/16/81      PAGE  1

ELEMENT  103-01      ALIGN IRON AND PLACE 4 RIVETS                                  LEARNING LEVEL 100
                                                                                        TOTAL MU - MANUAL     1688
                PRACTICE OPPORTUNITY                               UPDATED                         - PROCESS        0

                                                                                                     PROCESS                  NET
             LH MOTIONS                                               RH OR BODY MOTIONS       FREQ.      TIME     LH    RH    MANUAL

010 REACH    ALIGNING HANDLE        G01 -6                                                                          70            70
020 MOVE     ALIGNING HANDLE        P01 -2                                                                          36            36
030 GET      RIVET FROM TRAY        G43 -7          G43 -26    GET      RIVET FROM TRAY                            247   363     406B
040 PLACE    RIVET IN HOLE          P311-6      N   P311-6     PLACE    RIVET IN HOLE                                               B
050 GET      RIVET FROM TRAY        G43 -6          G43 -6     GET      RIVET FROM TRAY                            505   505     747B
060 PLACE    RIVET IN HOLE          P311-6      N   P311-6     PLACE    RIVET IN HOLE                              266   266     429B

                                                                                  TOTAL MU                        1124  1134     1688

MAI  67 PERCENT          RMB  47 PERCENT   GRA  25 PERCENT   POS  29 PERCENT          PROC   0 PERCENT
```

Figure 11–11. 4M Element Analysis. © MTM Association for Standards & Research, Inc., Fair Lawn, NJ.

NOTES OR OTHER COMMENTS
MISCELLANEOUS TIME ENTRY — MISCELLANEOUS DESCRIPTION — MISC TIME
RETRIEVAL OF SUB-ELEMENT — ELEMENT CODE
PROCESS TIME ENTRY — PROCESS DESCRIPTION — PROCESS TIME
DESCRIPTION CONTINUATION — LH — RH

M	X	LH DESCRIPTION	LH NOTATION	DIST.	WT.	LH	NV	DC	RH NOTATION	DIST.	WT.	X	RH DESCRIPTION	FREQUENCY	P	B/C
M			SM						P112	12			JACK			
P										POLISH				5000	O	
M			SM						G13	16			COVER		P	
M			SM						P02	12			TO WORK AREA		P	

			RH OR BODY MOTIONS	FREQ.	PROCESS TIME	LH	RH	NET MANUAL
P112-12	P112-12	PLACE	JACK			229	231	326
	POLISH				5000*			
G13 -16	G13 -16	GET	COVER			215	207P	280
P02 -12	P02 -12	MOVE	TO WORK AREA			134	134P	134
			ELEMENT SUBTOTAL		4586			740
				PROC 86%				

Figure 11–12. Overlapped and unabsorbed process time. © MTM Association for Standards & Research, Inc., Fair Lawn, NJ.

Methods Improvement Indexes

There are five of these powerful aids to the industrial engineer's efforts to accomplish or design the optimum method of production. Only a system like 4M can make the development of such indices a practical reality.

In order to set the scene and understand this discussion, it is first necessary to define the ratios and list their symbols. These, in a brief summary fashion, are as follows:

TL Total Micro-Units for motions performed by the left hand (the right hand is covered by TR), after any computer reduction to lower-case motions. This will not include any waiting time increments while motions of the other hand occur.

G01 -20	REACH	CONTROL BUTTON		131	131	131
P01 - 1	MOVE	BUTTON		25	25	25
G11 -10	GET	AIR DRIVER			117	117
P02 - 8	MOVE	DRIVER FROM HOLDER			106	106
P215- 8	PLACE	DRIVER ON SCREW				B
				86	246	246B
DRIVER ACTION			260			
P215- 3	PLACE	DRIVER ON SECOND SCREW			195	195
DRIVER ACTION			260			
P03 -12	MOVE	DRIVER TO HOLDER		91	152	152
		ELEMENT SUBTOTAL	520			972
		PROC 35%				

Figure 11–13. Unabsorbed process time. © MTM Association for Standards & Research, Inc., Fair Lawn, NJ.

TN Net time in Micro-Units for a one line or continuous set of motions, after the computer has completed its analysis as to how the motions occur with respect to each other.

T When there is no TP, T = TN. If there is some TP, then T equals TN or TP, whichever is larger—it is really a printout of the time required for a line, or set of lines, with or without process time.

TP The process time as entered by the analyst on the analysis form.

TP_w That portion of the process time that is *not overlapped* by assigned motions. During TP_w the operator waits.

TR Same as TL, but for the right hand.

Motion Assignment Index (MAI)

By definition

$$\text{MAI} = \frac{(\text{Sum of TL}) + (\text{Sum of TR})}{2(\text{Sum of TN})}$$

This index shows how completely the hands have been assigned work. All body motions except WALK are assumed (for the purpose of this index) to represent full use of the hands (this seemed to be the most rational approach to the developers of 4M). This means that body motion time must be accounted for in the numerator. It also seemed best, in this situation, to credit the time to *only* the right hand. Also, if the industrial engineer can reduce WALK time, it will improve the MAI index.

If one hand holds a part and only the other hand performs a motion or function, such as POSITION or other such activity, the hand holding a part (or assembly) is *not* credited in the numerator, since properly designed fixtures would normally eliminate the need for "holding."

Periods of inactivity are not considered in the development of MAI.

If MAI shows 50 percent, it indicates the *equivalent* of only one hand being used (see the definition of the expressions in the numerator if this meaning is not clear); 50 percent indicates a very poor motion pattern. However, a ratio of 100 percent is unattainable because of the delay patterns imposed by the Simo Chart for complex simo motions. Experience shows that a satisfactory range is 59–85 percent.

What is an acceptable range for one type of work may be entirely different for another type. When a low ratio is found when compared to a standard for that type of work, it is the clue to try and utilize better methods so as to utilize the body members more effectively.

GRASP, RELEASE, and APPLY PRESSURE Index (GRA)

By definition

$$\text{GRA} = \frac{(\text{Sum of GRASP and RELEASE}) + (\text{Sum of AP})}{\text{TL} + \text{TR}}$$

A large percentage of grasp time would indicate the need to seek ways to eliminate "search and select" (the high time value GRASPs). Excessive APPLY

PRESSUREs might be due to the improper fit of parts, insufficient usage of power assists, etc.

POSITION Index (POS)

By definition

$$POS = \frac{\text{Sum of POSITION values}}{TL + TR}$$

POSITIONs are high time value motions; the higher the class, the greater the amount of time required. Often "chamfering" or the use of better fixtures and/or parts feed (such as preorientation) can lower a POSITION classification.

REACH, MOVE, and BODY MOTION Index (RMB)

By definition

$$RMB = 100\% - (\%GRA + \%POS)$$

A higher percentage of RMB than normal could be due to *excessive* distances moved—the obvious solution is to try and minimize them. A high amount of BODY MOTION time obviously represents a work simplification opportunity. This index also includes time spent in performing such miscellaneous motions as CRANK, TURN, DISENGAGE, and Eye action.

Waiting-on-Process-Interval Index (PROC)

By definition

$$PROC = TP_w/T$$

Unutilized waiting time during process time obviously represents a methods improvement opportunity.

Some Possible Conclusions

A little thought would make clear that these indexes should make the job of performing *good* industrial engineering easier. They obviously can be of substantial value to the person performing the work. However, further thought will also disclose that these indexes will make the job of supervising such industrial engineering easier, the poorly performing industrial engineer will have Index Ratios consistently above the norm.

Error Messages

4M aids in the proper application of MTM-1 by not only simplifying "pattern writing," but also by such actions as automatic application of simo rules (including the automatic adjustment of case of REACH and MOVE), automatic report development and printing (there are many more available than the few mentioned in this chapter)—and also by calling attention to *errors*. There would be little gained by identifying each error detection signal. Suffice it to say that there are now 57 such messages. Examples would be as follows:

PRACTICE OPPORTUNITY INVALID (No. 5)—The Practice Opportunity Field was not "Y," "N," or blank.

ELEMENT LINE NUMBER KEY INVALID (No. 9)—Element and Line Number Fields on an analysis detail transaction were both blank.

RECORD TYPE INVALID (No. 10)—Record type on analysis detail transaction was not "M," "P," "T," "L," or "R."

Standard Data Maintenance

The 4M DATA System has three primary function.

1. The generation of elements based upon the proper application of MTM-1.
2. The application of these elements to form part-operation standards.
3. The maintenance of elements and standards through the use of mass updating procedures.

Each element with its heading and analysis detail is stored only once. They can be retrieved easily for the development of operation standards. Transaction codes are used on a daily basis to keep elements current. Every line of input to the 4M DATA System is prefixed by two numbers, one of which addresses the particular file to be accessed and the other the type of transaction. The transaction codes are 1, for delete; 2, for add; and 3, for change. (In the interactive version the corresponding codes are D, A, and C.) The file codes are 1, for Element Master; 2, for Analysis Detail; 3, for Operation Master; and 4, for Operation Elements.

For example, to place a new Element Master on the Element Master file, the entry begins with the digits 12. The first digit specifies the Element Master file and the second digit, the transaction code, instructs the computer to place the entry on the file. By the same token, the digits 11 instruct the computer to remove the entry from the file, and the digits 13 are used when an entry already on the file is to be changed in some manner. Consequently, since there are four files and three types of transactions, each line of data is prefixed by one of the following sets of two digits: 11, 12, 13, 21, 22, 23, 31, 32, 33, 41, 42, 43. Each report of an Operation Standard indicates whether it is updated and when such updating occurred.

Operation Standards are calculated upon request for screen display and/or printout in report format.

There is a master file of these data and it is divided into two sections. The first section contains heading information for each covered operation and is accessed through a 1 to 28 character part-operation number. The second section contains a listing of element codes which compose each part-operation located in the first section. Naturally, these files are accessed upon command to calculate a part-operation standard.

The following listing of reports completes this very brief review of the essentials of this 4M function:

R0820	(FULL) & R0830 (ABBREVIATED) 4M Operation Analysis
R0840	Operator Instructions
R0850	Summary Operator Instructions
R0860	MTM Analysis
R0815	Operation Standard Report

These reports were previously described and, in some cases, illustrated.

Where-Used, Search, and Updating

The third 4M DATA MOD II function is to update the above data files. Normal update has already been partially described. However, it can be done on a mass change basis, albeit this only occurs upon request by the analyst.

Associated reports are the following:

R0920	Element Where-Used Report
R0130	Element Master File Report
R0330	Operation Master File Report
R0530	Study Index File Report

The day to day transactions which deal with deleting, adding to, or changing the files results from the use of the standard input formats. The records generated are placed in the data files through the use of Job Control Language (JCL) No. 1.

The calculation of 4M element analyses and the issuance of the analysis reports are accomplished through the use of JCL 2.

JCL 3 is used to request operation standards calculations and reports dealing with this activity. The involved reports are:

R0815	Operation Standard Report
R0820	Operation Analysis—Full
R0830	Operation Analysis—Abbreviated
R0840	Operator Instructions
R0850	Summary Operator Instructions
R0860	MTM Analysis

JCL 4B is used for mass updating, but an Element Where-Used Report of a complete listing or up to 10 elements per run may be obtained with JCL 4A.

JCL 5 is used when a partial or complete listing of the Element Master File is desired.

To get a partial or complete listing of the operation file, JCL 6 is used.

JCL 7 is used to request a partial or complete cross-reference of work studies to part-operation numbers.

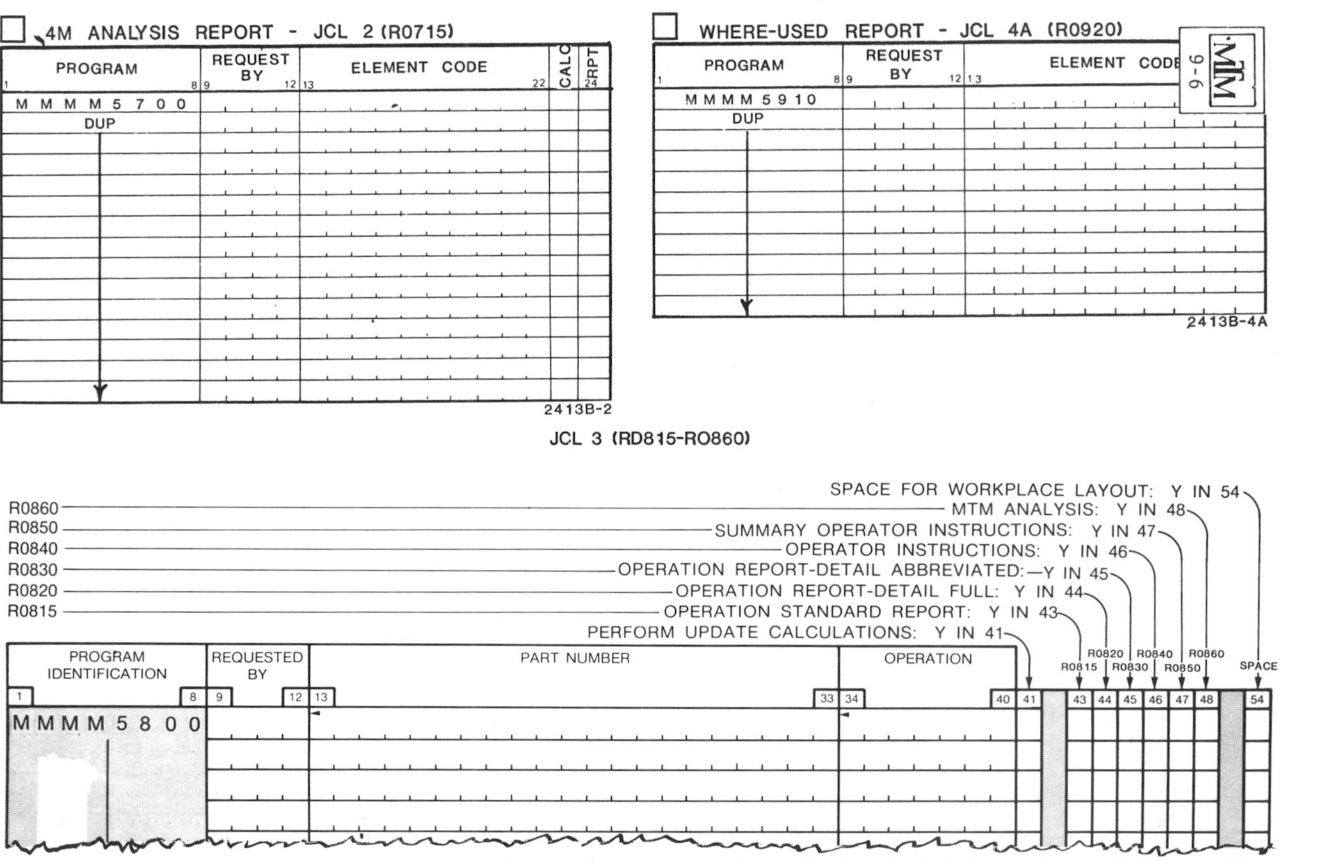

Figure 11–14. The essentials of JCL 2, JCL 3 and JCL 4A. © MTM Association for Standards & Research, Inc., Fair Lawn, NJ.

In case the reader thinks that this means great complication, Figure 11–14 illustrates the essentials of JCL 3, JCL 2, and JCL 4A.

Concluding Remarks

This somewhat abbreviated discussion of 4M DATA MOD II does not tell nearly all of the facts or information required to operate it. Naturally, purchase of the system will provide all needed information, program decks, *User's Manual,* and personal help in installation. However, this discussion should convey to the reader all of the essential information needed for a general understanding of the system. 4M DATA is really "the way to go" for the larger firm utilizing work measurement techniques.

CHAPTER 12

System Selection

During the last decade the MTM family has expanded to include a number of generic systems that can provide standards for the same operation. Other functional systems have been developed that have limited areas of application. Of course, the generic system can also be used in the functional areas, so now the MTM user has to decide which system is the most appropriate to use in any given circumstance.

The primary motivation for the development of the family of systems was economic. The users of the original MTM system, now known as MTM-1, felt that the system did a very good job, especially where detailed methods analysis was important, but that it was very time-consuming to use. This high time consumption was expensive not only for the engineers who were writing the standards, but also for the organization which wants quick standards coverage.

MTM-1 applicators observed that, for many operations, the same element sequences kept occurring over and over again.[1] Frequently, the only changing factors were the distances involved and the cases of the motions. For example, the element sequence—Reach, Grasp, Move, and Position—occurs with high frequency. These same elements (not necessarily in the sequence given) comprise a large percentage of the MTM-1 elements used. The percentage varies, but a figure of 80 percent is a good estimate for bench-type operations.[2]

System Evaluation Based on Accuracy Prediction

Investigations concerning motion sequence and the frequency of use of the MTM-1 motions resulted in the development of MTM-2 and, eventually, MTM-3.

The objective of minimizing the MTM application time was accomplished; Fig. 12–1 gives the results for each of the MTM systems.

[1]U. Åberg, "Frequency of Occurrence of Basic MTM Motions," *Research Committee Report No. 1*, Svenska MTM–Foreningen, Stockholm, Sweden, August, 1962. Also published in *Journal of Methods Time Measurement* **8** (5) (July–August, 1963).

[2]U. Åberg, "Sequence of Motions in Manual Workshops Production," *Research Committee Report No. 2*, Svenska MTM–Foreningen, Stockholm, Sweden, July, 1963. Also published in *Journal of Methods Time Measurement* **9** (2) (November–December, 1963).

System	Approximate Relative Time
MTM-2	Two times faster than MTM-1
MTM-3	Seven times faster than MTM-1

Figure 12–1. Speed of application of MTM-2 and MTM-3 as compared to MTM-1.

Once the systems were developed, the question arose as to which system should be used for a given situation. If application time were the only consideration, then the MTM-3 system should always be used. However, questions were raised about the accuracy of the quicker systems. There was an intuitive feeling that, if a system were faster to use, it would probably be less accurate than the more time-consuming system. As a result of these questions, a methodology was developed to predict the accuracy of the various systems.[3]

System accuracy is dependent on a number of variables. A brief description of these is as follows:

1. *System accuracy.* System accuracy is the precision of the system as reflected in how the data presented are to be used. For example, in the MTM-1 system, the time value for an R6B is used for all B reaches that are between 5 ½ and 6 ½ inches.[4] Thus there is an error introduced, which is called systems error. In MTM-2 the distance classes are much fewer than in MTM-1, thus producing a higher systems error.[5]
2. *Application accuracy.* When an applicator uses a MTM system he/she may make an error by selecting the wrong motion, distance, or class; this is called application error. For example, suppose a reach was actually 6 inches, but the applicator used the 7-inch category; he/she would have made an application error. It is interesting to observe that the *magnitude* of the error is inversely proportional to the number of cells in the data table, but the probability of making an error is directly proportional to the number of cells. For example, an applicator is more likely to make a distance error using the MTM-1 system than he/she is with the MTM-2 system because of the larger number of cell values in the MTM-1 system. However, the consequences of an error given that it occurs in the MTM-1 system are much less than in the MTM-2 system because the differences in the cell values are much less.
3. *The number of nonrepeated elements which occur during the review period time (NRT).* Prediction accuracy depends very much on certain characteristics of the standard and on the elapsed time over which performance is measured. The methodology is as follows:
   A. The review period is defined as the time period over which performance is reviewed. For example, if the performance of the worker is measured daily, then the review period is daily.

[3]"MTM System Accuracy and Speed of Application," distributed by the MTM Association for Standards & Research, Fairlawn, NJ. (No date.)

[4]W. Hancock, "The System Precision of MTM-1," *Journal of Methods Time Measurement,* **12** (2) (March–April, 1967).

[5]W.M. Hancock, J.A. Foulke, and J.M. Miller, "Accuracy Comparisons of MTM-1, GPD, MTM-2, MTM-3 and the AMS Systems," *Journal of Methods Time Measurement,* **18** (May, 1973).

B. The number of nonrepeated elements is determined by:
   (a) Determining the number of different jobs the worker does during the review process.
   (b) Reducing the frequencies in the standards of the jobs in (a) to one. (Exactly how to do this and what is meant are presented later in this chapter.)
   (c) Summing the TMUs of all jobs done during the review period after performing the frequency reduction in (b).

4. *Determining the confidence interval that one desires to use in the determination of the accuracy*. The typical interval is 95 percent. If 95 percent is used, then the true standard will lie within the confidence interval 95 percent of the time. If accuracy comparisons are being made with other work measurement systems, such as time study, it is more appropriate to use the 50 percent confidence interval, because the time study errors are usually averaged to get an average error.[6]

Analytic techniques have been developed to determine the systems error (see footnote 4) and experiments have been run to determine application error of the various MTM systems (see footnote 5). The net result is a series of graphs of which Fig. 12–2 is the most recent. Important points to observe when examining the graphs are as follows:

1. Both axes are in log units. They are used so that the accuracy lines will be straight lines and will be parallel to each other.
2. Comparing the hand-used systems, MTM-1 is the most accurate over the range of NRT displayed. MTM-2 is less accurate than MTM-1, and MTM-3 is less accurate than either MTM-2 or MTM-1. Thus for these systems, the faster the system is to apply, the less accurate the system is for a given NRT.
3. If we choose a certain percent accuracy ($\%A$) such as ± 5 percent, then all of the lines representing the various systems will intersect the 5 percent point. Thus all of the systems are capable of achieving a certain accuracy providing there is a sufficient TMU in the NRT time. This is an important observation because if the accuracy necessary is stated, then the minimum application time is obtained by having the applicators trained in all of the MTM systems.
4. The 4M Data System has higher accuracy than MTM-1, because the applicator error is greatly reduced by using a computer-aided system. Many of the decisions are made by the computer rather than being made by the applicator. Improvement in accuracy is a major stimulator for using computer-aided systems particularly

[6]For those wishing to derive the 50 percent confidence intervals, the lines of Fig. 12–2 were obtained using the equation:

$$\pm\%A = \frac{K_i\, S_j}{\sqrt{NRT}} \times 100$$

$\%A$ = Accuracy Level
K_i = the number of standard deviation at confidence level i
S_j = the standard deviation for MTM system j.

Figure 12–2 is the 95 percent confidence level. Thus for every MTM system (j), S_j can be determined using the above equation. K_i can then be set at whatever confidence level is desired. For the 50 percent confidence level, $K_i = 0.68$.

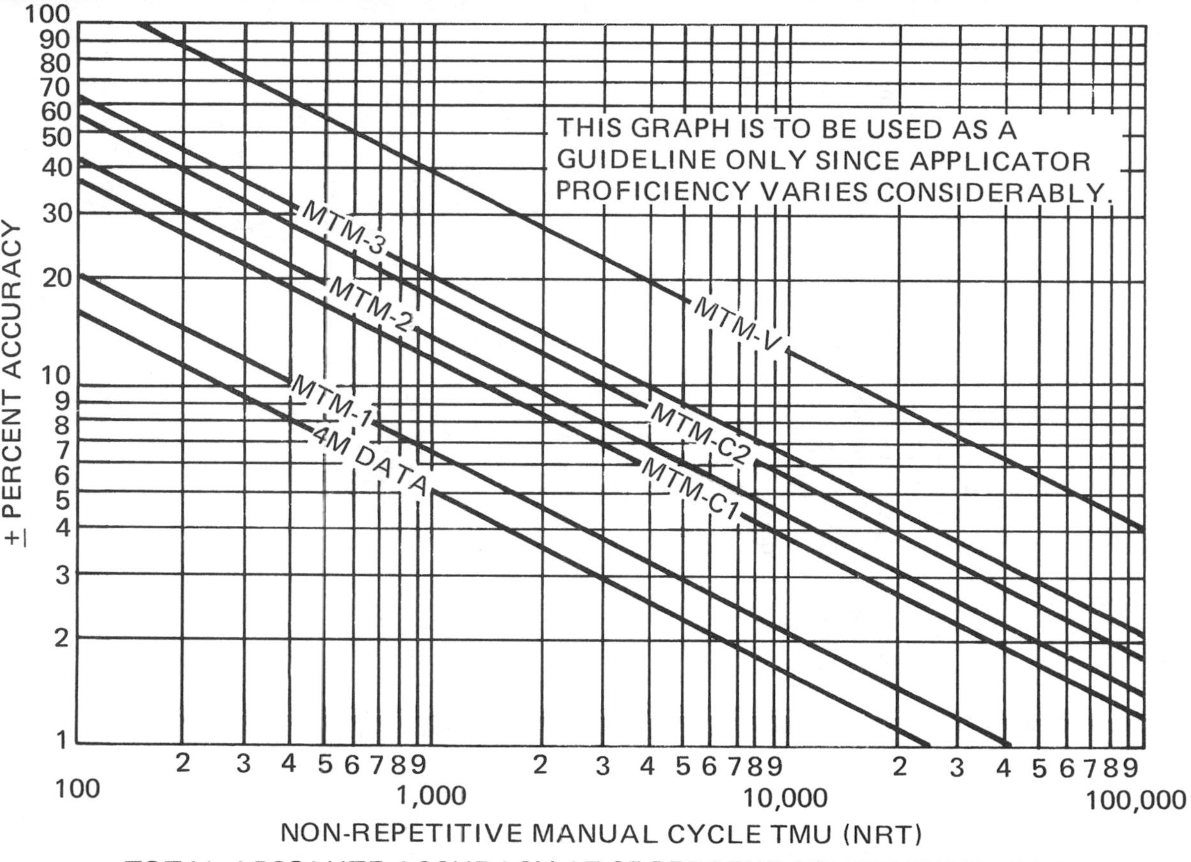

Figure 12–2. Total absolute accuracy at 95 percent confidence limits. Reproduced from MTM Accuracy and Speed of Application, *MTM Association for Standards and Research, Fair Lawn, NJ, with permission.*

where short cycle highly repetitive operations are involved or where high productivity losses or excessive labor complaints are incurred. (More details about productivity losses and later complaints will be given in this chapter.)

5. MTM-V, MTM-C1, and MTM-C2 are functional systems that are used for special purposes. Accuracy may be one of a number of considerations when determining whether to use one of the functional systems or one of the generic systems.

Examples of uses of Fig. 12–2. A frequent situation is that the applicator wants to use the fastest system consistent with the accuracy policies of the organization. A typical policy is that accuracies of ± 5 percent with 95 percent confidence are acceptable. Let us assume that we wish to use either MTM-1, MTM-2, or MTM-3. In order to determine the NRT TMUs in the review interval, suppose that the review interval is one day and the worker does the same job everyday. We then estimate the TMUs in his/her job and NRT by reducing the frequency of any repetitions within the job to 1. For example, if the worker is assembling a number of parts which take an approximate cycle time of 2.5 minutes ($2.5 \times 1666.7 = 4167$ TMUs) and, of the 2.5 minutes, six identical nuts have to be started and torqued with an estimated cycle time of 0.20 minutes per nut; then $2.5 - [(6 - 1)\ 0.20] = 1.5$ minutes or 2500 TMUs, which is the estimated NRT. We then enter Fig. 12–2 at 2500 TMUs. We find that MTM-1 will give us ± 4 percent accuracy; MTM-2, ± 7.5 percent; and MTM-3, ± 12.5 percent. Therefore, MTM-1 should be used to meet our policy of ± 5 percent.

Suppose that the review interval in the above example is for one week and the worker does an average of three jobs which have the following approximate NRT : 4000 TMU, 1200 TMU, and 2000 TMU. What system should we use? First sum the NRTs : $4000 + 1200 + 2000 = 7200$ TMU. Entering Fig. 12–2 at 7200 TMU, reveals an accuracy of ± 2.4 percent for MTM-1, ± 5.0 percent for MTM-2, and ± 7.5 percent for MTM-3. Thus MTM-2 can be used for this situation providing that none of the system restrictions are violated![7]

Suppose the NRT in the review interval is 500 TMU. Then MTM-1 will not meet the specification of ± 5 percent. The graphs of Fig. 12–2 are based on the frequency distributions of Åberg's Industry B. Any deviation from the frequency distribution of Industry B will result in changes in the levels of the curves. This is critical where short-cycle, highly repetitive operations are concerned. Also, the applicator variation can be reduced greatly by careful auditing of standards by a second applicator. Then, in the example of 500 TMU NRT, the following should be done to determine the percent accuracy:

1. Determine the standard using MTM-1.
2. If MTM-1 is used, have the standard audited by another applicator. This should reduce the applicator error to zero.
3. Using the data of footnote 4, compute the variance of the NRT (system variance). This is done by summing the variance of each of the elemental motions of the NRT. If σ_i^2 is the variance of each of the motions, then

[7]There are restrictions on using MTM-2 and MTM-3 that should be observed. See Chapters 6 and 7 for a discussion of these restrictions.

$$\sigma^2{}_T = \sum_{i=1}^{N} \sigma_i^2$$

where $\sigma^2{}_T$ is the total variance of the NRT.

4. Compute the 95 percent confidence interval by

$$\pm\%A = \frac{1.96\sqrt{\sigma_T^2}}{\text{NRT}} \times 100$$

Evaluation Based on Accuracy, Productivity, and Labor Complaints

After the accuracy methodology was introduced, questions arose concerning what accuracy would be most appropriate to use. Also, questions were raised concerning the desirability of reducing applicator time when the real objective was to maximize productivity. Attempts to respond to these kinds of questions lead to a methodology to determine the accuracy necessary to provide for minimum cost output.[8] Three factors were considered: the cost of developing the standards, the cost of settling the labor complaints due to standards that are perceived as unfair by the workers, and the productivity achieved. A brief discussion of the factors and how they are combined follows:

1. The cost of developing the standards. Figure 12–3 is the MTM Association's estimate of the time it takes to develop standards using the various MTM systems (see footnote 3). For example, if we needed to set a standard whose total TMU was estimated to be approximately 1 minute (1666.7 TMU), and if we were to use MTM-1, it would take an average of 250 minutes (416,675 TMU) to set the standard.
2. The cost of settling labor complaints. In this analysis, it is assumed that labor complaints are likely to occur if a worker is asked to work at a pace exceeding the standard rate. For example, if we have 1500 NRT TMUs in the standard for the review period and decide to use MTM-2, the 95 percent confidence limits are ± 11 percent. This means that the standard the worker is asked to perform against could be between 1335 and 1665 TMU 95 percent of the time. If the standard was established "loose," that is, at a level greater than 1500 TMU, then the worker will not complain. It will be easy for him/her to perform against the standard. However, if the standard requires the worker to perform at a rate faster than normal, i.e., the time allowed is less than 1500 TMU, then there is a chance that the worker will complain. There are two situations that may cause him/her to complain: he/she cannot attain the acceptable rate because he/she does not have the capability (see Chapter 2, Fig. 2–2 for the capability curve) or the worker perceives that the work is harder than "normal." The "harder" the worker perceives the standard to be, the more likely he/she is to complain. At any given level of "hardness" some work-

[8]W. Hancock and G. Langolf, *Productivity, Quality and Complaint Considerations in the Selection of MTM Systems,* MTM Association for Standards and Research, August, 1973.

System	Average Time to Set Standard
MTM-1	250 *X*
MTM-2	100 *X*
MTM-3	35 *X*
MTM-V	10 *X*
MTM-C1	125 *X* (tentative)
MTM-C2	75 *X* (tentative)
4M DATA	(15–60) *X*

Figure 12–3. The average time to set standards using the various MTM systems where X is the cycle time and all units are the same.

ers will complain and some will not. Figure 12–4 is an example of a Complaint Threshold Curve.

With the Complaint Threshold Curve, if the standard is 10 percent "tight" the probability of a worker complaining is 0.5. The curve is arbitrary, but it is felt to be most representative of the complaint threshold of employees currently working under low-task standards.

Since we have two sources of potential complaints, one where the worker simply cannot do the work and the other where he/she perceives difficulty, we can combine the two conditions into an overall complaint function. This is:

$$P_{\text{Complaint}} = \int_{a=-\infty}^{a=+\infty} f(a)p(a)\,da \quad (1)$$

where

a is the performance rate as dictated by the standard, $f(a)\,da$ is the density function for system accuracy $[N(100), 0.51\%A]$.[9]

$p(a)$ is the conditional probability of a complaint at performance rate a.

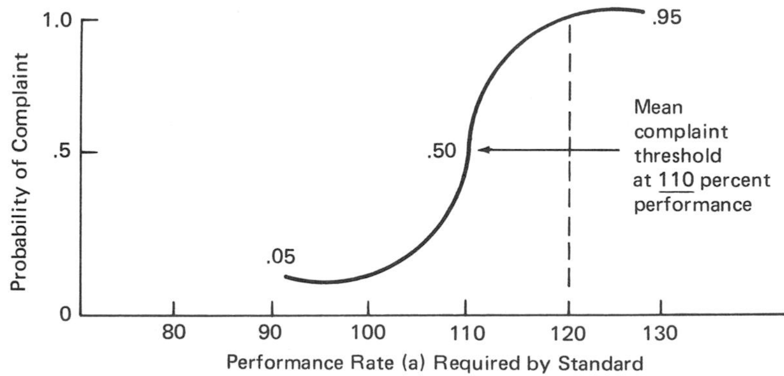

Figure 12–4. A complaint threshold curve. W. M. Hancock, and G. Langolf, Productivity, Quality and Complaint Considerations in the Selection of MTM Systems, *Aug. 1973, p. 9. Reproduced by permission of the MTM Association for Standards and Research.*

[9]The notation is for a normal distribution with a mean of 100 and a standard devition of 0.51%A. (A is the percent accuracy.)

If a complaint occurs, then we have to assume that the complaint will be corrected. "Tight" standards will be reestablished so that the performance rate is at 100. This correction process, because it will only occur when an employee complains and the employee will only complain if the standard is "tight," will cause productivity to decrease.

The average productivity level after correction of complaints is:

$$\text{MPL} = \int_{a=-\infty}^{a=+\infty} a\, f(a)[1-P(a)]\; da \; + \; 100 \int_{a=-\infty}^{a=+\infty} f(a)P(a)\; da \qquad (2)$$

where

a is the performance rate

$f(a)\, da$ is the system error density function $[N(100), 0.51\%A]$

$P(a)$ is the conditional probability of a complaint at performance rate a.

A computer program has been written to compute the equations for $P_{\text{Complaint}}$ [Eq. (1)] and MPL [Eq. (2)]. Two graphs (Figs. 12–5 and 12–6) have been developed, one for the 110 percent threshold and one for the 105 percent threshold. Each graph contains curves for $P_{\text{Complaint}}$, which is called "Complaint Probability," and MPL, which is called "Percent Productivity."

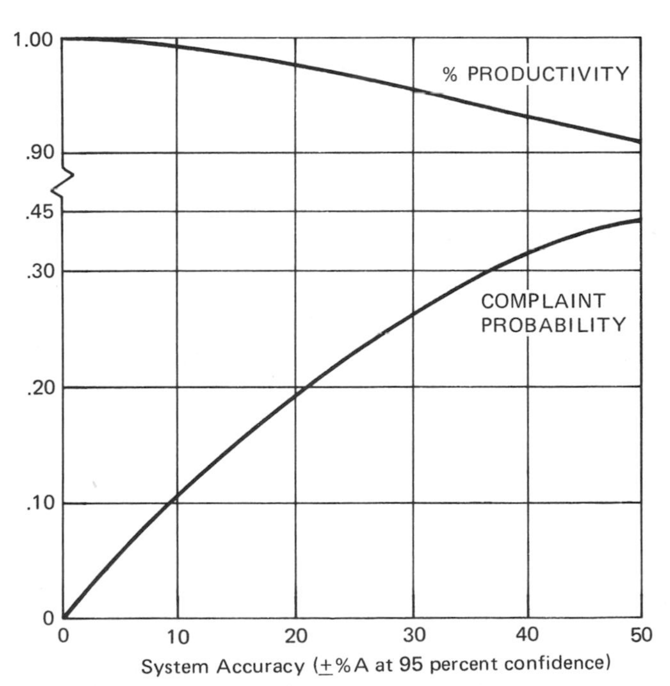

Figure 12–5. A plot of the $P_{Complaint}$ (complaint probability) and MPL (percent productivity) as a function of ± %A for the 110 percent mean complaint threshold case. F. Bayha, W. Hancock, and G. Langolf, "More Evaluation Parameters for MTM Systems" Journal of Methods Time Measurement, ***2(1)****, 1975. Reproduced by permission.*

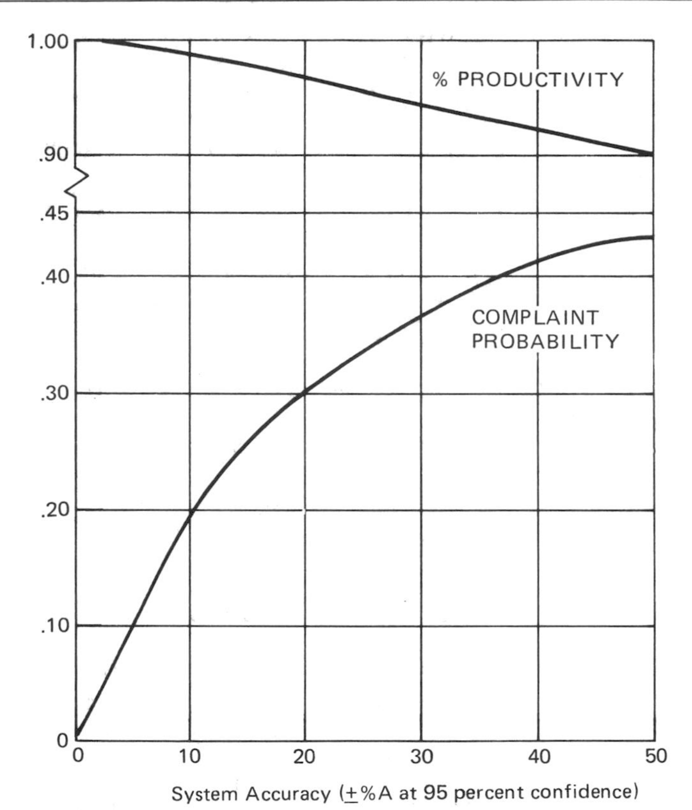

Figure 12–6. A plot of $P_{Complaint}$ (complaint probability) and MPL (percent productivity) as a function ± %A for the 105 percent mean complaint threshold case. F. Bayha, W. Hancock, and G. Langolf, "More Evaluation Parameters for MTM Systems," Journal of Methods Time Measurement, *2(1), 1975. Reproduced by permission.*

Examples of the Use of the Methodology

1. Operators are performing work where the NRT is an average of 5000 TMU and the standard time is 6000 TMU during the review period. Assuming a 110 percent complaint threshold, which system would be the most economical to use: MTM-1, MTM-2, MTM-3, or 4M DATA?

A. First, determine the *%A* for each system. From Fig. 12–2, the following percent values were obtained:

System	*%A*
MTM-1	2.9
MTM-2	6.0
MTM-3	9.0
4M DATA	2.2

B. Referring to Fig. 12–5, an estimate of the complaint probability and percent productivity is obtained:

System	Complaint Probability	% Productivity
MTM-1	0.04	99.9
MTM-2	0.07	99.9
MTM-3	0.10	99.5
4M DATA	0.03	99.9

Now suppose that there are:

a. Twenty new standards to be set on the average per month.
b. A labor rate, including fringe benefits, of $10.00 per hour.
c. An MTM applicator cost of $14.00 per hour to set new standards.
d. An average cost of $1000.00 to settle a grievance including reestablishing the standard.
e. An average of 250 employees on standards.

C. The costs of using each system can then be computed. When using the MTM-1 system as an example:

a. Figure 12–3 gives the time/TMU for setting MTM standards. The MTM-1 system takes 250 times the TMUs to be set. Thus, the cost to set the standards will be: [20 jobs/month × 6000 TMU/job × 250] × (1/10^5 TMU/hour) × $14.00/hour = $4200.00/month.
b. The costs of grievances (complaints) will be: 0.04 × 20 standards/month × $1000.00/grievance = $800/month.
c. The number of parts/day at 100 percent performance (assuming that the NRT and the standard are the same) will be: 8 hour/day × 1 part/6000 TMU × 10^5 TMU/hour = 133.3 parts/day/employee.
d. The cost/part assuming 100 percent productivity will be:

$$\frac{\$10/\text{hour} \times 8\ \text{hour/day}}{133.3\ \text{parts/day}} = \$0.60/\text{part}$$

e. The percent productivity using the MTM-1 system will be 99.9 percent. Thus, the increased price/part will be:
(1 − 0.999) × $0.60 × 133.3 = $.08/day/employee
f. The productivity loss/month will be:
0.08/day/employee × 250 employees × 21 days/month = $420.00.

The total cost of using the MTM-1 system would then be $4200.00 + $800.00 + $420.00 = $5420.00/month.

The following is a summary of the systems cost per month by category:

System	MTM Applicator	Complaints	Productivity Loss	Total
MTM-1	$4200	$ 800	$ 420	$5420
MTM-2	1680	1400	420	3500
MTM-3	588	2000	2099	4687
4M DATA	504[10]	600	420	1524

[10]Figure 12–3 gives a range of 15–60: 30 is assumed here.

The lowest total cost/month would be the 4M DATA, with MTM-2 being second. However, the computer costs for 4M DATA are not included in this analysis.

2. In this example, let us assume all of the conditions are the same as number 1 above, except that the 105 percent threshold is used. The only factors that will change are the complaint probabilities and the percent productivity. Using Fig. 12–6 and the *%A* from example 1, we get the following:

System	Complaint Probability	% Productivity
MTM-1	0.06	99.8
MTM-2	0.125	99.5
MTM-3	0.18	99.0
4M DATA	0.05	99.9

The following is a summary of the systems cost per month by category:

System	MTM Applicator	Complaints	Productivity Loss	Total
MTM-1	$4200	$1200	$ 840	$6240
MTM-2	1680	2500	2099	6279
MTM-3	588	3600	4200	8388
4M DATA	504[11]	1000	420	1924

Here, as in example 1, the 4M DATA has the lowest cost, however, computer costs would have to be added. If 4M DATA were not available, MTM-1 would be the least-expensive system to use, although it has by far the highest applicator costs.

Of course, the results of both examples are highly dependent on the costs of the various factors. To use the methodology, the various costs have to be obtained and the calculations made.

Note that it is assumed that the MTM applicator knows how to use all four systems so no cost is included for applicator training. Sufficient applicator staff is also assumed so that coverage can be maintained. The costs of lack of coverage are not included.

The methodology just presented is for unpaced operations where the worker does not have to complete the cycle within a fixed period of time. Typical examples are bench operations. The methodology gets much more complicated whenever the operation is "paced." Operations such as automotive assembly lines with no float between stations are typical paced operations.

As previously defined, paced operations are job assignments where the operator must get the work done within the cycle time allotted. This additional requirement causes the selection of the proper MTM system to be more complex because:

1. We must take into account two additional sources of variation: the variation among

[11] 4M DATA applicator time assumed as 30X.

workers and the variation exhibited by a given worker as he/she performs successive cycles.

2. The quality that is needed is affected by the number of operations that have to be completed correctly in order to have a good product, and the cost of repairing and/or completing the product if the worker cannot finish the work during the allotted cycle time.

Appendix A of footnote 7 contains the deviations of the between and within subject variances used here. The equations that are used in this methodology are as follows:

1. System error variance:

$$\sigma_S^2 = \left(\frac{\%A}{1.96} \times \text{NRT}\right)^2 \quad (3)$$

where $\%A$ and NRT are the same as used previously. The units of σ_s^2 are TMU^2.

2. The within subject variance:

$$\sigma_w^2 = S \quad (4)$$

where S is the sum of the elemental motion times for the work station or operation involved and can be considered the "unadjusted" cycle time. The units for σ_w^2 are TMU^2.

3. The between subject variance:

$$\sigma_b^2 = 0.02225S^2 \quad (5)$$

where S is the same as in Eq. (4).

The normal cycle time, which is the time the operator has to do his/her work with stated probabilities is:

$$\text{NCT} = S + Z\sqrt{\sigma_s^2 + \sigma_w^2 + \sigma_b^2} \quad (6)$$

substituting:

$$\text{NCT} = S + Z\sqrt{\sigma_s^2 + S + \sigma_b^2} \quad (7)$$

where Z is the number of standard deviations related to the probability of completing the job correctly.

The percent loading, which is the percent of the cycle that the standard time should be, is:

$$\%L = \frac{S}{\text{NCT}} \times 100 \quad (8)$$

where NCT is the cycle time (TMU) for people working against normal (low-task) standards.

Figures 12–7 through 12–12 give the relationship of NRT to percent load for different probabilities of completing the work (P) where the performance

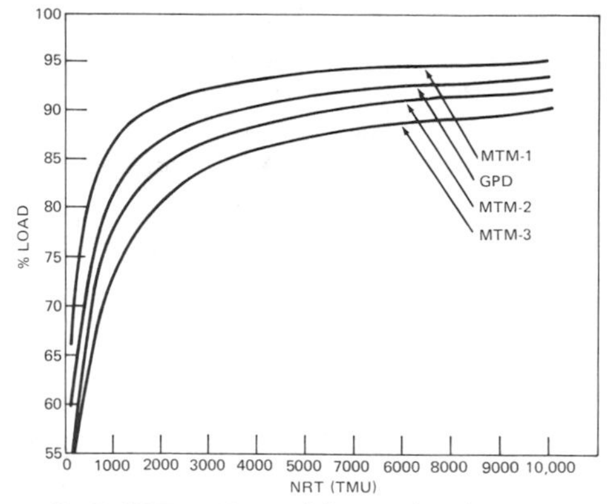

Figure 12–7. NRT vs % Load (normal task, P = *0.9993).*

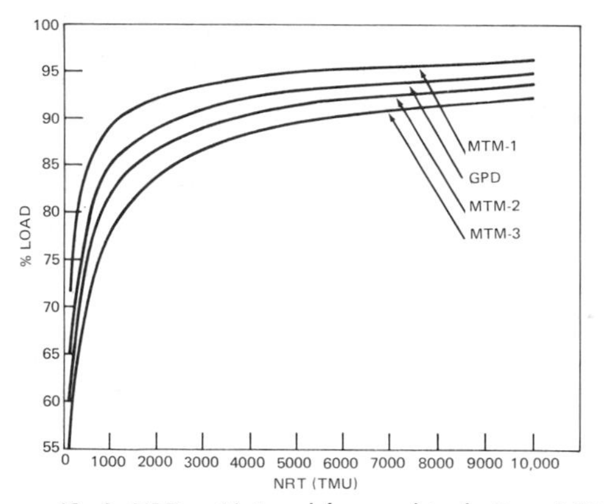

Figure 12–8. NRT vs % Load (normal task, P = *0.995).*

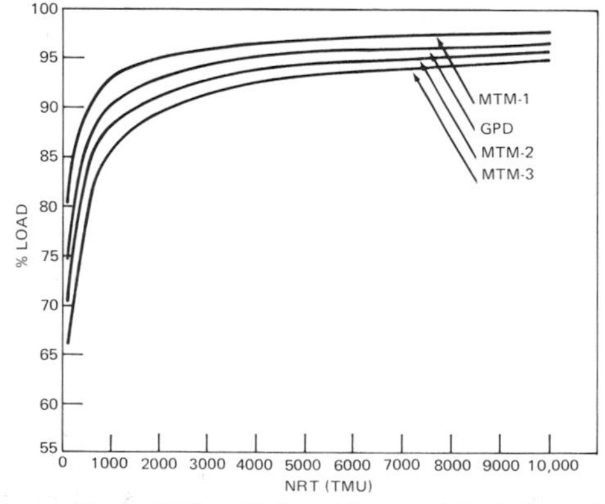

Figure 12–9. NRT vs % Load (normal task, P = *0.95).*

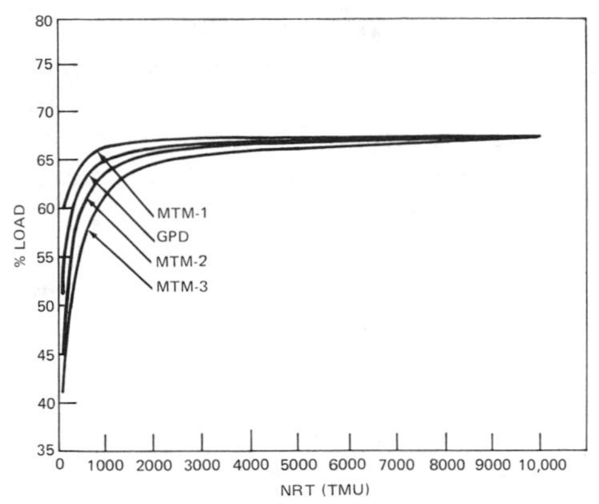

Figure 12–10. NRT vs % Load (high task, P = 0.9993).

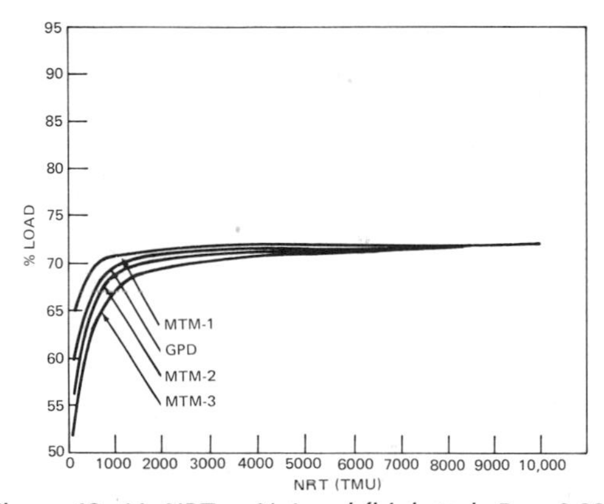

Figure 12–11. NRT vs % Load (high task, P = 0.995).

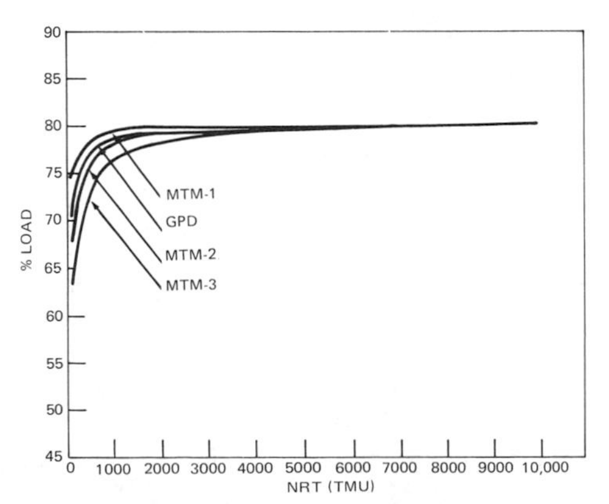

Figure 12–12. NRT vs % Load (high task, P = 0.95).

rate is set at normal task (100 percent rate) or high task (120 percent rate).[12] The high-task rate is included, because some organizations multiply the MTM values by 0.833 to produce high-task standards. σ_b^2 [Eq. (5)] is assumed to be zero when low-task standards are used because practically everyone can work at the 100 percent level or higher. When high-task standards are used, $\sigma_b^2 \neq 0$ because the variation is worker capability has to be taken into account. Also, please note that the GPD (General Purpose Data) system is illustrated along with MTM-1, MTM-2, and MTM-3. GPD is included because the graphs were reproduced from footnote 3. 4M DATA is not included because no data of this type is available for it.

Three levels of P—0.95, 0.995, 0.9993—are given in Figs. 12–7 through 12–12. All of these levels appear at first glance to be high; however, if we have several operations to be done in a series where all of them have to be performed to get a good product, the situation may change. Let:

$$P_n = P_1^n \tag{9}$$

where P_1 is the probability of getting a cycle done correctly and n is the number of serial operations involved. Then we can compute P_n for a given situation. For example, Figure 12–13[13] is a tabulation of the value of P_n for the three levels of P_1. Note that the probability of getting a good product falls off quickly unless P_1 is set at a very high level. For example, if $P_1 = 0.950$ and there are 20 operators,

Number of Operations	Value of P_n at three levels of P_1: P_1 = .950	P_1 = .995	P_1 = .9993
1	.950	.995	.9993
2	.903	.990	.9986
3	.857	.985	.9979
4	.815	.980	.9972
5	.774	.975	.9965
6	.735	.970	.9958
7	.698	.966	.9951
8	.663	.961	.9944
9	.630	.956	.9937
10	.599	.951	.9930
20	.358	.905	.986
30	.215	.860	.979
40	.129	.818	.972
50	.077	.778	.966
60	.046	.740	.959
70	.028	.704	.952
80	.017	.670	.946
90	.010	.637	.939
100	.006	.606	.932

Figure 12–13. Values of P_n *for different levels of* P_1 *and* n *in the examples given.*

[12] F. N. Bayha, W. M. Hancock, and G. D., Langolf, "More Evaluation Parameters for MTM Systems," *Journal of Methods Time Measurement,* **2** (1), 1975. Reproduced by permission.

[13] *Ibid.*

a good product will only be realized with a probability of 0.358 or 35.8 percent of the time!

In the following examples, the calculations can be shortened by introducing a calculation called BLC for basic line cost. Thus

$$\text{BLC (\$/month)} = \text{Labor Rate/Hour} \times \text{No. of Employees} \times 173.3 \text{ hour/month/employee}$$

Example: What prediction system should be used for an assembly line with 300 employees (100 per shift, three shifts) with a pay rate of $14.00 per hour, including fringe benefits? The cycle time is 1.0 minute (1666.7 TMU), the average NRT is 0.75 minute (1250 TMU), grievances cost $1000 per grievance to settle, 35 new standards are set/month, the MTM applicator rate is $14.00 per hour, and turnover is 4 percent per month.

1. *Complaint Costs.* For completeness, we will assume that there are two aspects to complaint costs: Those arising because employees are hired who cannot perform at 100. Of course, if these employees have already been selected out, this cost can be ignored. This cost would be:

$$0.04 \times 300 \frac{\text{employees}}{\text{month}} \times \frac{0.05 \text{ complaint}}{\text{employee}} \times \frac{\$1000}{\text{complaint}} = \$600/\text{month}$$

The other complaint cost occurs because the other employees will complain if they feel that the standard is unfairly set. Assume that $P = 0.995$ for the example. Then,

$$(1 - 0.995) \frac{\text{complaint}}{\text{employees}} \times [300 - 0.04\,(300)] \frac{\text{old employees}}{\text{month}} \times \frac{\$1000}{\text{complaint}} = \$1440/\text{month}$$

The total complaint cost will then be $600 + $1440 = $2040/month.

2. *Repair Costs.* The costs to repair items is a function of the number of items needing repair times the cost per repair. Suppose that the average repair takes 2.0 minutes, then the repair costs will be:

$$2\,(1 - 0.995) \times \text{BLC} = \$7278.60/\text{month}$$

where BLC = $14 × 300 × 173.3 = $727,860.

3. *Productivity Losses.* The productivity losses will differ with the choice of the MTM system used, which in turn affects the loading. The losses, as compared to doing the work on unpaced work stations, is:

$$\left(1 - \frac{\%\text{ loading}}{100}\right)\text{BLC} \tag{10}$$

For the MTM-1 system at $P = 0.995$ and an average NRT of 1250 TMU (see Fig. 12–8) the %loading will be 90 percent, therefore:

$$\left(1 - \frac{90}{100}\right)(\$727{,}860) = \$72{,}786$$

4. *MTM Applicator Costs.* The applicator costs will be the time to develop the new

standards. For MTM-1, with a standard development time of $250X$ (see Fig. 12–3), the costs will be:

$$35\ \frac{\text{standards}}{\text{month}} \times 1666.7\ \frac{\text{TMU}}{\text{standard}} \times 250 \times \frac{1\ \text{hr}}{10^5\ \text{TMU}} \times \$14.00 = \frac{\$2041.71}{\text{month}}$$

The total monthly costs, using MTM-1, would then be:

$2040 + $7278.60 + $72,786 + $2041.71 = $84,146.31/month

Using the same conditions as above, the costs for MTM-1, MTM-2, and MTM-3, would be:

System	Complaint	Repair	Productivity	Applicator	Total
MTM-1	$2040.00	$7278.60	$ 72,786.00	$2041.71	$ 84,146.31
MTM-2	2040.00	7278.60	118,277.25	816.83	128,412.68
MTM-3	2040.00	7278.60	145,572.00	285.84	155,176.44

Using the total costs as the criteria, the MTM-1 system is the best. Note that the productivity losses using MTM-2 and MTM-3[14] far exceed the reductions in MTM applicator time.

Since the percent loading is a function of the $\%A$ only and since the 4M system, according to Fig. 12–2, has a lower $\%A$ than MTM-1, one could conclude that the 4M system would have even lower productivity losses than MTM-1. Since productivity losses dominate the costs, at least in this example, one could conclude that the 4M system would probably be the most desirable system to use.

[14]There are additional constraints on using MTM-2 and MTM-3. See Chapters 6 and 7 for a discussion of them.

CHAPTER 13

Computerized Work Measurement Systems

by Karl Eady

Regional Director, MTM Association for Standards and Research

Introduction

Computer technology has advanced to the stage where there are a variety of sizes and complexities of computers to fit a multitude of industrial functions. There are mainframe computers for the large multidivision, multidepartmental organizations where accessibility must be universal. There are minicomputers, which can be considered scaled-down models of the mainframes, suitable for single and multiple plant organizations. The main difference is that they do not have the storage capacity of the mainframe computers.

Becoming increasingly popular for departmental operation are the desk-top or microcomputers. Many of these units have storage capacities which are entirely adequate for handling extensive standard data work measurement systems. The data storage device is usually a 5-inch square "floppy" diskette, but audio tape cassettes can also be used in many situations.

Finally there is the hand-held programmable calculator which can function as a personal "computer" in a limited way. This chapter describes the most popular work measurement systems designed for each of these categories of computers.

Mainframe and Minicomputer Systems[1]

MOST

This system was developed first as a manual technique by the Swedish Division of the H. B. Maynard Company, Inc., in 1967–1972 and was introduced in the United States in 1974. The MOST work measurement system is applicable for any cycle length and repetitiveness, as long as there are variations in the motion pattern from one cycle to another. Derived from the structure and theory of MTM-1 and MTM-2, MOST systems can be applied to direct productive work, such as machining, fabrication, and assembly, as well as to material handling, distribution, maintenance, and clerical activities.

All necessary procedures and principles for development of suboperation data in any work area are available as MOST application systems. The major features are:

[1]The 4M DATA system falls within this classification. (See Chapter 11.)

1. Compilation of work conditions in work management manuals.
2. Filing of suboperation data in a central data bank, based on a uniform coding system.
3. Construction of data application forms based on statistical principles.

The system employs a small number of predetermined models of fixed activity sequences which cover practically all aspects of manual activity.

The differences between the levels are the multipliers. Identical index numbers are applied on all levels. The multipliers are as follows:

1. Basic sequence models (basic MOST) = multiplier 10.
2. Bridge cranes and wheeled trucks = multiplier 100.
3. Job preparation, etc. = multiplier 1000.

General Move. The first sequence model—"general move"—is defined as moving an object from one location to another freely through the air. This can account for as much as 35 percent of the work of a machine operator and even more for an assembly worker. This activity is represented by the following sequence of letters or subactivities:

A B G A B P A

A = action distance (mainly horizontal hand or body motions)
B = bend (mainly vertical body motions)
G = grasp
P = position

The variations for each subactivity are indicated by an index figure, for instance,

$$A_6\ B_6\ G_1\ A_1\ B_0\ P_3\ A_0$$

A_6 = walk three to four steps
B_6 = bend and rise
G_1 = grasp one light object with one hand
A_1 = move within reaching distance
B_0 = no bend
P_3 = place objects with adjustments
A_0 = no return move

The index values representing various elements are memorized from an index table. The developers state that all values on this and other index tables are derived from MTM-2 or MTM-1.

The time value for the sequence is obtained simply by adding together the index numbers and multiplying the sum by 10. For instance, the MOST standard time for the preceding sequence is

$$6 + 6 + 1 + 1 + 0 + 3 + 0 = 17, \quad 17 \times 10 = 170 \text{ TMU}$$

Controlled Move. "Controlled move" is a movement of the object that is restricted in at least one direction by contact with or an attachment to another object.

Tool Use. "Tool use" covers not only conventional hand tools such as wrenches, screwdrivers, gauges, and writing tools, but also fingers and mental processes.

The MOST system is designed for consistency and, with its table, subdivisions are selected so that they will span typical small variations in method and layout. The design meshes well with the instinctive "rounding" tendency practiced by some analysts and avoids the "fine-tuning" in areas where they may be too detailed for variations encountered in practice.

It appears that there is a large category of operations for which the consistent behavior of MOST will produce standards with accuracy approximately equivalent to that of higher-level systems, but with far less effort and in shorter time.

Computer Configuration. Through the use of a minicomputer, quite often located in the industrial engineering department, or a mainframe IBM in the data processing department, with on-line interactive access for the industrial engineer, MOST time standards can be calculated from workplace data and a method description. The computer thus will do all the work measurement, relieving the industrial engineer of practically all paperwork.

The MOST computer systems, which are generic and second- and third-level, include all necessary steps and procedures to develop a complete time standard for an operation, such as:

1. Development of suboperation data based on MOST work measurement systems.
2. Calculation of time standards, including manual times, process times, and allowances.
3. Printing of method instructions and route sheets.

The core of the program consists of comprehensive filing and editing procedures for retrieval, revision, and mass updating of time standards. The output formats can be tailor-made to customer conditions and requirements.

The H. B. Maynard Company asserts that, by using the computer, one hour of work can be measured with one to five hours of analysis, and a time standard can be calculated in an average of five minutes.

Process Time. The computer can generate a series of optional modules to calculate process time for such operations as machining, welding, line balancing, and labor reporting. These times can be integrated at the proper point to give a complete standards calculation.

Simulation. Finally, the computer gives the user simulation capabilities. Proposed work standard changes can be displayed to determine whether the new method or equipment will add or subtract time from the standard. Hypothetical conditions can be created for the method, workplace, and process.

Although the MOST developers state that the system is derived from MTM techniques, the documentation is not readily available because of the proprietary nature of the system. Information about MOST is contained in the book, *MOST Work Measurement Systems,* by Kjell V. Zandin, published by Marcel Dekker, Inc., New York, 1980.

Figures 13–1 and 13–1a show the output of a MOST operation.

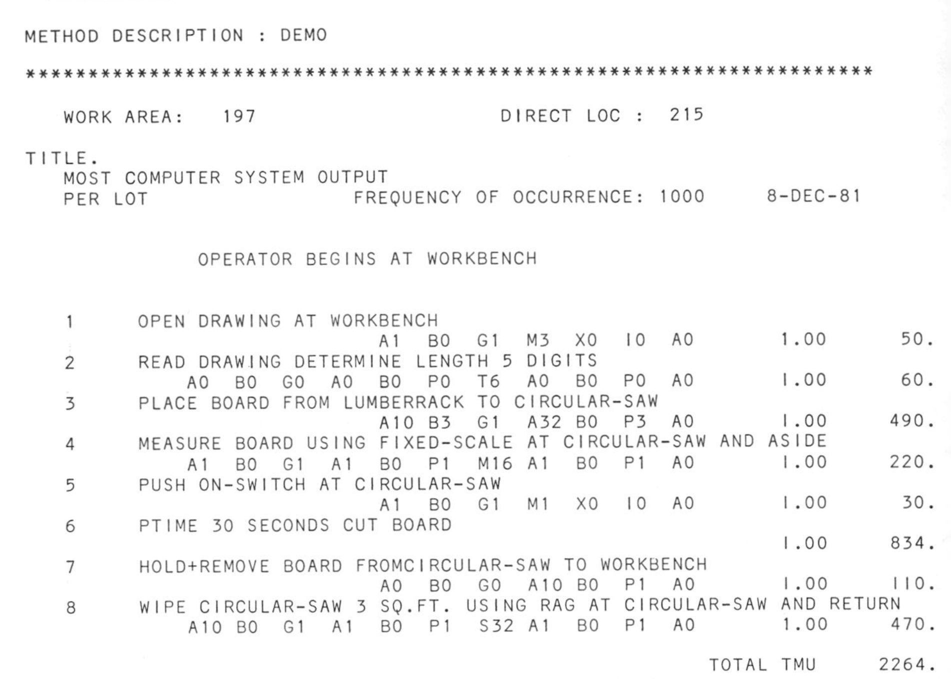

```
METHOD DESCRIPTION : DEMO
******************************************************************

  WORK AREA:    197                    DIRECT LOC :  215

TITLE.
  MOST COMPUTER SYSTEM OUTPUT
  PER LOT                FREQUENCY OF OCCURRENCE: 1000      8-DEC-81

            OPERATOR BEGINS AT WORKBENCH

  1     OPEN DRAWING AT WORKBENCH
                       A1   B0   G1   M3   X0   I0   A0        1.00      50.
  2     READ DRAWING DETERMINE LENGTH 5 DIGITS
           A0   B0   G0   A0   B0   P0   T6   A0   B0   P0   A0  1.00      60.
  3     PLACE BOARD FROM LUMBERRACK TO CIRCULAR-SAW
                       A10  B3   G1   A32  B0   P3   A0        1.00     490.
  4     MEASURE BOARD USING FIXED-SCALE AT CIRCULAR-SAW AND ASIDE
           A1   B0   G1   A1   B0   P1   M16  A1   B0   P1   A0  1.00     220.
  5     PUSH ON-SWITCH AT CIRCULAR-SAW
                       A1   B0   G1   M1   X0   I0   A0        1.00      30.
  6     PTIME 30 SECONDS CUT BOARD
                                                               1.00     834.
  7     HOLD+REMOVE BOARD FROMCIRCULAR-SAW TO WORKBENCH
                       A0   B0   G0   A10  B0   P1   A0        1.00     110.
  8     WIPE CIRCULAR-SAW 3 SQ.FT. USING RAG AT CIRCULAR-SAW AND RETURN
           A10  B0   G1   A1   B0   P1   S32  A1   B0   P1   A0  1.00     470.

                                                        TOTAL TMU     2264.
```

Fig. 13–1. MOST computer output.

UNIVEL

In 1964 Willared L. Kern organized Management Science, Inc., of Appleton, Wisconsin, for the purpose of developing and marketing computer software for industrial engineering applications utilizing new concepts. The systems utilize universal formulas to generate precise elemental standard times. The system allows the formulas to be resident in the computer core to eliminate the need to develop, store, and retrieve standard data. Because values were developed through the use of mathematical formulas, only numerical input data are needed in the system. This eliminates the task of developing standard data stored and coded through alphanumerically organized reference tables and of maintaining the data as well as the significant alphanumeric coding structures as future changes are required.

The totally integrated systems are named "Univation Systems" and are made up of modules for various manufacturing controls in which predetermined time systems are employed using UnivE1 as a data base. Output information automatically generated or mass changed through the use of the UnivE1 system consists of elemental methods instructions and standards and a complete "network data base" of all manufactured products and operations.

The base of UnivE1 is unknown to the writer. Since it is a proprietary system, the documentation is not part of the system.

```
          WORK AREA LOCATER NUMBER   197

          -------   -----------   -----------------
          GRINDER   '         '   '               '
          -------   '         '   '               '
                    CIRCULAR-SAW  '               '
                    '         '   '   WORKBENCH   '
                    -----------   '               '
                                  --------(x)-------   -------
                                                       LUMBERRACK
                                                       '     '
                                                       -------

                        -------------
                        '           '
                        'TOOLCABINET'
                        '           '
                        -------------
**

 NAME                     LOCATION                      BODY/FRAG/PT
---------                ------------                  ----------------

)RKPLACES:
'ORKBENCH                 35,15     20,5
'OOLCABINET               25,1      12,4                 PBEND
:IRCULAR-SAW              20,16     10,4
.UMBERRACK                60,10     5,5                  PBEND
RINDER                    10,18     5,2

OLS:
CREWDRIVER               WORKBENCH
DJUSTABLEWRENCH          WORKBENCH
AMMER                    WORKBENCH
OXWRENCH                 WORKBENCH
AWBLADE                  TOOLCABINET
ENCIL                    KARL

JECTS:
OOLBOX                   WORKBENCH
OARDS                    LUMBERRACK
*

ERATORS:
ARL                      WORKBENCH                       45,15  B

   FROM                     TO                          STEPS
------------             ----------                   ---------

RKBENCH                  TOOLCABINET                     10
RKBENCH                  CIRCULAR-SAW                     5
RKBENCH                  LUMBERRACK                       6
RKBENCH                  GRINDER                         20
OLCABINET                CIRCULAR-SAW                     8
OLCABINET                LUMBERRACK                      12
OLCABINET                GRINDER                         10
RCULAR-SAW               LUMBERRACK                      20
RCULAR-SAW               GRINDER                          4
MBERRACK                 GRINDER                         30
```

Fig. 13–1a. MOST computer output.

WOCOM

WOCOM is the registered trademark of a work measurement system developed by Science Management, Inc., Moorestown, New Jersey, the holding company for the Work-Factor Foundation.

The generic basic-level WOCOM system provides automated application of the two most recognized systems of predetermined times: MTM and Work-Factor. It is designed for simple operation and flexibility in meeting varying individual company needs and does not require previous experience in computer equipment or work measurement methods. Programs are available through the national time-sharing network or for operation on in-house computers. Information on the system is available from the Work-Factor Foundation.

The system consists of eight modular computer programs that can be used to analyze human and machine work, alternate work methods, and assembly line operations and to maintain labor standards. A test installation in a division of a large corporation produced the following results:

1. The system proved to be accurate and faster than the existing method of establishing new Detailed Work-Factor standards.
2. Training time for personnel on the new system was less than one-half of that required for manual application of Detailed Work-Factor.
3. The system provided the standards group with more time to devote to industrial engineering projects, since most of the effort devoted to the establishment and maintenance of standards could be accomplished by personnel other than industrial engineers.
4. A 50 percent reduction in the clerical phase of establishing Detailed Work-Factor standards was achieved.

Microcomputer (Desktop) Systems

In recent years the microcomputer has seen increasing use in the industrial environment. These computers, which have been designed as self-contained desktop units, are comparatively inexpensive, are easy to use, and are approaching the flexibility and capacity of mainframe systems. In keeping with the development and increased availability of this hardware, software packages have been developed and are available for a great variety of industrial engineering applications, including work measurement.

One of these software packages has been developed under the direction of the MTM Association for utilizing some of the generic and functional MTM systems in the development of standards and standard data with full storage, retrieval, and mass updating capabilities. At this writing it is the only work measurement system known by the author which has been developed for the micro computer and which utilizes the predetermined time concept. Time study and work sampling programs also exist, however.

ADAM

Automatic Data Application and Maintenance is the full name of the system which allows the user to operate a work measurement system with the aid of a computer to increase the speed of application while maintaining the accuracy of the specific technique being used. The ADAM system is available from the MTM Association for Standards and Research, Fair Lawn, New Jersey. It was introduced in 1980 as the first predetermined time system developed for the microcomputer designed to create and maintain labor standards derived from standard data.

ADAM is a fully integrated work measurement system. Just as a word processing system reduces effort and errors in composing and editing, ADAM does the clerical and computational work involved in creating and applying labor standards. Basic elements may be derived from any measurement technique, whether it be an MTM-based predetermined time system, time study, historical data, etc., through the use of appropriate modules. The user is able to combine the basic elements to form intermediate levels from which the standards themselves are ultimately set.

Modules already developed for the ADAM system are called ADAM-2, ADAM-C, and ADAM-V. ADAM-2 utilizes the notations and applications of MTM-2, while ADAM-C contains all of the first- and second-level elements of MTM-C. Likewise, ADAM-V utilizes the elements of MTM-V. Furthermore, ADAM has a formula application capability available for all modules.

The basic ADAM system, written in the PASCAL programming language, can be run on any microcomputer which has the capability of compiling PASCAL. It can be purchased without modules as self-contained software for the application and maintenance of company-developed data bases, whether MTM-based or not. The modules can be purchased in addition to the basic system as desired in any single or multiple combination.

ADAM-2

ADAM-2 can be used wherever MTM-2 can be applied manually. The advantages of easy input, high speed of application, consistency, and accuracy are retained. All of the nine elements and two weight factors of MTM-2 are present in the system.

ADAM-2 analyzes simultaneous motion combinations and applies the proper overlap values when required. It also analyzes combined motions (those done by the same body member at the same time) and "limits out" the shorter motion(s).

Elements developed with ADAM-2 may be stored as standard data and retrieved for use within studies or other higher-level elements.

Inputting any of the ADAM modules is easy to do. Inputs are of "free format," releasing the analyst from the restrictions of the 80-column card with designated field sizes and locations. Figure 13–2 illustrates an ADAM-2 input. The "&" notation separates left-hand motions from right-hand motions.

```
/FILE-HAND TOOL-CIRCULAR MOTIONS
GB2 & GB6
GB6 & PC6
& PB6 × 18
& A × 18
R × 9
PA6 × 9 & PB6 × 18
& PA6
```

Fig. 13–2. ADAM-2 input.

Figure 13–3 is the output prepared by ADAM-2 for the input of Fig. 13–2. For each MTM-2 notation, ADAM-2 supplies an appropriate "action" word, while the analyst need only input object information. Of course, the automatic action word can be suppressed, allowing the analyst to supply a more descriptive one if desired.

ADAM-C

Using ADAM to apply the MTM-C clerical standard data system provides the ultimate in speed of application and quality of documentation. As in ADAM-2, the ADAM-C modules contain all of the elements of MTM-C on both levels. Level 1 consists of approximately 370 elements and Level 2 approximately 250 elements.

ADAM-C will accept and apply compound frequencies of occurrence. This is especially valuable in clerical situations when frequency relationships between the work measurement unit and final product unit must be indicated. For example, a frequency of 6 × 1/4 may be entered to note an occurrence of six times in an element which occurs once for every four units being produced.

Figure 13–4 shows the free format input of an ADAM-C analysis and Fig. 13–5 shows the resultant output computed by ADAM.

ADAM-V

The data base of MTM-V covers a wide range of machine shop operations including measuring, marking, moving levers and buttons, rotating, part handling, tool handling, and hoist loading. The entire data base of MTM-V is contained in the ADAM-V module. The same input and output features as those

```
1 FILE-HAND TOOL-CIRCULAR MOTIONS
2 GET COMPLEX              GB2) 10   GB6         GET COMPLEX
3 GET COMPLEX              GB6) 33  *PC6         PUT W/ 2 CORR'NS
4                              270   PB6    18   PUT W/ 1 CORR'N
5                              252   A      18   APPLY PRESSURE
6 REGRASP            9     R    54
7 PUT EASY           9     PA6)270   PB6    18   PUT W/ 1 CORR'N
8                                6   PA6         PUT EASY

                   TOTAL TMU:  895
```

Fig. 13–3. ADAM-2 output.

```
/REMOVE OLD SHEETS
G5A2 O1 /ACCOUNT BOOK
RN × 2 LC12 I30 × 6
G1A2 /OLD SHEET
/REPLACE WITH NEW SHEETS
HI14 C4
C1 G5A2 /UPDATED ACCOUNT BOOK
```

Fig. 13–4. ADAM-C input.

found in the ADAM-2 and ADAM-C modules are also present in the ADAM-V module.

Figures 13–6 and 13–7 illustrate the input and output of an ADAM-V study.

Formula Application using ADAM. ADAM contains a flexible system for defining and evaluating formulas. The system provides room for saving 100 formulas, each of which may include up to 10 variables. To include a formula in

DEPARTMENT: INSURANCE
FUNCTIONAL AREA: FILING
TASKNAME: UPDATE ACCOUNT BOOK
TASKCODE: IFA-1
DATE: MARCH 12, 1981 REVISION DATE:
UNIT OF MEASURE: BOOKS UPDATED
BY: IMA GOODWIN APPROVED: KE
REMARKS: INCLUDES GET & ASIDE ACCOUNT BOOK

NE	DESCRIPTION	ELEMENT	TIME	FREQ	TOTAL
	REMOVE OLD SHEETS				
	GET & ASIDE ACCOUNT BOOK	G5A2	29	1/1	29
	OPEN ACCOUNT BOOK	O1	29	1/1	29
	READ NUMBERS	RN	7	2	14
	LOCATE SHEET(S)	LC12	129	1/1	129
	IDENTIFY THREE WORDS	I30	22	6	132
	GET & ASIDE OLD SHEET	G1A2	32	1/1	32
	REPLACE WITH NEW SHEETS				
	INSERT ON BINDER-RINGS	HI14	84	1/1	84
	CLOSE BINDER-RINGS	C4	35	1/1	35
	CLOSE UPDATED ACCOUNT BOOK	C1	27	1/1	27
	GET & ASIDE UPDATED ACCT BK	G5A2	29	1/1	29

TOTAL TMU	540.000
PF&D ALLOWANCE	0.1500
STND. TMU	621.000
STND. MINUTES	0.3726
STND. HRS./UNIT	0.0062

Fig. 13–5. ADAM-C output.

```
/ REMOVE MILLED PART TO BENCH
MAA4 HO3
/ GET PARTS PLACE IN CHUCK
HO3 /NEW PART HO2 /NEW PART MAA4
/ STRIKE WITH HAMMER
FLA30 FLA10
/ START MACHINE
MAA4 F 11
/ STOP MILLING MACHINE
MAA4 MPT X 112
```

Fig. 13–6. ADAM-V input.

a study, the analyst supplies the formula number along with the values of the variables. ADAM substitutes the values into the stored formula, computes the result, enters the result into the study, and inserts the names and values of the variables into the study. ADAM saves the formula permanently (until the ana-

LINE	DESCRIPTION	ELEMENT	TIME	FREQ	TOTAL
1	REMOVE MILLED PART TO BENCH				
	OPERATE TRIGGER	MAA4	50	1	50
	GET OR PLACE OBJECT	HO3	90	1	90
2	GET PARTS, PLACE IN CHUCK				
	GET OR PLACE NEW PART	HO3	90	1	90
	GET OR PLACE NEW PART	HO2	40	1	40
	OPERATE TRIGGER	MAA4	50	1	50
3	STRIKE WITH HAMMER				
	FASTEN/LOOSEN W/STRIKING TOOL	FLA30	160	1	160
	FASTEN/LOOSEN W/STRIKING TOOL	FLA10	70	1	70
4	START MACHINE				
	OPERATE TRIGGER	MAA4	50	1	50
	MILL	F-11	1136	1	1136
	— CUT LENGTH	—	0	12	0
	— CUTTER DIA.	—	0	4	0
	— CUT WIDTH	—	0	.125	0
	— RPM	—	0	95	0
	— FEED/TOOTH	—	0	.025	0
	— TEETH	—	0	8	0
5	STOP MILLING MACHINE				
	OPERATE TRIGGER	MAA4	50	1	50
	PROCESS TIME	MPT	1	112	112
			TOTAL TMU		1898
			PF&D ALLOWANCE		0.1500
			STND. MINUTES		1.3096
			UNITS/MINUTE		0.7636

Fig. 13–7. ADAM-V output.

```
F11 MILL                        (Formula number and name)
TIME = 1667 × (V1 + SQRT((V2 × V3) − (V3 × V3)) + .25)/(V4 × V5 × V6)
      Cut Length
      Cutter Dia.
      Cut Width
                                (These are the variable names)
      RPM
      Feed/Tooth
      Teeth
```

Fig. 13–8. Define and save the formula (one time only).

lyst explicitly elects to change it) and computes its result wherever it is included in an analysis.

Figure 13–8 demonstrates the input required for defining and saving a formula. This is done one time only. It can then be retrieved whenever desired for insertion into a study.

Figure 13–9 illustrates the input necessary to call out the formula defined in Fig. 13–8 with appropriate values for the variables. The first number is the number of the stored formula, and the following numbers are the values for the variables in sequence.

Figure 13–10 is the output from ADAM inserted as a line in a study.

ADAM Standard Data Storage and Retrieval. Once an element has been created which occurs in many operations, it is never analyzed again. A single keypress will save the element under an operation code and name. The assignment of the code is up to the analyst. ADAM will recognize up to 11 characters for this code. The element is saved with the current date and the analyst's name. Then, anytime the analyst encounters this element in an operation, he or she simply enters its code. ADAM will retrieve it and enter it into the current study. The analyst may decide if ADAM should retrieve all the detail or merely its name and time.

A good manual system will also accommodate this building-block approach. However, the greatest advantages of a computer approach are realized when changes must be made. Because a method has changed, for example, an element may require revision. Very quickly, the analyst retrieves the element from the mass memory of the computer, changes it as required, and presses a key to save the revised element.

All operations using a changed element may also be revised. The analyst presses a key to get a listing of all standards using the revised element. He presses another key to update all the standards listed. Or, if appropriate, he may update just those standards which he chooses, either on an individual basis or by function.

Figure 13–11 is a representation of how elements are stored on several levels and combined to created still higher levels or labor standards.

```
⋮
F 11   12   4   .125   95   .025   8   (Formula number followed by values
⋮                                       for the six variables)
```

Fig. 13–9. Call out the formula in any study.

LINE	DESCRIPTION	ELEMENT	TIME	FREQ	TOTAL
	.				
	.				
7	MILL	F-11	1136	1	1136
	— CUT LENGTH	—	0	12	0
	— CUTTER DIA.	—	0	4	0
	— CUT WIDTH	—	0	.125	0
	— RPM	—	0	95	0
	— FEED/TOOTH	—	0	.025	0
	— TEETH	—	0	8	0
	.				
	.				

Fig. 13–10. Output.

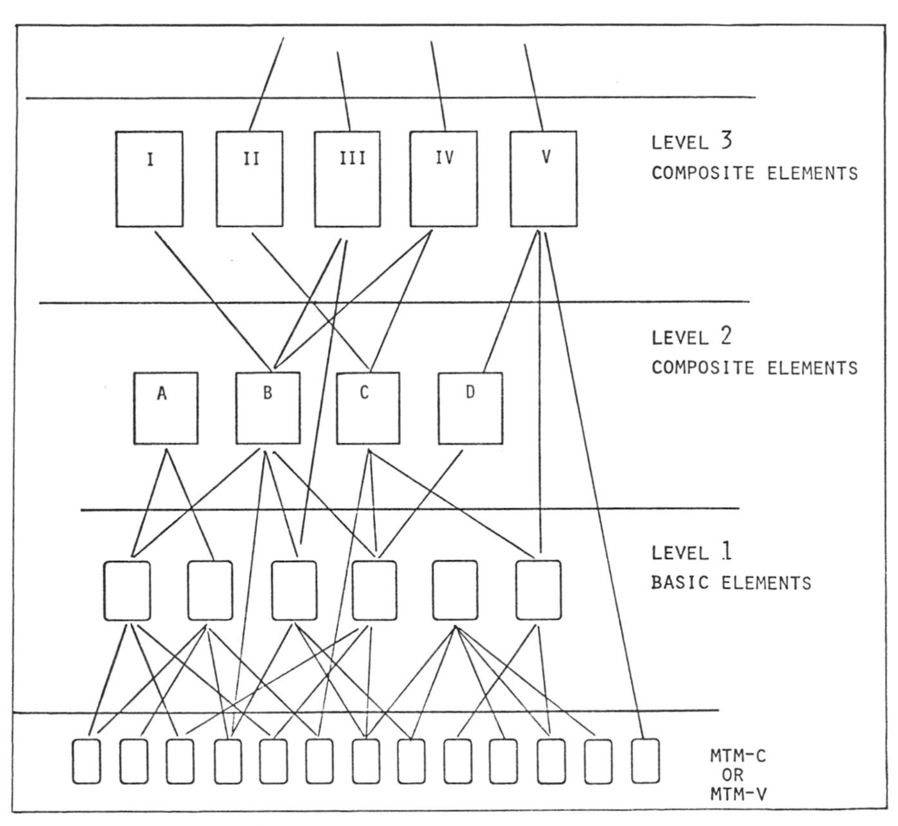

Fig. 13–11. ADAM data base structure.

ADAM STANDARDS CREATION & MAINTENANCE ACTIVITY LOG

ACTIVITY DATE: 2/17/82

ABX-15R7 128 PLACE PART IN TEST SET
CREATED BY: R. V. JONES PRIOR ACTIVITY DATE: ORIGINAL
REMARKS: —
FORMULA 9 VERIFY SORT
MODIFIED BY: PAUL SMITH PRIOR ACTIVITY DATE: 1/18/80
REMARKS: REVISED TO REFLECT ADDITIONAL PROCESS TO CHECK SECTORS

RT9-13BO 1549 PART HANDLING—CLAMPS
MODIFIED BY: GERRY SMITH PRIOR ACTIVITY DATE: 3/04/79
REMARKS: ELIMINATION OF TWO CLAMPS BY NEW FIXTURE

Fig. 13–12. Standards maintenance log.

ADAM Standards Maintenance Activity Log. More frequently than not, a company's standard data base tends to lose traceability over time. Changes in procedures, personnel, and policies can combine with imperfect recollections and hastily produced documentation to yield a maze of confusion and missing facts. To deal with this phenomenon, ADAM establishes and enforces the procedures for creating and maintaining standard data. Moreover, it automatically records every standards maintenance activity into a log which may be printed daily. This facility guarantees that every standard data element can be traced from its current status back through all previous modifications to its original development, providing the dates, changes in time, persons responsible, and reason for change. An example of such a log is shown in Fig. 13–12.

Hand-Held Programmable Calculators

The state of the art has advanced to such a degree in the area of hand-held programmable calculators that it is possible to develop rather sophisticated work measurement programs for them. Although these programs do not supply method description in great detail, they are somewhat faster than manual systems and reduce the possibility of application error to an appreciable degree. Two MTM-based programs have been developed for these programmable calculators. They are known by the acronyms CUE and MATE.

CUE

Cue was developed by Douglas M. Towne, General Analysis, Inc., Redondo Beach, California. It consists of data derived from MTM-1 by the process of averaging, elimination, and simplification. It is designed for the Texas Instruments TI-59 calculator and printer. The original version, CUE-I, does not provide for the analysis of simultaneous motions, but uses the predominant hand concept, in which the analyst determines which hand requires the most time to perform its motion.

The TI-59 has five function keys which will handle ten functions when used in conjunction with a shift key. The keys are labeled A, B, C, D, and E, and, when used with the shift key, A′, B′, C′, D′, and E′. The MTM motions and motion aggregates have been assigned to each of these function keys. A, for example, is the GET aggregate, while B is the MOVE motion. A 16-inch GET would be inputted by pressing 1-6-A, and 8-inch MOVE is 8-B, and APPLY PRESSURE (AP) 7-E.

The complete listing of motions for the CUE system are

A	GET
B	MOVE
C	POSITION
D	Frequency Multiplier
E	Miscellaneous Motions
A′	TURN
B′	Process Time
C′	Finger Motions
D′	Horizontal Body Motions
E′	Vertical Body Motions

The operation description is hand written but the output is printed on the printer to which the calculator is attached, as shown in Fig. 13–13.

OPERATION DESCRIPTION	CUE CODE	CUE	OUTPUT
Obtain & Pull Down Lever	12A	12.	GET
	6B	6.	MOVE
Get Next Casting to Machine	20A	20.	GET
(10 lbs.)	20.10B	20.1	MOVE
Wait For Machine Cycle	.05C′	0.0550	MIN
Loosen Holder	4A	4.	GET
	7E		AP1
	90A′	90.	TURN
Remove Finished Casting	4A	4.	GET
	A	0.	GET
To Bin	15.10B	15.1	MOVE
Place New Casting in Fixture	15A	15.	GET
	4B	4.	MOVE
	9C	9.	POSN
Turn & Tighten Holder	6A	6.	GET
	90A′	90.	TURN
	7E		AP1
		329.	TMU
		0.1973	MIN
		15	% PFD
		0.2269	TOT

Fig. 13–13. Annotated CUE output.

MATE

MATE is a system of MTM-2 developed for the Hewlett-Packard HP41C or HP41CV hand-held calculator by the author and is presented to show the potential of this device as a work measurement tool. The program is contained on four magnetic strips and the data base on seven magnetic strips.[2]

The keyboard for the MATE system is shown in Fig. 13–14. The MTM-2 motion aggregates are entered through the numeric keys as shown. Both left- and right-hand motions may be entered as well as simultaneous motions. Seven calculate and print functions are performed by the program as follows:

1. RH Right-hand motion
2. LH Left-hand motion
3. SM Simultaneous motion not requiring overlap
4. SM+ Simultaneous motion requiring overlap
5. PT Process time entry
6. EL Calculate element standard
7. STD Calculate operation standard

The RH and LH functions calculate single-handed motions, applying frequencies when required, and print out the result in conventional MTM-2 format. The SM function calculates two-handed motions not requiring overlap by selecting the limiting motion and printing its time value. The SM+ function calculates two-handed motions requiring overlap in the same manner, and applies the proper overlap motion as a separate line entry.

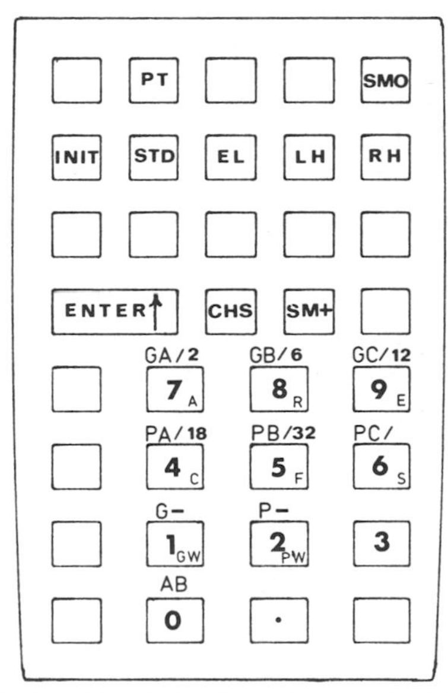

Fig. 13–14. The HP-41C keyboard customized for MATE.

[2]Further information on availability may be obtained from the author at the MTM Association Midwest Office.

	Motion	Distance	Enter	Motion	Distance	Frequency	Function	OUTPUT		
RH GB12:	8	9	↑			1	RH		14	GB12
LH 2PC6:	6	8	↑			2	LH	2PC6	52	
SM PB6 & PA12:	5	8	↑	4	9		SM	PB6	15	PA12-R
SM+ GC6 & PC18:	9	8	↑	6	4		SM+	GC6 GC2	36 14	PC18-L
Steps 8S:	6		↑			8	RH		144	88
Arise from Bend:	0		↑			1	RH		0	-

Fig. 13–15. Key press sequences.

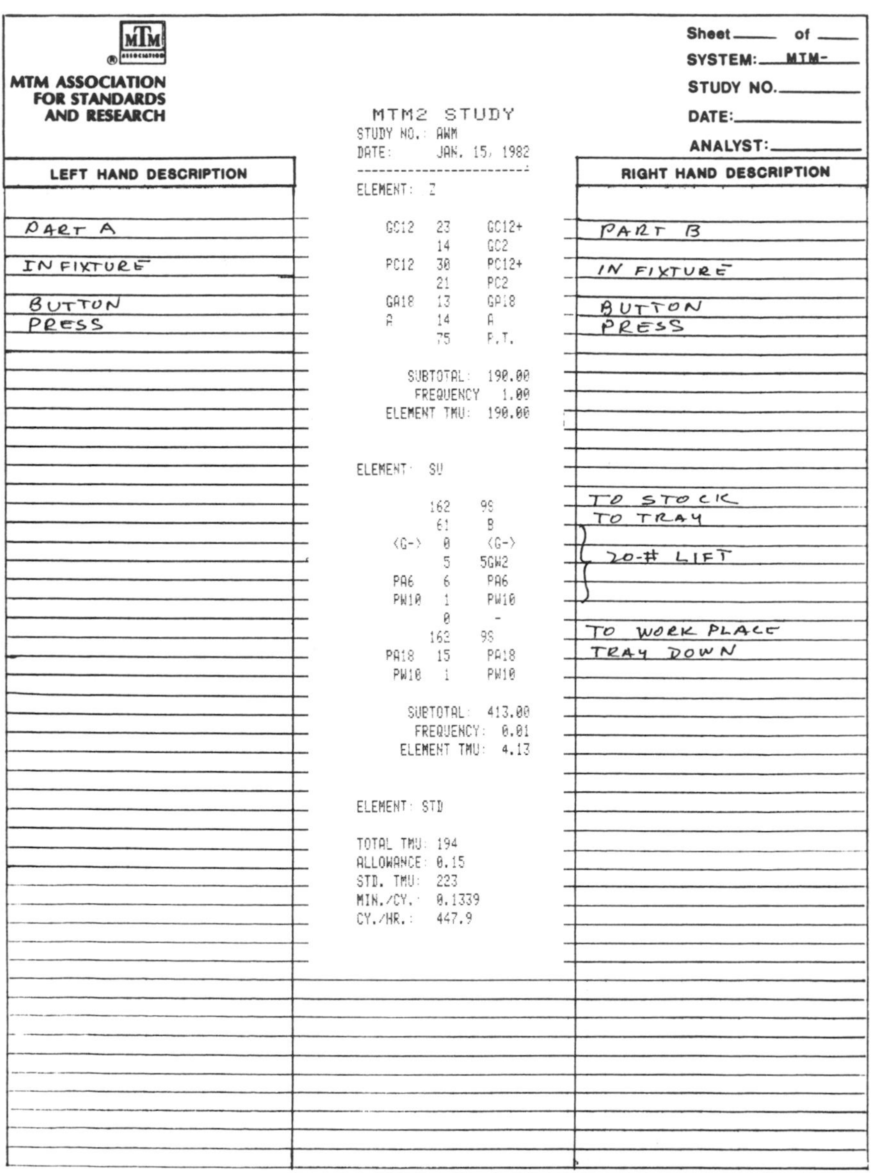

MTM ASSOCIATION FOR STANDARDS AND RESEARCH

Sheet ____ of ____
SYSTEM: MTM-
STUDY NO. ________
DATE: ________
ANALYST: ________

LEFT HAND DESCRIPTION

PART A
IN FIXTURE
BUTTON PRESS

```
MTM2 STUDY
STUDY NO.: AWM
DATE:      JAN. 15, 1982
------------------------
ELEMENT: Z

    GC12   23     GC12+
           14     GC2
    PC12   30     PC12+
           21     PC2
    GA18   13     GA18
    A      14     A
           75     P.T.

       SUBTOTAL:  190.00
        FREQUENCY   1.00
    ELEMENT TMU:  190.00

ELEMENT:  SU

          162     9S
           61     B
     <G->   0     <G->
            5     5GW2
     PA6    6     PA6
     PW10   1     PW10
            0     -
          162     9S
    PA18   15     PA18
     PW10   1     PW10

       SUBTOTAL:  413.00
        FREQUENCY:  0.01
      ELEMENT TMU:  4.13

ELEMENT: STD

TOTAL TMU: 194
ALLOWANCE: 0.15
STD. TMU:  223
MIN./CY.:  0.1339
CY./HR.:   447.9
```

RIGHT HAND DESCRIPTION

PART B
IN FIXTURE
BUTTON PRESS
TO STOCK
TO TRAY
20-# LIFT
TO WORK PLACE
TRAY DOWN

Fig. 13–16. An MTM-2 study using MATE.

Process times may be entered into the element by keying in the process time in TMU and pressing the PT function key. The EL function totals the motion times for the element, applies the frequency factor, and calculates the element normal time.

The STD function totals the element times, applies an allowance and prints the standard in TMU, hours per cycle, minutes per cycle, and cycles per hour.

MATE Data Entry and Output Report. The key presses for data entry are simple as illustrated by Fig. 13–15. Key presses and resultant printouts are shown for several typical MTM-2 motion patterns.

Figure 13–16 is an example of a two-element MTM-2 study which illustrates the functions of the MATE program.

CHAPTER 14

Miscellaneous Topics in Work Measurement

The purpose of this chapter is to present several subjects which are important to the work measurement practices but which do not require separate chapters. The topics presented in this chapter are:

1. Suggested solutions to staffing problems where the demands for the service have high uncertainties and where service has to be supplied within a certain time. Examples are the staffing of maintenance departments, fire stations, gasoline stations, bank teller operations, tool room attendents, and sales clerks.
2. Suggested solutions to staffing problems where the demands for service have high uncertainties and where the service has to be completed before the end of the shift with stated probabilities. An example is the determination of the number of nurses necessary to care for patients with varying degrees of illness.
3. Response to the critics of predetermined time systems. The industrial engineer frequently has to respond to the critics of predetermined time systems, especially in arbitration proceedings, and it is important that he/she be aware of the technically correct responses.

Staffing Problems where the Demand Has High Uncertainty and where the Work Must Be Done within a Stated Time

There are many situations that occur in service organizations and indirect labor applications where staff must be provided to take care of a work load that may vary widely from day to day or hour to hour. The usual work measurement approach of determining how long it takes to do a job is not sufficient, because we also have to cope with the probability of whether the job will need to be done and the average amount of time the receiver of the service will have to wait before service is begun.

Fortunately, there is a discipline called queueing theory that has concerned itself with analytic solutions to these types of problems. The purpose of this section is to present the solutions which have been found to be most useful for work measurement specialists.

In the introduction, several types of working environments are listed as having properties that require the approach to be described. Note that all of the examples have the common properties that:

1. The service is being performed.
2. The service to be supplied is random.
3. The service must be supplied in a reasonable time. What is reasonable may vary from example to example, but must be stated.

Let

A = the average number of times/hour that the work has to be done. (It is assumed that the distribution of arrivals is Poisson, which is a reasonable assumption if the arrivals are independent of each other.)
W = the number of workers that are available to do the work.
S = the standard time to do the work in hours. The distribution of the time is exponential. Telephone conversations, many repair operations, and various other service operations have been found to be reasonably approximated by the negative exponential distribution.[1]

The percent of time the workers will be busy during the work is (since each worker included in W provides 1 hour of available manpower):

$$B = \frac{AS}{W} \times 100 \tag{1}$$

and the average time that will elapse before the service (like at a tool crib window) can begin (in hours) is:

$$T = \frac{L}{A} \tag{2}$$

where

$$L = \left(\frac{(AS)^W (AS^{-1})}{(W - 1)!\,(WS^{-1} - A)^2}\right)P \tag{3}$$

L is the expected number of people waiting in line to be served and

$$P = \left[\sum_{M=0}^{W-1} \frac{1}{M!}(SA)^M + \frac{1}{W!}(SA)^W \frac{WS^{-1}}{WS^{-1} - A}\right]^{-1} \tag{4}$$

P is the probability that there is no one waiting for service or being served. M is the number of customers in the system. The average total time (waiting to begin plus the service) in hours is:

$$TT = T + S \tag{5}$$

As an example of the use of the equations, suppose we have people arriving at a tool crib at an average rate of six per hour. The standard time to provide the service requests (fill the tool requests in our example) is 0.333 hours and

[1]Morse, P. M., *Queues, Inventories and Maintenance,* John Wiley, New York.

there are three tool clerks. Thus, $A = 6$, $S = 0.333$, and, using Eq. (1), $B = \frac{(6 \times 0.33)}{3} \times 100 = 66.7$ percent = percent utilization of the clerks. Using Eq. (4),

$$P = \left[1 + \frac{1}{1}[0.333(6)]^1 + \frac{1}{2 \times 1}[0.333(6)]^2 + \frac{1}{3 \times 2 \times 1}[0.333(6)]^3 \frac{\frac{3}{0.333}}{\frac{3}{0.333} - 6}\right]^{-1} = \frac{1}{9}$$

Using Eq. (3),

$$L = \left(\frac{[(6)0.333]^3 \frac{6}{0.333}}{(2 \times 1)\left(\frac{3}{0.333} - 6\right)^2}\right)\frac{1}{9} = \frac{8 \times 18}{2(9) \times 9} = \frac{8}{9} = 0.889$$

Using Eq. (2), $T = \frac{0.889}{6} = 0.148$ hours = the average expected time waiting to be served. (In our example, the average time a person will wait before a tool crib attendant begins to provide the service.)

Using Eq. (5), $TT = 0.148 + 0.333 = 0.481$ hours = the average total expected time to be served. (In our example, the average total time to get the tools.)

Many work measurement people attempt to load workers subject to random demands, as in this case, to 100 percent as they would production line workers. If we let B in Eq. (1) = 100 percent and solve for the number of workers (W), we get:

$$100 = \frac{6(0.333)}{W} \times 100,$$

$$W = 2$$

Thus, two clerks would be fully utilized. The problem is that if you solve for T [Eq. (2)], you will find that the average waiting time before being served is infinitely high! This is a well-known result of queueing theory and is presented to demonstrate the tradeoffs that have to be considered where demands on a work force are random.

To pursue the problem further, the usual case is where the organization must pay not only the clerks, but also the people being served. In the example, if there are three clerks, then

$3 \times \frac{(100 - 66.6)}{100} = 1$ clerk hour/hour that is wasted per hour due to idle time.

Since there are an average of six people needing service per hour,

$6 \times T = 6 \times 0.148 = 0.89$ hour/hour that would be lost waiting by the people needing service.

The average total time that would be lost would be:

1 + 0.89 = 1.89 hours/hour with three clerks.

If four clerks are used, we would provide better service because T would decrease to 0.029 hour using Eq. (2), but we would lose 2 hours/hour due to lower utilization of the clerks' time. Thus the average total time lost would be 2.17 hours/hour. If it were possible to give the clerks other assignments, such as stocking shelves when there were no customers waiting, then one could increase the number of clerks to the extent that 100 percent work load could be provided for them. This would provide minimum waiting for the people needing service. If no additional work could be provided, three clerks would be the minimum cost solution.

For a more thorough discusison of the equations given, see footnote 2.

Staffing Demands for Different Types of Work where the Work Has to Get Done by the End of the Shift with Stated Probabilities

In this section we are concerned with getting the work done with stated probabilities, but the work can be performed at any time during the shift. Like the previous section, whether or not the work will need to be done is subject to random variation. Also, there may be several types of work to be done. (Also true, in most cases, in the first example.)

The determination of the amount of nursing staff to care for a group of patients is the best example. The nurses have to care for as many as 40 patients on a given floor. Some of the patients can feed themselves, others cannot; some need to be isolated, others need dressings changed. How much staff is needed on each shift?

If we use the more classical work measurement techniques, then we might determine the average amount of care required by each patient and multiply it by the number of patients to be cared for. For example, if the average time is 2.2 hours/patient on the first shift and there are 40 patients to be cared for, $\frac{40 \times 2.2}{8} = 11.0$ nurses needed, assuming each nurse works 8 hours. What can we say about the probability that the work can get done? We can say that on the average, the work will get done, and we can say that if the nurses work their normal pace, they will get the work done about 50 percent of the time. Fifty percent does not sound like an acceptable figure for something as important as caring for sick patients. Actually, as we shall see, 50 percent is too low a figure if we assume that the nurses will work the limits of their capability if the particular patients that they are caring for require it. At first glance, this assumption seems severe, but it is used in practically all measured day work situations. Generally, workers are asked to work at an average normal pace, but when the occasion demands it, they are expected to work faster. Likewise, there

[2]Gross, D. and Harris, C. M., *Fundamentals of Queueing Theory*, John Wiley, New York, pp. 95–103.

will be occasions when work requirements are less than normal, and they can take it easy.

The expected or average time to care for a patient is:

$$E(T) = \sum_{i=1}^{m} p_i f_i t_i \tag{6}$$

[t_i and thus $E(T)$ can be in any time units; usually minutes or hours are used], where

i = the type of work that the nurse has to do.

Example: change bed linen, give medications, change dressing.

p_i = the probability that the work will have to be done. (Since there are "m" procedures, we need a p for each procedure.)

Example: medications may only be required on the average by 33 of the 40 patients. Thus $p_i = 0.825$.

f_i = the average frequency that the ith procedure has to be done during the work day *given* that it has to be done.

Example: of the people needing medications, they are required 2.37 times on the average. Thus $f_i = 2.37$.

t_i = the average time to perform i.

Example: the time to give medications is 4.25 minutes. Thus $t_i = 4.25$ minutes.

m = the number of procedures (types of work) involved.

The variance of $E(T)$ is:

$$V(T) = \sum_{i=1}^{m} p_i f_i \left[\sigma_i^2 + (1 - p_i) f_i t_i^2 \right] + \sum_{i=1}^{m} p_i t_i^2 \phi_i^2 \tag{7}$$

where $V(T)$ is in time units squared, usually expressed as minutes squared or hours squared and where

σ_i^2 = the statistical variance of the average time to do t_i.
ϕ_i^2 = the statistical variance of f_i given that $f_i > 0$.

As an example, if we assume the value of $E(T) = 82.553$ minutes, the average time for 40 patients would be:

$$\begin{aligned} E(T') &= 40[E(T)] = 40 \times 82.553 \\ &= 3302.12 \text{ minutes or } 55.04 \text{ hours.} \end{aligned} \tag{8}$$

$$\begin{aligned} V(T') &= 40\,[V(T)] = 40 \times 882.188 \\ &= 35{,}287.52 \text{ minutes squared or } 9.80 \text{ hours squared,} \\ &\quad \text{where } V(T) \text{ is assumed to be } 882.188 \text{ minutes squared.} \end{aligned} \tag{9}$$

Then the critical level (CL) of nurses needed is:

$$CL = \frac{E(T') + Z\sqrt{V(T')}}{P} \tag{10}$$

CL is that level of nurses that can do the work with the probability level associated with *Z*, which is the number of standard deviation units of the normal distribution associated with the probability of getting the work done. *P* is the ratio of the average amount that the nurses can exceed the normal rate; 1.33 is the suggested figure. For example, if we want the work completed 95 percent of the time, Z = 1.645, then

$$CL = \frac{\frac{55.04}{8} + \frac{1.645\sqrt{9.80}}{8}}{1.33} = 5.7 \text{ nurses}$$

where *Z* is obtained from a cumulative normal probability table. Thus, 5.7 nurses can do the work 95 percent of the time. This is the minimum number of nurses needed. *E*(*T*′), which the average amount of staff needed, is $\frac{55.04}{8}$ = 6.9 nurses.[3]

Also,

$$\left(\frac{6.9 - 5.7}{6.9} + 1\right) 100 = 117.4 \text{ percent}$$

is the average pace compared to normal (100 percent) that the nurses will have to work if they have 5.7 nurses. In this example, 6.9 nurses can get the work done more than 95 percent of the time, working at an average pace of 100 percent. However, it is not always true that *CL* is less than the average number of nurses needed, so if one is concerned about making sure there are enough personnel to get the work done, the computation of *CL* is suggested. Furthermore, there exists at the present time a shortage of nursing personnel in parts of the country. If the hospital is short of nurses, *CL* is important to know in order to provide for more uniform assignment of nurses between floors.

If nurses would agree to work at the *CL* staffing levels rather than the *E*(*T*′) level, then they could be paid 17.4 percent more with savings to the hospital (fringe benefits).

Discussions of Predetermined Time Systems

Predetermined time systems were introduced from approximately 1920 until 1960. Throughout this period but more especially during the 1950s and 1960s, there were a number of technical criticisms of predetermined systems. The industrial engineer that uses predetermined time systems should be aware of the criticisms and of the efforts to respond to the criticisms so that if asked, he/she can be prepared.

The following are presented primarily for users of the manual MTM-1 sys-

[3]Hancock, W., Segal, D., Rostafiniski, M., Abdoo, Y., DeRosa, S., and Conway C., "A Computer-Aided Patient Classification System Where Variation Within a Patient Classification is Considered," in *System Science in Health Care,* edited by C. Tilquin, Pergammon, London, 1980, Vol II, pp. 1639–1650.

tem which was introduced in the late 1940s and now enjoys widespread use both as a manual and computer-aided system (4M).

1. The independence of the elements. The major criticism is that if the elemental times are not independent of each other, then the sum of the elemental times will not equal the total time. Several authors [4–6] have criticized predetermined time systems on this point. The major argument is that the motion times preceding and following a particular motion are affected by the particular motion. Thus, the preceding and following motions are not independent of the particular motion. The work of Smith and associates, [7–10] is frequently cited as evidence that the motions are not independent. These authors used a piece of equipment called the Universal Motion Analyzer that provided time values for various reach, move, and position motions. In particular, they found that the travel motion times were affected by the type of "manipulative" motions that preceded and followed the travel.

The independence criticism can be refuted in several ways: First, one can question the assumption that independence is necessary. Second, one can admit the dependence exists and is part of the rules of the MTM-1 system. And third, there are several studies that found that independence exists and the conditions under which independence is obtained.

Is independence necessary for the elemental motion times to be summed to obtain the standard time? The answer is that as long as we are summing mean (average) times, which is what the elemental motion times are, the sum of mean times is equal to the mean of the sum, whether or not independence exists. Buffa[11,12] discusses this issue and makes the point that the question of independence is only of importance if the variance of the elemental time is being used. This is rarely done. The only known potential application is contained in Chapter 12, where paced operations are discussed. Thus, when one is summing the elemental times to obtain the standard and the variances are not used, it is immaterial whether or not the elemental times are independent of each other.

The second response concerning independence is that the MTM-1 system

[4]Davidson, H. O., *Functions and Bases of Time Standards,* The American Institute of Industrial Engineering, Columbus, Ohio, 1952.

[5]Gomberg, W., *A Trade Union Analysis of Time Study,* 2nd ed., Prentice Hall, Englewood Cliffs, N.J., 1955.

[6]Goltlieb, B., *Predetermined Motion Time Systems in the USA,* Department of Research, AFL-CIO, Washington, D.C., 1959.

[7]Wehrkamp, R., and Smith, K. U., "Dimensional Analysis of Motion: II," *Journal of Applied Psychology,* Volume 36, pp. 201–206, 1952.

[8]Smader, R. C., and Smith, K. U., "Dimensional Analysis of Motion: VI," *Journal of Applied Psychology,* Volume 37, pp. 308–314, 1953.

[9]Harris, S. J., and Smith, K. U., "Dimensional Analysis of Motion: VII," *Journal of Applied Psychology,* Volume 38, pp. 126–130, 1954.

[10]Davis, R. T., Wehrkamp, R. F., and Smith, K. U., "Dimensional Analysis of Motion: I," *Journal of Applied Psychology,* Volume 35, pp. 363–366, 1951.

[11]Buffa, E. S., "The Additivity of Universal Standard Elements," *Journal of Industrial Engineering,* Volume VIII, No. 5, pp. 217–223, Sept.–Oct., 1956.

[12]Buffa, E. S., "The Additivity of Universal Standard Data Elements, II," *Journal of Industrial Engineering,* Volume VIII, No. 6, pp. 327–333, Nov.–Dec., 1957.

has "rules of thumb" that indicate that dependence does exist. For example, a M_C move is required before a position. Thus, the position is dependent on the use of a M_C move.

The third response is that there have been studies that conclude that independence does exist. For example, Nanda[13,14] concludes that independence existed for a high percentage of subjects tested. He also found that the mean elemental times are additive. Muckstadt[15] also concluded that independence exists, but that the elemental motions become independent of each other only after there is sufficient practice opportunity. This is an important finding because it may supply the reason why Smith and associates' work on the Universal Motion Analyzer produced the dependence results. In many cases, the practice opportunity of the subjects prior to collecting the data does not appear to be sufficient to ensure that learning to the level of an experienced person occurred.

2. *Predetermined time systems are not universally applicable.* The criticism here is that predetermined time systems are not universally applicable to all types of human endeavor. This criticism is true. They are not. For example, the MTM-1 system predicts the low-task times for experienced healthy people where members of the body are being used. Also, until recently, there was inadequate coverage of decision times where no body member motion was involved (see Chapter 5).

Over time, the use of the system has been expanded to include not only experienced people, but inexperienced people (see Chapter 3). Also, Chapter 2 gives the relationships between the time values and the capabilities of the working population where it is asserted that the MTM-1 value can be achieved or exceeded by approximately 95 percent of the working population. Thus, approximately 5 percent cannot achieve the values.

A second aspect of the universality is whether or not the data on which the systems are based are a "statistically valid sample." The words "statistically valid" have several meanings. The inference is that the types of work studies did not represent a wide enough range of work situations to be able to assume that the predetermined data apply to all types of work.

Responses to questions of this type of universality have to be system-specific. The original MTM-1 data were compiled over a number of years and published in 1948.[16] The data were primarily collected from bench-type operations, where machine tools such as drill presses were being used. The authors made a number of suggestions concerning the need to do more studies to determine if the data could be more widely used. The MTM Association for Standards and Research was formed in the early 1950s to encourage the further development of the MTM system. The major emphasis of the Association during the 1950s and

[13]Nanda, R., "Developing Variability Measures of Predetermined Time Systems," *Journal of Industrial Engineering,* Volume XVIII, No. 1, pp. 120–122, Jan., 1967.

[14]Nanda, R., "The Additivity of Elemental Times," *Journal of Industrial Engineering,* Volume XIX, No. 5, pp. 235–242, May, 1968.

[15]Muckstadt, J. A., "Independence of MTM Elements," Proceedings—12th Annual MTM Conference, MTM Association for Standards and Research, New York, NY, Sept., 1964.

[16]Maynard, H. B., Stegemerten, G. J., and Schuab, J. L., *Methods-Time Measurement,* McGraw-Hill Book Company, New York, 1948.

the early 1960s was to determine whether or not the MTM elemental motions could be used more broadly and whether the time values were correctly established. Laboratory and industrial studies were made and numerous reports published giving the findings. As an example of the nature of the work that was done during this period, the following is a list of the research topics that were published:

1. The Disengage Motion[17]
2. Standards for Reading Operations[18]
3. Simultaneous Motions[19]
4. Analysis of Short Reaches and Moves[20]
5. Research Methods to be Used in Collection and Analysis of Rate[21]
6. Studies of Arm Movement Involving Weight[22]
7. Studies of Positioning Movements[23,24]
8. Apply Pressure Research[25,26]

This research effort resulted in a number of changes to the MTM-1 data card as well as important clarifications of the application rules.

After the early 1960s, the questions coming from users about the validity of the data diminished. At the same time, computer techniques were developed[27] that greatly expanded the ability to collect data both in the laboratory and from industry. Sample sizes in the thousands were obtained. This technology also made it possible to collect very large sample sizes of data and at the same time explore other aspects of work measurement. Thus, the emphasis of the research shifted to investigations that would broaden the use of MTM. Typical projects during the period of the 1960s and early 1970s were:

---

[17]Lang, A. M. and Biel-Nielsen, H. E., *Preliminary Research Report on Disengage,* Research Report 101, MTM Association for Standards and Research, Feb., 1951.

[18]Lang, A. M., *A Research Report on Standards for Reaching Operations,* Research Report 102, MTM Association for Standards and Research, April, 1952.

[19]Raphael, D. L. and Clapper, G. C., *A Study of Simultaneous Motions,* Research Report 105, MTM Association for Standards and Research, Sept. 1952.

[20]Raphael, D. L., *An Analysis of Short Reaches and Moves,* Research Report 106, MTM Association for Standards and Research, Sept., 1953.

[21]Raphael, D. L., *A Research Methods Manual,* Research Report 107, MTM Association for Standards and Research, April, 1954.

[22]Raphael, D. L., *A Study of Arm Movements Involving Weight,* Research Report 108, MTM Association for Standards and Research, March, 1955.

[23]Raphael, D. L., *A Study of Positioning Movements I. The General Characteristics II. Special Studies Supplement,* Research Report 109, MTM Association for Standards and Research, 1957.

[24]Raphael, D. L., *A Study of Positioning Movements III. Application to Industrial Work Measurements,* Research Report 110, MTM Association for Standards and Research, Nov. 1957.

[25]Goodman, B. E., *A Laboratory Study of Apply Pressure,* MTM Association for Standards and Research, July, 1959.

[26]Foulke, J. A., and Hancock, W. M., *Industrial Research on the MTM Element Apply Pressure,* Research Report 111, MTM Association for Standards and Research, December, 1961.

[27]Hancock, W. M. and Foulke, J. A., *A Description of the Electronic Data Collection and the Methods of its Application to Work Measurement,* Research Information Paper No. 1, MTM Association for Standards and Research, 1961. Also published in *Journal of Industrial Engineering,* Volume 13, No. 4, July–August, 1962.

1. Learning Curve Research[28–30] (see Chapter 3)
2. Factors in Manual Skill Training[31]
3. Personnel Selection Test Development[32] (see Chapter 2)
4. Decision Time Research[33–36] (see Chapter 5)
5. Accuracy Prediction Research[31–39] (see Chapter 12)
6. System Selection Studies[40] (see Chapter 12)
7. Human Performance While Using Microscopes[41] (see Chapter 9)

Most of the studies that were done during the 1950s and 1960s were done under contract by students, employees, and faculties of universities, and most contain the data as well as the conclusions and recommendations, thus partially meeting the criticism that the data upon which the system was based were not made publicly available.

A third aspect of universality is concerned with the concept of operational validity. Operational validity means that if the system produces a desired result, it is valid operationally. The MTM family of systems appears to meet this criterion in an ever-increasing number of applications. The use of the system has been expanded by the users, so that now it is widely used for industrial and clerical operations in the United States, Europe, Japan, and Australia. The potential users did not blindly adopt the system without preliminary investigations.

[28]Hancock, W. M. and Sathe, P., *Learning Curve Research on Manual Operations: Phase II, Industrial Studies,* Research Report 113A, The MTM Association for Standards and Research, 1969.

[29]Hancock, W. M., "The Prediction of Learning Rates for Manual Operators," *Journal of Industrial Engineering,* Volume 18, No. 1, Jan., 1967.

[30]Hancock, W. M. and Foulke, J. A., *Learning Curve Research on Short Cycle Operations: Phase I. Laboratory Experiments,* Research Report 112, MTM Association for Standards and Research, 1963.

[31]Chaffin, D. B. and Hancock, W. M., *Factors in Manual Skill Training,* Research Report 114, MTM Association for Standards and Research, August, 1966.

[32]Poock, G. H., *Prediction of Elemental Motion Performance Using Personnel Selection Tests,* Research Report 115, January, 1968 (PhD dissertation, University of Michigan).

[33]Neter, M. A. *Critical Data Analysis of Repetitive Man Machine Systems Operators,* Research Report 117, MTM Association for Standards and Research, 1970. (PhD dissertation, University of Michigan).

[34]Raouf, A. *Study of Visual Performance Times for Inspection Tasks* (PhD dissertation, University of Windsor, Ont. Canada, 1970).

[35]Glen, T. M., *The Prediction of Work Performance Capabilities of Mentally Handicapped Young Adults* (PhD dissertation, University of Michigan, 1964).

[36]Thomas, M. U., *Some Probablistic Aspects of Performance Times for a Combined Manual and Decision Task,* Research Report 116, MTM Association for Standards and Research, 1972. (PhD dissertation, University of Michigan).

[37]Åberg, U., *Frequency of Occurrence of Basic MTM Motions,* Research Committee Report No. 1, Svenska MTM-Foreningen, Stockholm, Sweden. August, 1962. Also published in *Journal of Methods-Time Measurement,* Volume 8, No. 5, July–August, 1963.

[39]Magnusson, K.-E., "The Development of MTM-2, MTM-V and MTM-1," *Journal of Methods-Time Measurement* Volume 17, No. 1, Jan.–Feb., 1972.

[39]Hancock, W. M., Foulke, J. A., and Miller, J. M., *Accuracy Comparisons of MTM-1, GPD, MTM-2, MTM-3 and the AMAS Systems,* MTM Association for Standards and Research, May, 1973.

[40]Hancock, W. M. and Langolf, G. D., *Productivity, Quality and Complaint Considerations in the Selection of MTM Systems,* MTM Association for Standards and Research, August, 1973.

[41]Langolf, G. D., *Human Motor Performance in Precise Microscopic Work Development of Standard Data for Microscopic Assembly Work* (PhD dissertation, University of Michigan, 1973).

Thus, at least for the MTM systems, the use continues to grow, the systems appear to be operationally valid, and universality is being attained over time.

3. Predetermined time elements must have irreducible elements, i.e., they cannot be reduced further or factors cannot be found that can affect the values. The criticism here is that if the elemental values are reducible, the system is not universal. The comment that the elemental values are reducible is correct, but the comment that they must be irreducible in order to be universal is not correct. Humphreys[42] describes an experiment by Gerald Nadler where the angle of movement to a dial caused the reach motion to vary by as much as 16 percent. Similar studies can also be found in McCormick,[43] where the direction of motion affects the reach and move times.

The MTM-1 motion reach, for example, does not take direction into account. The mean values for a particular class and distance are obtained by determining the elapsed time of motions in all directions found in industrial operations. Direction is treated as a random variable. Thus, when one chooses a particular reach value, the direction of the reach can be regarded as a random variable. Because the elemental motion values are very small compared to the standard, and each of the motions of the job may be considered random samples of the particular motions involved, the sum of the mean elemental motion values and the sum of the actual time values will converge as the number of elements increases. The popular expression for this is "averaging out." The extent to which the convergence occurs is partially expressed by the accuracy equations of Chapter 12. In these questions, the random errors due to class, distance, and the other variables are taken into account. The random errors due to direction of motion are assumed to be confounded with the systematic error.

4. Predetermined time systems require the use of leveling by the developer. The criticism here is that although the applicator does not have to level, leveling was used to develop the elemental motion values, thus, there is no difference between predetermined times and Time Study. The response is that it is true that leveling was used, at least in the MTM-1 system. However, the leveling was done under carefully controlled conditions and frequently by more than one person. Thus, the variance in leveling has been minimized and the same values apply to all workers when the system is used. The importance of doing the leveling as part of the system development is that the standards produced by the system will be much more consistent than if leveling were done at the point of use, such as in time study. Consistency is an important attribute of a predetermined time system because humans (workers) are quick to perceive different work pace requirements if they are working in proximity of each other. This characteristic is known as "small difference comparisons" or in psychology as "just noticeable differences."

Under certain conditions, even without leveling as a factor, the use of predetermined time systems will produce results which will cause labor complaints.

[42]Humphreys, R. W. *Work Measurement, A Review and Analysis,* Institute for Labor Studies, Division of Social and Economic Development, West Virginia University, Morgantown, WV, pp. 142–143.

[43]McCormick, E. J., *Human Factors Engineering,* 3rd ed., McGraw-Hill, New York. 1970, pp. 325–326.

The extent to which this will occur and the economic consequences are covered in Chapter 12.

5. *If predetermined time systems produce elemental values that are correct, then the elemental values of different predetermined time systems should all be the same.* This statement is one of the hypotheses of Davidson.[44] He proceeds to demonstrate that the time values are different for a number of the basic motions of the more popular systems. The problem with his arguments is that he does not sufficiently consider the differences in definitions of the beginning and end points of the motions. Differences in these points can and do produce values that are not comparable.

6. *A check time study and a predetermined time standard do not agree. Therefore, the predetermined time standard is incorrect.* There are many reasons why the two may differ, but assume the more obvious ones have been taken care of, such as, in the case of the MTM-1 system, whether or not the worker was experienced, in good health, working at the normal pace, and the motion pattern of the check study and the predetermined time standard were the same. If there still remains a difference, then the comparison should be made using the method suggested at the end of Chapter 5 (Decision Times). Differences will exist for short-cycle studies, which are frequently used to "prove" that predetermined time standards are not accurate. Since the accuracy of predetermined time systems is less for short cycles, the confidence interval of the prediction should be taken into account utilizing well-known statistical techniques when using time study to verify predetermined time standards.

7. *The MTM-1 values are not the minimum energy values.* This criticism was made by Schmidtke and Stier[45] with the apparent assumption that the basis for establishing the MTM values was the time required to do the work involved at the lowest level of energy expenditure. The response is that minimum energy expenditures were not the basis of setting the elemental motion times. The basis was the normal rate of working. Thus, although the MTM-1 values are surprisingly close to the minimum energy expenditures, the criticism is based on a misunderstanding of the design of the system.

[44]Davidson, H. O., *Functions and Bases of Time Standards,* The American Institute of Industrial Engineering, Columbus, Ohio, 1952.

[45]Schmidtke, H. and Stier, F., "An Experimental Evaluation of the Validity of Predetermined Elemental Time Systems," *Journal of Industrial Engineering,* Volume XII, No. 3, pp. 182–204, May–June, 1961.

CHAPTER 15

Work Measurement Applicable to Professional Employees

The proper establishment and application of job or performance standards for professional workers (engineers, accountants, managers, etc.) produces as large (in many cases significantly larger) a positive economic impact on the overall performance of the organization as does the establishment and application of standards for factory workers. This is in spite of the fact that the work standards cannot be nearly as precise or exact with respect to time as the measurement of blue collar and clerical workers performing routine clerical tasks. It must be recognized that professional worker performance must be judged over a long term—like 6–12 months; rarely, and only for a few less important standards, can the measurements occur over a shorter period.

That the work of professionals needs to be measured and performance against such standards evaluated is attested by the explosion in the numbers of such workers and the associated salary and support costs. As in the factory, this alone is reason enough for measurement, which, in turn, is required for control. However, in this case, there is another and perhaps more important reason for the measurement. These people have such a large effect on the performance of the organization that this fact really should be the overriding reason for measurement and control.

Many managements recognize the need, but a majority (especially those in the smaller and/or less-sophisticated managements) prefer to use a "softer" and much less effective approach—job or staff appraisal. Usually such "soft" measurements involve using a form specifying specific measurement criteria such as personality, production, relationships with others, etc. These measurement attempts are general (not specific) and are virtually useless except in the eyes of idealists. Seldom do such evaluations have a significant impact on performance, and they seldom affect promotions, salaries, or job assignments except at very low levels of the organization, or when the worker is *very bad*. It is better than no measurement, but it is not an adequate measure.

Regardless of semantics and approach, it behooves management to establish work standards that directly relate to the work and work tasks and that specify required performance in measurable terms. These standards must also be realistic, relevant, and achievable. The resultant measurements are generally viewed as goals or objectives. Such goals must also be timely and in consonance with overall organization goals or objectives. Since they are both organization goals and measurements, they must be stated in measurable terms.

It is time for industrial engineers to expand their horizons and move into this functional or work segment of organizations since it is a logical extension of their now normal pervue of factory and office work measurement. However, since it represents a very high level work that intimately involves top level professionals, the activity should possibly be restricted to industrial engineers holding both the IE degree and the MBA—the IE does not cover the managerial considerations involved in planning. The reasons for this statement should become obvious as this subject is discussed further. What has been said so far merely "sets the scene"; it is an introduction.

More Specific Information and Facts

There are a number of significant differences between manual work measurement and the work measurement of professional workers. Perceiving such differences should substantially help the reader's understanding, learning, and success in this area of work measurement.

The time span of the measurement was previously mentioned. For production workers and the supporting indirect employees the work involved almost always produces an immediate result, a result that can immediately be examined for volume produced and for defects caused by errors—committed by the worker or by the machine. Just visualize such workers as janitors, machine operators, assemblers, stock pickers, and inspectors performing their jobs and this "immediacy of result" will be obvious. The same holds true for most routine paperwork and even for a large portion of the routine activities associated with computer operations. When measuring the work of professionals and many semiprofessionals, the time span of the measurement must be lengthy since the ultimate effect of the actions (mentally oriented or mentally "driven" actions) of these workers often is 12 or more months in the future. Often it is 2–5 years in the future for the top management team of an organization with regard to the long-term objectives or goals to be perceptible (capable of being identified).

The greatest attention must be paid to the actions and their effects within a 1-year time span, since anything longer ceases to have more than a very minor motivational effect. This means that those actions which cause very-long-term effects must be so specified that partial measurements can be made within a 12-month period based on short-term actions directly related to the long-term goals. It is a complicating factor.

Even in the case of 1-year measurements, they should be so stated (whenever possible) that some degree of progress can be judged in the 3–6-month range. All this latter discussion of motivation and the perception of performance effects is owing to the fact that the manager/supervisor should check progress against the standards or goals and discuss such performance with the concerned professional(s). Such discussions should occur every 3–6 months, with a major review at the end of each 12 months. The 12-month review can occur on any calendar date, but it should always occur essentially every 12 months (approximately).

Other facets of professional work measurement include a recognition of the fact that we are dealing with a very complex output, which has an important impact upon the organization, yet which is affected by such aspects and factors as the following:

1. Complex mental processes
2. Motivation
3. Personalities
4. Objectives
5. Knowledge
6. Political aspects and implications
7. Social impacts
8. Experience
9. Governmental regulations and/or laws
10. Moral implications
11. Intelligence
12. Perceptions
13. Effects (especially on personal life)
14. General economic condition

Not only do the above apply to the worker being measured, but also to those affected by the actions of the measured worker.

It should be obvious that, when one takes into account all of the possible variables, any performance standards established will be somewhat imprecise, not only with respect to the time element, but also with regard to actual output result—especially since there often is no "one best way" to accomplish the goal. However, one simple example may make clear that the measurement and control process is not impossible. Most readers know professionals (nonmanagerial and managerial) that seem to move from one job success to another, whereas others continually fail to perform at a satisfactory level, and education and/or experience seems to have only a somewhat minor effect upon results. Obviously, the high performers have somehow succeeded in turning in superior performance. Good standards do make it possible to monitor performance and make changes where required. Later, empirical research results related to goal achievement will be cited that should fully convince any readers who doubt that all this is possible.

Finally, it should be recognized that the quality and variety of results produced by a person often cannot be determined except through considerable unbiased study and analysis—the exact opposite of factory work measurement.

No knowledgeable person has ever seriously considered measuring professional work with the techniques of time and motion study or by any of the directly associated techniques. A few have suggested work sampling, but this only tells us the kinds of activities in which the worker engages and the percent of time spent on each—but even in this area the observers cannot really distinguish between such things as deep meaningful thought and daydreaming or between idle conversation and purposeful business-related conversation (all by mere *observation*). By now the reader probably suspects that both objective and subjective measurements are involved; it is a correct deduction.

Management by Objectives (MBO)[1]

About the only useful measurement involves the setting of meaningful and important objectives or goals expressed in measurable terms and providing an estimated time required for performance. These must be identified and defined by the worker *with* his/her superior. They are then used to measure the performance of the worker. The procedure is commonly referred to as Management by Objectives. The objectives will not produce the desired results if they are established by the supervisor and arbitrarily imposed upon the professional.

Even more importantly, many MBO programs have failed, some totally and others partially, because the MBO program was not preceded by properly integrated and formalized long- and short-range planning. Without the long- and short-range plans as a guide, incorrect or unachieveable or otherwise unsatisfactory objectives are often set. Even if the general thrust is correct, the degree of movement and the time factor will be incorrect without proper planning.

Some of the reasons why MBO programs fail without being preceded by formal planning are discernible with a little study. First, the MBO objectives will likely focus on the wrong objectives, pointing the individual and/or the organization in the wrong direction. Another, but related, phenomena is that the participants tend to be buffeted by both the external and internal environments. MBO without formal and proper planning cannot properly take these various environments into account—economic, legal, governmental, social, demographic, industry trends, etc. The participants and the organizational entity become somewhat like a motorboat that has lost its rudder—there is a lot of action, but much of it in wrong directions. Finally the participants, without proper formal planning, do not really believe the objectives are real, and they tend to not be fully committed to their achievement. If personal objectives relate to overall objectives established in a formal planning effort of the proper sort, there is both belief and commitment. Additionally, with planning present, the organization's morale tends to be high—another plus.

Many people labor under the erroneous belief that most corporations plan and, therefore, the above argument is likely invalid. The actual fact is that the vast majority of organizations do *not* engage in proper planning. At the most, no more than about 30–40 percent of the nation's largest corporations properly plan. Less than 10–15 percent of the small ones, even up to $150,000,000 sales, engage in proper overall formal planning. The nonplanners have budgets for the year ahead and some of them even engage in a form of product planning spanning 1–3 years. However, budgets should be the result of planning, and product planning is only an expression of *one* aspect of proper overall planning. The budgeting process is never a proper planning process.

Before proceeding to the next section it should be clearly understood that the comments apply to both profit-seeking and non–profit-seeking organiza-

[1]An excellent short history of MBO is found in Greenwood, R. G., "Management by Objectives: As Developed by Peter Drucker, Assisted by Harold Smiddy," *The Academy of Management Review*, Volume 6, Number 2, April, 1981.

tions, including governmental organizations. Much of the discussion references business and industry, but it does apply equally well to nonprofit organizations. About all that needs to be done by the nonprofit organization reader is to substitute "Cost Benefits" for "Profits."

Since proper planning must be considered an integral part of Work Measurement for professional (and often semiprofessional) employees, proper planning processes and procedures will be discussed in considerable detail.

What Is Proper Planning?

The American National Standards Institute (ANSI) in its Industrial Engineering Terminology Standard[2] defines planning as follows:

> PLANNING—is the process whereby an individual or an organization identifies opportunities, needs, strategies, objectives and policies that are used to guide and manage the organization through future periods. All planning consists of a) accumulation of information, b) sorting and relating bits of information and beliefs, c) establishing premises, d) forecasting future conditions, e) establishing needs, f) identifying opportunities, g) establishing objectives and policies, h) structuring alternative courses of action, i) ranking or selecting total systems of action which will achieve the best balance of ultimate (future) and immediate objectives, j) establishing criteria and means for measuring adherence to the selected program of action and k) so managing the organization to achieve the objectives. Some kinds of planning include short and long range planning, product planning, financial planning, etc.

A somewhat related ANSI definition is

> POLICY—the verbal, written, or implied general plans of action that guide the members of the organization in the conduct of its operation, and incorporating in them broad premise and limitations within which further planning activities take place.

Other definitions,

> Planning is the formulation of the thoughts and ideas to guide the strategy and tactics of an organization[3] [or of an individual].
>
> It is the exercise of foresight to adjust the company and its activities in advance of events which will influence them.[4]

A very simple definition is "thought in advance of action."

Another and much better way to describe or define planning for our purpose is to cite what it involves by means of definitive statements of the major elements or steps in Formal Integrated Long- and Short-Range Planning. These

---

[2]American National Standards Institute, *Standard Z94, Industrial Engineering Terminology* (the Organization Planning and Theory section), published by the American Institute of Industrial Engineers and the American Society of Mechanical Engineers, 1972, 1981.

[3]Karger, D.W. and Murdick, R.G., *Managing Engineering and Research,* 3rd ed., The Industrial Press, New York, 1980.

[4]Karger, D.W. and Murdick, R.G., *New Product Venture Management,* Gordon and Breach Science Publishers, New York, 1972.

were defined by Dr. Z.A. Malik in his PhD research, which investigated the benefits of planning.[5–7] In order for an organization to be considered as a planner in Dr. Malik's study and in our discussion, all of the following elements have to be present, otherwise the organization should be classified as a nonplanner:

1. The resultant long- and short-range plans are in writing and in the possession of top management.
2. The plan is the result of all elements of the management team working together.
3. The long-range plans must have at least a 5-year time horizon.
4. The various internal and external environments, present and future, are taken into account when developing the plans.
5. The long-range plan identifies goals and objectives, and stresses taking advantage of opportunities. It also should identify the principal strategies and tactics to be pursued in order to achieve the plan.
6. The short-range plan (SRP) is created immediately after the long-range plan (LRP) is completed—it really is an expansion of the near-term portion of the long-range plan, *not the reverse*.
7. The organization is managed in accordance with the plans—meaning that progress is measured against planning goals and objectives and corrective action is taken when significant deviations are identified.

The above are the elements that must be present for the organization to be considered a planner in Dr. Malik's research and in the context of this discussion.

The eighth step, when exploiting MBO, is to establish agreed upon objectives for each professional worker. These are developed by each employee and his/her manager—both must be involved. For this effort to be successful there must be a *free* exchange of ideas, and the manager cannot simply impose his/her thoughts upon the worker.

There are three main kinds of MBO objectives. The first kind is related to company long- and short-range plans. The second variety is related to departmental plans, and the third class or variety of individual professional goals is only partially related to the organization—like a goal to enhance personal professional development actions. A fourth kind, which is not a part of MBO and which may not be discussed by the professional and his manager, is those personal objectives that are not even indirectly related to the organization.[8] These are all later discussed in detail.

The ninth step with MBO present would be to measure performance against the professional's goals and objectives and to reward good performance as well as taking corrective action when poor performance is detected.

---

[5]Malik, Z.A., *Formal Long Range Planning and Organizational Performance* (doctoral dissertation), Rensselaer Polytechnic Institute, Troy, NY, Dec., 1974.

[6]Karger, D.W. and Malik, Z.A., "Long Range Planning and Organizational Performance," *Long Range Planning*, Dec. 1975.

[7]Karger, D.W., "What Effective Planning is All About," *Industrial Management*, July–August, 1976.

[8]More information on this can be found in Karger, D.W. and Albrecht, D.D., "The Measurement of Professional Work," *Industrial Engineering*, August, 1971.

Why Do It?

Why exert all the effort and spend all the money and energy required to accomplish the above-described procedures? The answer is simple: One engages in long- and short-range planning in order to significantly increase the overall performance of the organization—the effect is well documented.

MBO, as an adjunct to proper planning, is practiced in order to increase the performance of the organization's professional members as individual performers—which is naturally related to "team" performance. Individual performance, in addition, is substantially influenced by incentives and/or rewards tied to goal achievement (the achievement or meeting of standards). This is as true for professional employees as for direct labor employees. All this is well documented in the literature. For example, 80–90 percent of all salesmen are on some form of incentive pay, simply because organizations long ago found that it improves performance. Most top managements receive some form of incentive pay, even in the U.S.S.R. Why would not the same action be appropriate for individual professional employees?

It was mentioned that documentation exists concerning the effect of proper planning upon organizational performance. It does exist, but the number of well-documented cases is small since the companies that do it like to keep the benefical effects somewhat concealed—also they often do not clearly perceive the effect.

One of the first documented research efforts was that of Thune and House,[9] who in 1970 performed a pioneer empirical study measuring the overall impact of planning on an organization. Their research concentrated on the chemical and drug industries, probably because the companies comprising this industrial grouping early found planning vital to success—therefore, a reasonably sized group of planners could be found to compare to the nonplanners. Profits and sales were significantly higher for the formal planners over about a 10-year study period.

David M. Herold[10] restudied these industries in 1972 and essentially substantiated the earlier result, sales and profits per year for the formal planners exceeded those of the nonplanners by about 60–150 percent depending upon the year and the factor.

The more recent and previously referenced study by Z.A. Malik was much more comprehensive and broad. It therefore sheds considerably more light on the benefits accruing to those organizations that engage in proper formal planning. It measured achievement against 13 commonly recognized corporate financially oriented performance measures over the 10-year period of 1963–1973 for New York Stock Exchange–listed firms in three industry groups. The industry groups were (1) mechanical, (2) electronics, and (3) chemicals and drugs—these last two were again combined to form one group in order to have enough

---

[9]Thune, S.A. and House, Jr., R., "Where Long Range Planning Pays Off," *Business Horizons,* August, 1970.

[10]Herold, D.M., "Long Range Planning and Organizational Performance, A Cross Valuation Study," *Academy of Management Journal,* March, 1972.

Index	Electronics Compare Means ($\bar{x}_1 > \bar{x}_2$)	Machinery Compare Means ($\bar{x}_1 > \bar{x}_2$)	Chemical & Drugs Compare Means ($\bar{x}_1 > \bar{x}_2$)
Annual Rates of Change (percent)			
	$\bar{x}_1 > \bar{x}_2$ by —%	$\bar{x}_1 > \bar{x}_2$ by —%	$\bar{x}_1 > \bar{x}_2$ by —%
1. Sales Volume	151%	127%	45%
2. Sales per Share	108%	68%	32%
3. Cash Flow per Share	447%	151%	32%
4. Earnings per Share	1009%	321%	64%
5. Book Value per share	147%	186%	52%
6. Net Income	410%	292%	65%
Mean Annual Rates (percent)			
7. Earnings per Capital	38%	48%	11%
8. Earnings/Net Worth	39%	86%	13%
9. Operating Margin	26%	55%	32%
10. Dividend/Net Income	$\bar{x}_2$ is 35% $> \bar{x}_1$	$\bar{x}_2$ is 13% $> \bar{x}_1$	$\bar{x}_2$ is 10% $> \bar{x}_1$
Mean Values per Year			
11. Capital Spending/Share ($)	106%	$\bar{x}_1 < \bar{x}_2$	$\bar{x}_1 < \bar{x}_2$
12. Stock Price ($)	23%	$\bar{x}_1 < \bar{x}_2$	$\bar{x}_1 = \bar{x}_2$
13. Price/Earnings Ratio	$\bar{x}_1 < \bar{x}_2$	13%	22%

***Figure 15–1.** The performance of the respondent planners was compared to that of the nonplanners for each industry group over a 10-year period (1963–1973). The ultimate sources of the data used for evaluation were the annual reports. $\bar{x}_1$ is the mean increase for a group of planners and $\bar{x}_2$ is the mean increase for a group of nonplanners in the same industry group.*

planners and nonplanners. Only firms having sales of $50,000,000 to $300,000,000 were considered so as to have a somewhat comparable size for each of the firms included in the study.

The planners exceeded the nonplanners by about 25–1000 percent in the *rate of growth or improvement,* depending upon the factor and the industry group. Figure 15–1 summarizes the results of the study.

Currently, a study of the effect of proper planning on small firms, $10,000,-000 to $50,000,000 in sales, is being conducted by Professors D. Karger, L. Cox, and R. Roberts of the University of West Florida. Preliminary results seem to indicate benefits similar to those experienced by larger firms, but the results seem to be erratic. Moreover, there appears to be some evidence that indicates that planning efforts which contain few of the necessary planning elements mentioned earlier actually do worse than firms which do not claim to adhere to formal planning.

All of the studies of benefits experienced by planners indicate that the financial benefits are so great as to not require a search for other reasons. However, there *are* other important reasons why organizations do (or should) plan—the kind of reasons that motivated the first planners, those who began planning before the financial results were disclosed by research:

Provides a unified sense of direction to the company. No firm can succeed when top executives are each trying to move the firm in different directions; sometimes these efforts are in direct opposition without planning.

Technological, social, economic, competitive, and legal environments are changing at an ever increasing rate.[11] Such changes must be anticipated and properly taken into account. Trying to achieve good organizational performance without doing this is an impossibility—and planning makes it possible to properly take the environment into account.

LRP provides direction and continuity to short-range plans. The cost detriment and other effects of not planning, such as year to year changes in organizational purpose and direction, make not planning unacceptable.

Planning forces the organization's consideration of its various internal and external environments.

Planning permits the better management of debts and of capital.

Planning substantially helps optimize an acquisition program.

Proper planning makes for logical divestments of marginal or otherwise unproductive activities.

Without proper planning the management team must guess what the Chief Executive Officer (CEO) will approve—and no manager can afford to be "wrong" very often—hence, the lack of plans inhibits effective performance.

An organization needs policies to guide the management team; these are virtually impossible to state clearly without proper plans.

Not only are the benefits of planning great, but planning also unlocks the door to another set of benefits, those associated with the measurement of the performance of individual professionals. Top management professionals are often essentially measured by the overall plans; this is not true for individuals lower in the organization, management or nonmanagement employees. The measurement and control of professional work cannot help but enhance the beneficial effects of planning since it assures greater dedication and effort by those who create and execute the plans. This great importance of planning is why the subject dominates this chapter.

The Basic "How To" Steps in Planning

The major steps in properly integrated long- and short-range planning are as follows:

1. Select and appoint the planning director.
2. Educate the planners regarding the planning process, the expected usage of the plans, and the potential beneficial results.
3. Create a standard view of the organization.
4. Define future horizons, external and internal, for the length of the planning period. It must be 5 or more years in length; more preferably, it should be about

[11]See Alvin Toffler's *Future Shock* and his latest book *The Third Wave*. He documents the ever increasing rate of change experienced in the world's societies. The books are in hard cover and paperback.

equal to the longest product life cycle of the firm's major products (or planned for products). Such forecasts must take into account rates of change in the factors. (See footnote 11.)
5. Select the planning site.
6. Define present business(es).
7. Define future business(es).
8. Decide on long-term strategic and tactical objectives.
9. Expand the near-term portion of the long-range plans into the organization's short-range plans.
10. Operate business(es) per plan; this includes the measurement of performance against the plans.
11. Revise plans yearly and when *really important* new data appear.

The Planning Director

This is a position to which Chief IEs might aspire because of planning's obviously important major positive impact upon the results achieved by the organization. The position normally carries vice-presidential status and is therefore a good upward position move for IEs—even just to move into a senior staff position in the planning activity.

The planning director is a very important person, but he/she does not do the planning. The director's task is to see that the right things get done correctly at the proper time, much like the director of an orchestra. The orchestral director is both important and essential, but so are the musicians—in our case the managers and the professionals, especially the top managers. Essentially, he/she is a sort of substitute for the CEO with regard to some of the planning activities.

The planning director could be the CEO, since planning and managing organizational results *are* his/her chief responsibility. However, it is usual that such men are too busy to handle the details involved. This is especially true in larger companies. In smaller companies, the expertise is often missing from the CEO's skills and knowledge. The result is that the organizing work, the detail, and some of the top level "thinking" is delegated to a planning director. However, the CEO still plays a vital role in the planning and its direction, even with a well-qualified planning director.

Ideally, the director is an existing highly placed manager, like a vice president. However, if the company has never before planned, then it is usually best to initially go outside for the director and have the future in-house director become trained by working with the outside director.

It is vital to planning success that the director not only be experienced, but also that he/she possesses enough stature and "presence" that all the planners will perform as requested. Sometimes the director must even curb the actions of the CEO, if he/she becomes too dominating.

Experienced outside directors can be obtained from management consulting organizations and from among the professors of some schools of business or management. However, make sure that the person who will serve as your outside director of planning has *actually* directed a planning effort or has, as a minimum, actively helped another person in leading such an effort. Do not accept merely having the Supervising Principal of the consulting organization possess the actual experience unless you are fully cognizant of the hazards and are satisfied that what you are buying is acceptable or is the best you can afford. Adequately experienced people actually do not cost significantly more than inexperienced people. Moreover, when possible benefits vs nonbenefits are considered, one is usually foolish not to procure the very best available.

Tasks to be performed by the planning director include:

Sees that the right things get done by the right people at the right time.

Selects, with the approval of the CEO, the planning site. Should be at a quiet and well-appointed establishment away from headquarters.

Obtains and distributes all general material and services required by the planners.

Educates the planners.

Creates a standard view of the organization.

Specifies the basic data required and arranges for their procurement and distribution.

Arranges and secures, with the approval of the CEO, the services of consultants, industry associations, etc., that may be desirable and helpful to the planning effort.

Sets meeting dates and makes all travel, eating, and living arrangements.

Acts as chairman of the planning sessions, which includes the development and distribution of agenda.

Publishes the official results and reports of all planning sessions.

Do not give the planning director the job of actually measuring performance against the plans. Such activity will close "doors" and greatly inhibit his/her performance in directing future planning efforts.

Educate the Planners

Unless all planners understand the planning process and the role of the planning director, no really beneficial planning can result. Everyone has heard of planning. *Everyone* claims they plan. To state that one does not plan seems almost immoral and/or very wrong. However, few managers have ever participated in meaningful and proper formal long- and short-range organizational planning. The same holds true for long- and short-range *individual* planning of a formal and structured nature. Therefore, before meaningful planning can commence, the planners must be educated regarding the following:

1. the uses and benefits of long-range planning (LRP) and short-range planning (SRP)
2. the actual steps in the planning process
3. the sequence in which the planning actions must occur
4. the roles of the planning director
5. the reasons why consultants are sometimes used
6. why the organization's strengths and weaknesses must be identified
7. why a strategy must be developed
8. the data requirements of LRP and SRP

9. how the plans are created
10. how the plans should be stated and used

"If all of the above are not common knowledge among the planners, each manager will go off in a different direction. Furthermore, needed information may not become available at the right time."[12]

Another very specific duty of the planning director, in cooperation with the CEO, is to create a Standard View of the Organization—more will be said on this subject later in the chapter.

The "Working" Planning Manual

The first time through the planning process there is no Standard System and Procedures document to follow, and each step is a "first-time" event. Incidentally, a standard procedure should be prepared as the first-time planning effort proceeds in order to guide the planning effort in future years. The content of a manual adapted to the needs of the "first-time" planner could logically consist of the following:

1. Correspondence related to the planning process. It will cover who does what and when. Everybody directly concerned should get copies of all correspondence and/ or notices. File sequentially by date, with the most recent date in the forward position.
2. Essential information on the planning process—the material required to educate the planners. (Such information is provided by the material referenced in footnotes 12 and 13.)
3. Environmental forecasts, at least 5–10 years into the future. These include economic, technological, social, demographic, ecological, political, legal, and competitive (by kind of industry). Scope is somewhat adjusted to the importance of the specific topic to the organization. Often such forecasts need to be by country. Most of these are available from governments, trade publications, professional publications, industry associations, and business publications. However, a specialist consultant can often be very helpful regarding a specific industry outlook and with regard to industry trends. A knowledgeable planning director will have many of these forecasts and will know how to get the others required. These must all be summarized as to important and possibly relevant aspects, and copies must be provided to each planner by the planning director. Everyone must plan for the same future.
4. Corporate and functional background material, including the identification of strengths and weaknesses—a facet later discussed in some detail. It also relates to developing a standard view of the organization.

[12]Karger, D.W. and Murdick, R.G., *Long Range Planning and Corporate Strategy* (a how-to manual and self-evaluation guide), Karger and Murdick, 506 Circle Dr., DeFuniak Springs, FL 32433, 1976, 1978.

[13]Another possible solution would be to give each planner a copy of this text. However, it was not designed for this particular purpose, hence, some information is missing. Also, this text contains much other information.

5. Financial background material presenting the past overall (and by division and product where appropriate) 10-year history and a simple forward projection for the next 5 years (or for the planning horizon, if longer). It is essential that this segment include a cash flow history and projection as a guide to determine what planning objectives can be financially supported *on the basis of historical patterns*. Eventually the financial function must also summarize and project the financial effects of (a) the proposed objectives and (b) the adopted objectives. These projections possibly include pro forma P&Ls, cash flow, pro forma balance statements, debt projections, capital investment projections, etc.
6. Definitions of the present business(es) and the business(es) the organization wants to be in 5–10 years in the future. This is one of the first tasks to be accomplished in a planning session. Modern and active boards of directors usually devote time to accomplishing this task with appropriate top level officers. Remember, planning is concerned with how to get from "where we are to where we want to be."
7. Long-range overall and sectional (division, subsidary, etc.) plans.
8. Short-range plans for the units considered in step 7.

Create a Standard View of the Organization

Each member of any organization always has a somewhat different idea or concept of the organization. Hence, each will be planning for a different organization unless a standard view is created.

Additionally, section 4 of the planning manual called for the presentation of the strengths and weaknesses of each function or major organizational element—because planners must take advantage of organization strengths and minimize the effects of weaknesses.

Strengths are immediately available for exploitation, whereas major weaknesses take 1–3 years to correct. The identification of these facts presents a major problem to the planning director. No one wants to admit to having a weakness, and strengths are usually exaggerated. After the first time, some of the fear of exposing weaknesses vanishes if the CEO can be restrained from taking too strong corrective actions. Therefore, the real problem is how to detect and bring forth these data for the first planning session—here is where it is advantageous to have a competent outside consultant act as planning director. This individual will know how to detect the weaknesses and also will usually be able to keep the CEO from becoming too excited about disclosed weaknesses, often a difficult to impossible task for the inside planning director, especially in a first-time planning effort.

One way the consultant discovers the strengths and especially the weaknesses is to talk with each manager about their area of responsibility and how they are affected by other functions. If this is properly done, the consultant will identify the major strengths and weaknesses—usually by indirect rather than by direct statements. For example, a manager might comment "we have a great team that is capable of superb work, but often we cannot get the data needed from Joe Green's department." Only thorough checking will determine who is really at fault.

The outside consultant director can usually present weaknesses in a manner that will not drastically upset the managers. The inside planning director, usually, could not afford to take such an action.

In addition to the functional or major organizational element data, there really must be a probing search for strengths and weaknesses in all of the following (from footnote 12):

1. Corporate Board of Directors
2. Management skills
3. Professional and craft skills
4. Financial position and potential
5. Marketing
6. Engineering
7. Production
8. Computer hardware, software, and data communications systems
9. Raw materials sources and vendor relationships
10. Facilities and locations
11. Organizational dynamics
12. Legal skills and legal positions
13. Company policies
14. Corporate image

Board of Directors

This is not a resource that can be discussed freely and openly with employees, it cannot even be openly and freely discussed with the top-level management team. However, the CEO with one or two key people, and sometimes with a senior consultant, will discuss board strengths and weaknesses and what can be done to improve the "makeup" of the board. The material which follows (taken from footnote 12) presents some of the questions that should be considered, and these in turn point the way to the possible strengthening of this asset of the organization.

> What is the size of the board compared with similar size companies and others in the industry? What are the ages of the directors?
>
> What is the proportion of outside directors? Are the outsiders really only former insiders who have retired from the company? Are they people who actually are in the employ of the company (like a legal firm)? Do they mainly represent your sources of money and, hence, really should not be privy to all company actions?
>
> In a similar manner, we might explore attitudes. How do the board members feel about mergers and acquisitions? What is their general perspective? Is the board dominated by an individual or group or by any set of biases or prejudices which may affect new product planning or any activity of the management of the corporation? What about experience? Is the executive experience of the board and of the management of the company as a whole broad enough to guide development and marketing of new

product lines or the entry of the company into a completely new field? This is much easier said than done.

Are they capable of admitting that in an acquisition program some of the acquisitions will likely turn out to be faulty and require divestment? Some people tend to hang on stubbornly when they should let go.

Will the board permit "betting the company?" Or contrariwise, is the board strong enough and wise enough to pull back on a CEO who is too enamored with a pet project?

What skills do the outsiders possess? What resources do they command? Are they being utilized to further the advancement of the organization?

What of the composition of the board in terms of strength of will and general management skills. Are the people on it the right ones to *aid* the organization in achieving its goals? If not, what skills, resources, experience, and training are needed? What kinds of people are likely to possess them? Can you find such individuals who would be willing to serve?

Do any of the board members represent negatives? If so, who and why? Can they be removed? If not, can they be neutralized by one or more specific individuals.

The Board of Directors represents an important resource of the company if it is properly structured as to represented member assets, such as skills, knowledge, influence, access to information, etc.

One important use of the board is that of a sounding board for the thinking and ideas of the inside members. If it contains only insiders, they are talking to themselves—a rather useless procedure.

All this is especially important since the cost of maintaining a well-structured board with outsiders is about the biggest bargain left to U.S. industry. In very small companies ($1–4 million sales) the outsider usually receives only $200 to $400 per meeting. In firms with $200 million or more of sales, a board member will cost approximately in total (expenses, insurance, and fees) $8000 to $24,000 per year (assuming 6 to 10 board meetings per year). [Now substantially higher, about $10,000 to $30,000.]

Other Areas to Consider

Available space does not permit discussing each of the 14 identified resource areas of an industrial organization. The key functional managers with the planning director should be able to draw up a "bill of particulars" for each area. However, some guidance will be provided by discussing two important and dissimilar areas, the kinds of considerations mentioned for these should indicate the detail and depth that should be pursued for the others.

Manpower (taken from footnote 12)

While the subject of manpower evaluation is somewhat common, the following can serve as a checklist of the items to be considered:

1. Managerial personnel. What are their qualifications on the basis of age, present ability, special talents, and potential growth?

2. Depth of management talent (or backup for present managers). A related question: Is there a manpower inventory system that includes management personnel?
3. Engineering and scientific talent. What is the company's engineering strength in terms of number and quality of engineers by field, depth in field, and specialties?
4. Professional talent other than engineers and scientists. What quality, quantity, and variety of talents, such as financial specialists, salesmen, marketing specialists, production and industrial engineers, etc., are available?
5. Technical supporting personnel (laboratory assistants, computer programmers, engineering assistants).
6. Dispersion of the above personnel as to subsidiaries, divisions, departments, geographic areas, fluency of language, etc.
7. Skilled, semiskilled, and unskilled labor. What quality and quantity are available within the company and its various plants? What potential reserves are there in the areas surrounding each plant and/or office? What is the labor history in the involved areas? Are there any union limitations?
8. Geographical dispersion of plants.
9. Compensation practices. Are they totally uniform within the company? Is there a written compensation policy and procedure covering present and expected enlargement of the organization? If there is no integrated plan for a dispersed company, how can this desirable and necessary attribute be achieved?
10. Performance measurement. Performance of employees when measured against realistic standards, especially when tied to rewards, is significantly greater than without this aid. What is the situation re this factor for *each* class of employee by division, subsidiary, plant, and/or department? Standards for professional and semiprofessional employees should be determined by planning coupled to an MBO program.
11. Morale and loyalty of employees. By subsidiary, division, and/or department as well as by geographical area.

The various subheadings are pretty much self-explanatory to an experienced manager and space does not permit a discussion of each. For example, it is obvious that the depth of management talent, or backup for present managers, becomes of great importance when embarking upon an aggressive new product and/or an acquisition program. If you do not have the necessary backup personnel, then you must attempt to acquire the extra people required for such a program well *before* the actual need. If you do not, the new people have to learn both the parent company and the problems of the new product and/or acquisition. This *never* works as well as first getting the new employee properly acclimated to the parent company.

We all know that integration of personnel acquired in an acquisition can be done. However, it is not often realized that the integration of such talent is far more successful when it is accomplished when the parent company has an adequate supply of skilled managers. This is even more important when new products are acquired with the merged company.

Marketing

The following analyses should be made available to the senior planners, usually by major country and/or geographical marketing area involved:

1. Total available market for each product and/or service line.
2. Fraction of market secured for each product and/or service—and the trend of the market proportion secured.
3. Why has not a larger market share been secured?
4. For each product and/or service line, identify market structure, customer characteristics, geographic location of markets, price–volume relationships, product and/or service substitutes, and life of products and/or services (life cycles and positions in the cycles).
5. The replacement parts–service market for each product. For service businesses, identify the facts concerning the associated product markets.
6. Cyclical factors.
7. Competitive position by product and/or service and for the total organization.
8. Channels of distribution for each product and/or service.
9. For products, the service policies, facilities, and service organization details should be established.
10. Brand and/or service-mark recognition if involved.
11. Price and discount structure.
12. Patent position by product.

Strategic Long-Range Planning

The strategy of a firm is the long-term set of objectives to be achieved by the actions and policies spelled out in the LRP. Strategy is the means by which the company survives in its environment. Many firms try LRP and become disillusioned, because they fail to set a specific strategy first. Arbitrarily setting generalized goals for LRP and mechanisticly developing plans does not mean that the company has tuned in competitive actions or economic, social, legal, and technological changes. Strategic objectives specify the direction which the company will take in the struggle against competitors and other external forces. The time for achievement of these objectives represents the span of time covered by the LRP.

. . . Definition of strategy requires the development of the following components:

1. *scope* of the business
2. *competitive edge*
3. *risk* level
4. *specifications*
5. *deployment* of resources
6. *dynamics*—the timing, monitoring, and adjustment of milestones of the LRP.

. . . The company should first establish a "base line," point of departure, or *strategic posture*. The strategic posture is simply the explanation of the current status of the six components of strategy. It is the result of past strategic planning or of drifting.

> When we specify and describe what these six factors will be like 5–10 years hence, we are specifying the new (future) strategy. LRP is the design to bridge the gap from strategic posture to strategic goals [from footnote 12].

As mentioned earlier, one of the first things to be accomplished in a planning session is to define the firm's present business(es)—which is much more than stating that the firm manufactures and sells widgets for a profit. Next one defines the future business(es) of the firm in very specific terms.[14]

Space is not available to cover all aspects of strategic planning and how to do it, but some mention must be made of competitive edge. This factor of a firm determines what advantage, if any, it has in the market place—and without such an advantage it will either fail or be a "me to" and faltering organization.

> The competitive edge is what every company must have to exist. It is the sum of all factors which make the company different in some way from every other company from the *customer's viewpoint*. This last point cannot be emphasized enough! A firm may have high morale, low production costs, well-educated management, good sources of raw materials, or any number of *internal* operating advantages. If these are not *translated* into services and products with distinguishing features which are *recognized by the potential buyers*, they do not constitute a competitive edge.
>
> The competitive edge may *stem* from such operating advantages as listed or from other sources such as:
>
> Unique product design
> Product reliability
> Favorable price/quality ratio
> High integrity of the company
> Strong financial position
> Industry leader
> Modern plant and facilities
> Skillful marketing
> Good Management Information System
>
> The competitive edge as viewed from the purchaser's side when comparing products is:
>
> Utility of the product function to the customer
> Price
> Quality of workmanship and appearance
> Reliability of the product
> Life of the product
> Total life operating and maintenance costs
> Delivery lead time
> Status of product ownership
> Personal relationships with the company personnel
> Service [from footnote 12]

[14]A good discussion of this subject can be found in Drucker, P., *Management*, Harper & Row, New York, 1975.

Product Planning

Product plans must be considered an integral part of LRP and SRP. However, it is too vast a subject to be included in this overview of planning and MBO. It is therefore suggested that the reader who may be involved in this aspect of the planning process study such sources as *New Product Venture Management,*[15] *Managing Engineering and Research* (also contains the information on the factors to consider in evaluating the technical function),[16] and, if a small firm, *Problems of Small Business in Developing and Exploiting New Products.*[17]

In the April 16, 1981, issue of the *Wall Street Journal* Booz, Allen & Hamilton are quoted as saying that 25 major consumer-products companies found that they expect internally developed new products to account for 25 percent of 1985 sales. These writers believe that the proportion is about as large for industrial products. Obviously, new products and product planning are a very important area of concern in planning and in the operation of a business.

Learning Curves

This topic was generally discussed in Chapter 2 with applications presented with regard to Work Measurement. There were no applications explained or cited relative to the specific use of learning and experience curves in the planning for an organization's future or for adaptation to pricing and general business analysis. The material which follows fills this need and was taken from *Long Range Planning and Corporate Strategy* (footnote 12).

> The Boston Consulting group has been proposing the concept of the Experience Curve as a planning tool for at least a decade.[18–20] . . . The essence of the idea is that unit costs decrease on the value-added portion of a product as accumulated volume (experience) increases—usually about 10–20 percent every time production experience doubles. Naturally, suppliers also are affected by the learning phenomenon and therefore should be able to extend lower prices as their learning experience increases.
>
> It should be obvious that different firms learn at different rates. Therefore, being a fast learner will give a firm a major advantage. What affects the rate of learning? The quality of the firm's management.

---

[15]Karger, D.W. and Murdick, R.G., *New Product Venture Management,* Gordon & Breach Science Publishers, New York, 1972.

[16]Karger, D.W. and Murdick, R.G., *Managing Engineering and Research,* Industrial Press, New York, 1963, 1969, 1980.

[17]Karger, D.W. and Jack, A.B., *Problems of Small Business in Developing and Exploiting New Products* (a published research report), Rensselaer Polytechnic Institute (and the Small Business Administration), Troy, NY, 1963. This book is out of print but can be found in libraries.

[18]*Perspectives on Experience,* The Boston Consulting Group, One Boston Place, Boston, Mass. 02106.

[19]*Experience Curves as a Planning Tool,* The Boston Consulting Group, One Boston Place, Boston, Mass. 02106.

[20]*The Experience Curve—Reviewed,* The Boston Consulting Group, One Boston Place, Boston, Mass. 02106.

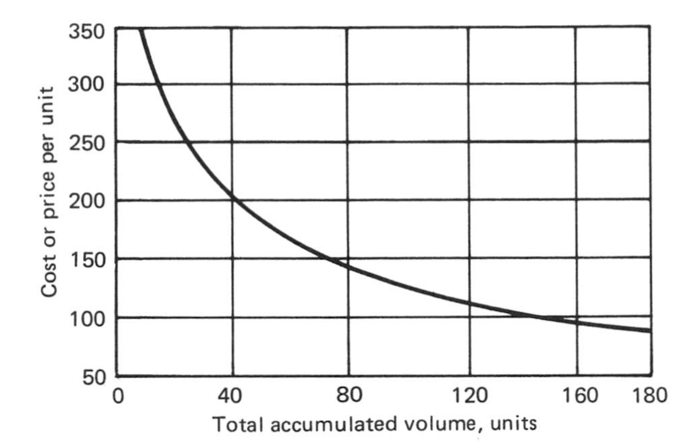

Representation of experience relationships graphically on linear scales.

Exhibit VI–2

The learning effect is caused by all elements (functions) of the firm learning, not just the production function. This is why high quality management in all functions and at all levels should foster a high rate of learning.

Exhibit VI-2 illustrates the learning phenomenon in its natural state. If plotted on log-log coordinates, it appears as a straight line as illustrated in Exhibit VI-3.

In studying the phenomenon it was noted that product price seems to become stable (not subject to *erratic* variation) when the slope of the price curve is parallel to that of the learning curve of the lowest cost producer (best noted when both are plotted on the same graph paper in log-log coordinates).

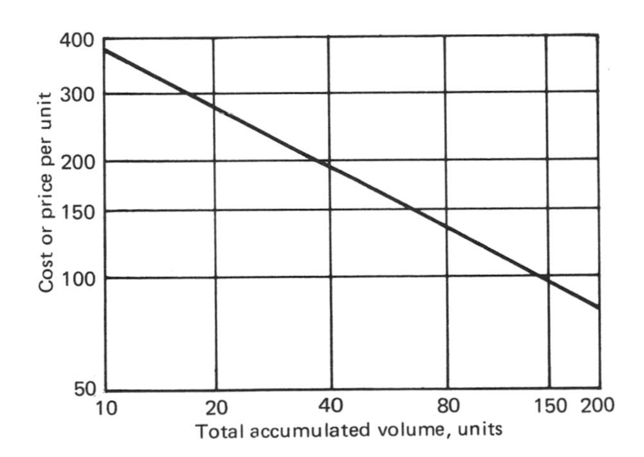

Representation of experience relationships graphically on log-log scales.

Exhibit VI–3

If you want to determine a competitor's rate of learning, you should try to plot the price they ask for a mature product. If it has assumed a typically stable shape (see *Perspectives on Experience* and the material on p. 49, including Fig. 3–11), then the slope of the bottom portion of the curve likely represents their rate of learning, especially if they have the major market share of the product.

Fundamentally, if a company has a market share of, say 60 percent, it almost always will be further down the curve in terms of price than its competitors unless some competitor has a much faster rate of learning and also has a substantial share of the market, like 40 percent.

The implication for companies holding small shares of the market is to "drop dead." There are, however, some arguments in favor of hanging in there if you are a small market-share holder. First, an examination of the learning curve shows that the unit cost differential between the large and small market-share holders continually decreases. Second, it is possible through efficient operation and selection of the strategy components to carve out a very profitable niche in the industry.

The tables below will give the reader a "feel" for the effect of market share. Also note that the rate of learning (a variable between companies) can make up for a smaller market share, especially in combination with superior strategic actions. Also, the superior strategic actions may have placed the firm into a desirable market niche.

One word of caution to those who would use learning curves. The abscissa is not TIME but UNITS PRODUCED. This does not equate to time since different firms produce at different rates. Also, they likely started producing at different times.

Comparative Performance Through 1976

Company	Sales ($ Millions)	Return on Equity (percent) (5-Year Average)
IBM	16,304	20.5
Burroughs	1,871	14.3
Sperry Rand	3,203	13.1
NCR	2,312	11.8
Honeywell	2,495	9.7
Control Data	1,331	4.7

Comparative Performance Through 1976

Company	Sales ($ Millions)	Return on Equity (percent) (5-Year Average)
Crown Cork & Seal	910	16.2
Continental Can	3,458	14.1
National Can	917	12.2
American Can	3,143	12.0

Realities: The CEOs Desires

This topic covers an important factor. Again we quote from *Long Range Planning and Corporate Strategy* (footnote 12):

> It is necessary to look at the realities of what might be an immovable object in a chosen path of strategic development—real support in terms of actions.
>
> Whether the reader is a consultant, a corporate planner, or even the CEO, the first step to be taken is a private one. If you are not the CEO, get with him on a one-to-one basis and find out what he or she really wants out of life at this time. If you are the CEO, think it over and communicate your thoughts to the planners. If you are the planning director, then you too must communicate your findings. However, be very careful what you say and how you say it, because you can scuttle the planning effort at this point with merely a few words. Why and how should be clear when one realizes that there are "accomplishment oriented," "don't rock the boat," and other "kinds" of CEOs. If the CEO really controls the company, and most do, then the firm must develop a strategy that will sustain the CEO's wishes.
>
> More on this subject can be found in "Changing Life Ways and Corporate Planning."[21]

Profit Impact of Market Strategy (PIMS)

One way to keep from falling into the trap of trying to rescue a business that cannot really be rescued or to attempt the rescue in an incorrect manner is to test the strategy and tactics to be attempted against the experience of other Strategic Business Units (SBUs)—other independent businesses (independent on the basis that the results achieved were independent of any company affiliation). This can be done by gaining access to ths ever growing computer model of over 3000 SBUs by subscribing to the Strategic Planning Institute (SPI), One Broadway, Cambridge, Massachusetts 02142. It has included in its computer model over 3000 manufacturing-type SBUs in a wide variety of SIC classified industries. By entering your data in a carefully structured and defined manner you can see how your SBUs and each of their strategies and tactics "stack up" based on the experience of the companies in the computer model.

The actual research and modeling began in 1960 by Dr. Schoeffler and associates of General Electric. It started as an investigation of marketing strategies and ultimately resulted in identifying the "Profit Impact of Market Strategy" (PIMS). Soon the effort expanded much beyond the original focus on marketing. The model and the effort are now managed by SPI. Dr. Schoeffler is its President. However, SPI has not completely shed its original "mantle," PIMS. Today, SPI publishes from time to time for its members SPI's "Findings" in a series of "so-called" *PIMS Letters*—they are numbered in a series.

---

[21]Mitchell, A., "Changing Life Ways and Corporate Planning," *Planning Review*, January, 1976.

At this point we continue by again quoting from *Long Range Planning and Corporate Strategy* (footnote 12).

> The basic model lumps all businesses, but this does not seem to blunt the results which are based upon 37 independent variables in a regression model. It predicts about 80 percent of the resultant ROI. This "lumping approach" is considered by some to be a weakness, but it seems to work and also is "the only game in town."
>
> PIMS probably achieves its primary usefulness in the analysis and diagnostic appraisal phases of policy and strategy formation. It permits comparing your present strategy to proposed actions. Also, you can measure the apparent correctness of each of the above to the model optimums. However, you must define your strategies in terms of the model variables.
>
> The PIMS variables are listed below since they also indicate what factors should be considered in strategy and product planning.

Independent PIMS Variables

Those largely controllable by management

- Change in advertising and promotion/sales
- Change in product quality
- Change in return on sales
- Change in sales force expense/sales
- Corporate payout
- Degree of diversification
- Growth of sales
- Inventory/purchases
- Manufacturing costs/sales
- Market position
- Marketing less sales force expense/sales
- New product sales
- Product quality
- Price relative to competition
- Receivables/sales
- R & D expense/sales
- Sales/employee
- Vertical integration

Factors partially controllable by management

- Capacity utilization
- Change in market share
- Change in selling price index
- Change in vertical integration
- Corporate size
- Instability of market share
- Market position impact
- Relative pay scale

Independent PIMS Variables

(continued)

Largely uncontrollable variables

Buyer fragmentation index
Change in capital activity
Competitive market activity
Fixed capital intensity
Industry exports
Industry long run growth
Investment intensity
Investment intensity impact
Sales direct to end user
Share of four largest firms
Short run market growth

Do not let defining the variables for your SBUs scare you since the degree of definition required is practical and achievable—but you will gain a better understanding of your businesses as you accomplish the task.

There are some generalized findings that can be mentioned, but it also should be cautioned that while ordinary attempts to take advantage of them will help, the best results are obtained when you can test your policies and strategies against the experience represented by the model.

The characteristics of the market(s) of the business itself and of competitors constitute 80 percent of the reasons for success or failure—operating skill or luck only accounts for about 20 percent. Being in the right business in the right way is 80 percent of "the game."

Investment intensity has a negative impact on net cash flow and on profitability, which includes ROI. SPI has developed a number of reasons why this is so and how the problem can sometimes be minimized.

Businesses producing high value added per employee are more profitable than those with low value added.

Market share naturally has a significant positive impact on profit and net cash flow. It relates to the discussion of corporate "learning."

Firms producing quality products (quality is defined as the customers' evaluation when compared to the offerings of competitors) experience significant favorable impact on all measures of financial performance.

Innovation (new product introduction, R & D, marketing effort, etc.) produces a positive effect on performance if the firm *has* a strong market position (market share). If it has not, then the effect usually cannot be seen or it is negative.

Vertical integration is good for the firm if it is in mature and stable markets. In other markets the opposite is usually true.

Actually, there are nine major strategic influences (some of these were identified in the above material). Sometimes they offset each other and sometimes they reinforce each other. Sometimes the effect of a strategic factor reverses; it depends on the situation regarding other factors. For example, in the area of innovation mentioned above, R & D effort tends to increase earnings if performed by a firm with a strong market position. The latter firms probably would better pursue a strategy of following and of being good operators.

All of the above is saying that policy and strategy formation is complex and difficult. It requires the dedicated attention of the best minds available.

The above are only a very few of their findings and data presentations.

The model affords most users a way of testing many strategies before actual implementation.

Short-Range Planning

Short-range planning (SRP) details the immediate-time portion of the long-range plan, usually the first year. The SRP usually coincides with the firm's fiscal year, which usually coincides with the calendar year for tax and reporting reasons. Another pressure moving planning to the calendar year is that many published environmental forecasts are in terms of calendar years. The SRP provides the flexibility for the firm to change tactics within its broad strategic plan. It also ties the functional and operational activities into a unified system.

The departmental objectives are subordinate to the functional plans and/or divisional plans, as are the objectives of the individual professionals. Moreover, these plans should support the firm's LRP

Strategy is defined in terms of objectives, and strategic planning is the establishment of the programs required to achieve the strategic objectives. The short-range plans for the company must provide for the best interests of the company as a whole, not for the best interest of subsidiaries, divisions, product lines, or an individual manager.

The SRP should consist of a system of activities which establish the specific objectives and responsibilities of each managed unit. The structure of the system will need to be in accord with the organization of the company. The development of the system must be a mutual and cooperative effort between the concerned managers, those concerned with each specific objective. Each short-range task, like the long-range objectives, must be supported by definitive task descriptions, time schedules, budgets, cash flow analyses, and/or cost benefit calculations.

Short-range tasks are (a) those longer-range tasks which were started earlier and which will be completed within the year or in succeeding years (in the latter event present or state each year's events separately) or (b) tasks being started that will be completed within about the next 24 months (again separate into specific years) or (c) tasks which are to be started and completed within the planning year (the more usual case for short-range planning objectives).

Typical overall company examples of such tasks are:

1. Specific mergers, joint ventures, acquisitions, and/or divestments.
2. Replacement and/or acquisition of major managers and/or professionals.
3. Restructuring one or more organizational elements.
4. Design and implementation of a performance measurement system and/or an incentive system for one or more organizational units.

5. Year-ahead capital budget items.
6. Year-ahead product improvements and/or new products.

Time

The initial planning effort will require an overall time span of about 3 months minimum for a relatively moderate size operation (sales of $10,000,000 to $100,000,000)—the degree of organizational complexity and dispersion will also affect the time required. Also, this short time span will only be possible if a very knowledgeable planning director is at the helm and a real strong effort, approaching a "crash" effort, is attempted. Six months would permit a relatively leisurely attempt in the general industry size range indicated above.

Once the initial planning has been accomplished, it is a year long effort.

MBO for Professional Employees

Professional employees for an MBO system are here defined as being all managers, all engineers, all accountants, and any other employees in managed units having essentially a "knowledge worker" type of job.

A "knowledge worker" is defined as having primarily a mentally oriented job (mental decisions, largely unstructured decisions), a job where results largely depend on relatively high level mental activity and where the quality of performance of the activity could have a discernible effect upon the organization. It is not routine clerical activity.

Pragmatically, MBO goals must be established at essentially three levels:

1. The very top executive level (represented by over-all organization goals)
2. Functional, divisional, departmental, and program goals (All can be involved. Some would argue that these represent two or more levels, and, if this makes it more real and correct to the reader, consider them as different levels.)
3. Individual goals for each professional

All lower-level goals must be in consonance with and supportive of higher-level goals—and they must not be in opposition to the welfare of the organization.

The overall organizational goals are used to measure the performance of the CEO and/or Chief Operating Officer (COO).

The functional goals are used to measure the performance of the functional managers.

The goals for divisions, subsidiaries, departments, and programs are used to measure the performance of the cognizant managers.

The goals for individual performers are used to measure their performance.

The goals of individual performers will consist of one or more of the individual goal segments illustrated in Fig. 15–2.

Each manager should meet individually with each of his/her professional and semiprofessional employees to help them develop a set of goals for the year ahead. The emphasis should be on *help,* since the manager should not set the goals for the employees. If the manager perceives that a goal suggested by an

Goals related to organizational objectives	Goals related to functional, divisional, and/or subsidiary goals	Goals related to departmental goals	Goals only indirectly related to the welfare of the organization	Goals not related to the organization—private goals that often are not expressed before the individual performer's supervisor

Figure 15–2. Possible segments of an individual performer's goals.

employee is not suitable, then the reason for such unsuitability must be explained in terms that the employee can understand and accept.

One of the big flaws in the typical MBO program is that the goal negotiations and establishment are not between *equals*—one man is the boss and the other is a subordinate. It is a handicap that must be recognized and strong efforts must be made to neutralize the problem. Some managers and workers have no problems, whereas with others the problem is of major magnitude.

Immediately after the negotiations, one person should record "the understanding" in writing. These writers suggest it be done by the worker and, if the manager disagrees, the misunderstandings should be immediately corrected by mutual action.

Possible reasons for employee-suggested goals being unsuitable are:

1. Impossibility of achievement, or very likely impossible of achievement. Goals should be achievable with a reasonable expenditure of energy. In the first year of an MBO program it is *vital* to the ultimate success of the plan that all goals or objectives be achievable without excessive effort.
2. A goal is in opposition to one or more of the company's, functional, divisional, subsidiary's, and/or departmental goals.
3. The goal is too easy to achieve.
4. A professional development or personal goal is contrary to the best interests of the organization. Employees may have goals, such as "acquiring a different job and/or a higher rated job," but if it means leaving the company, the goal should be private to the employee and not a part of the official MBO program.

The manager and the employee should be sure to adequately cover the employee's work area responsibilities in setting goals.

All goals—LRP, SRP, and MBO—should be stated in measurable terms. A later subsection of this chapter will discuss how this can be accomplished and provide examples.

The same starting date for measuring MBO goal achievement is usually desirable and should be set by the company for all involved employees. Generally, this should coincide with the starting date for the LRP and SRP goal measurement. If all of the individual goals are not established by the starting date, it is not very important so long as the "miss" is not more than about 30 calendar days.

Few individuals, or even few managers, will have goals directly and strongly related to one or more overall organizational goals, except for the top officers.

Functional, divisional, subsidiary, and departmental managers are usually essentially measured by truly SRP goals and only to a minor degree by any of the LRP goals established for their responsibility area. This has also been largely true for CEOs and the top officers.

This short-term emphasis of executive performance measurement is what has largely been responsible for the lamented decrease in productivity of the American worker. With emphasis on short-term performance measurement (from the board of directors down) and the frequent job hopping by executives so as to escape the bad consequences of emphasizing short-term decisions (neglecting product development, product improvement, capital investment in modern machinery, etc.) are what really caused the phenomena. It is really largely a creature of governmental laws, regulations, and an improper outlook (coupled to actions) by the board and top management. Japan, for example, has long been pursuing the long-range view with the result that they *are* highly competitive.

> Managers helping professionals set objectives must realize that the more personal and professional development objectives of the individual (such as "acquire an MBA through evening graduate study") need not be immediately or directly supportive to those of the overall organization or of the department.
>
> Generally, the professional worker wants mainly three people to be involved in specifying his work objectives. They are: himself, his immediate supervisor, and possibly a higher-level professional supervisor. He prefers only nominal influence from colleagues, nonprofessional executives, and subordinates.
>
> Conflict often arises between managers and professionals because the manager does not understand what the individual is working toward (his objectives as *he* perceives them), or the individual does not understand what the company is working toward. Both sets of objectives must be written and discussed so that the meanings and intents of the statements are clear to each person involved.
>
> A common measure of alignment of objectives is a determination of degree of contrast between the type of work the professional prefers to do and the type of work he actually does. Dissatisfaction because of undesired jobs is the first indication that there may be a mismatch between the individual and the department. Mismatch does not mean that the individual necessarily belongs elsewhere. Instead, it usually means that the individual and the manager must get together and smooth out differences. Without this adjustment, departmental and organizational goal achievement is hindered.
>
> Ambitious professional personnel are interested in achieving recognition from their peers in professional associations. It is, therefore, important that a company encourage the building of personal reputations in the outside world through the publication of papers, presentations before professional groups, advanced professional education, etc. This is a type of personal goal that does not hinder departmental achievement of objectives (from footnote 12).

Incentive Systems for Professionals and Managers

A whole book could be written on this subject, so what is said here is merely an introduction.

A common and long used scheme for lower-level managers, professionals, and semiprofessionals is to set aside a portion of the yearly profits (sometimes by class of employee) and then divide the sum(s) in proportion to the departmental salary budgets. The "pot" for the cost center is then divided, more or less, in accordance to performance based on the manager's judgment, with or without the benefit of MBO objectives. Guidelines for the manager to follow should be formulated by top management, and these should be applied by all managers in the same manner. However, the benefits of such incentives will be lost unless a "connection" is made by the manager between achievement and the amount of the award—this is true for any incentive system tied to MBO.

With managers responsible for the *profit performance* of an organizational unit, the problem is more complex. First, the pay of such managers is usually a combination of salary and perquisites (perks). The bonus or incentive payout cannot, or at least should not, simply be keyed to profits and/or Return on Investment (ROI) or a similar measure. If it is, the manager will be strongly tempted to maximize the measurement(s) in the short term (the usual case in the United States), since he/she will likely be in another unit or in another company by the time the short-term strategy of maximizing profits and/or ROI produces an adverse effect—but we have seen cases where such players were caught by the ultimately falling profits. This is not a consolation to the company involved, since it has suffered real and lasting harm. What is needed is a complex usage (complex system design) of long- and short-term measurements in such a fashion that it places the desired emphasis on long-term vs short-term measures. This latter remark also applies to the first-mentioned system and even to nonincentive MBO systems. If the reader does get involved in the design of one of these incentive systems, it is recommended that the devised system be tested by pursuing various strategies to determine whether the manager can evade the constraints and "milk" the system for rewards that are not properly due him/her. In simple terms but slightly restated, simulate the persuit of various strategies to make sure that the system constraints upon a manager's performance will hold him/her in the desired direction and that excessive rewards cannot be obtained by other kinds of performance activities.

This apparent conflict between long- and short-term actions is why many CEOs opt for acquiring new products and/or businesses via the acquisition or merger route. Here success is more sure than developing and growing your own business. There usually are not large short-term detriments in pursuing this route—in fact, there often are positive and desirable short-term effects.

It should be obvious that using compensation to motivate professionals is a complex matter. It can be used to increase performance on the job with respect to either current or long-term objectives or for almost any mix of these. It is used to attract people and to hold people. Less popular is its use to move people out of the organization. It should be a management tool, not a thing governed by staff specialists. Line management should have a strong voice in determining compensation—more than is usual. Only they know what is going on and what needs to be done.

It was recommended that good performance should be rewarded. Typically, this is less than a 15 percent reward. If one gives 15 percent (often less) to the top performer and 8–10 percent to an average performer, the difference is rather meaningless at almost any level when you factor in a reasonable inflation rate, even a small one. Another usual fault is that when the employee gets above the midpoint of his/her salary range, the amount of the increase is decreased.

The increase should be related to the degree of achievement. For very superior performance, the reward probably needs to be at least in the 20+ percent range (especially when inflation is in the 7–12 percent range)—an inflation adder would appear appropriate.

All of the above remarks apply to executives and professionals receiving bonuses.

Writing Measurable Objectives[22]

It is difficult to phrase goals and objectives in measurable terms. However, it can be done if one gives the job enough thought and consideration.

First, it is suggested that the individual objectives, especially those for the nonmanangers, be written under categories such as:

1. Related to the overall organization and organizational unit objectives
2. Related to specifically assigned responsibilities
3. Problem solving
4. Related to administrative and other ongoing matters
5. Motivation
6. Professional development
7. Personal

While the above suggestions are logical and should be followed, there is a task crucial to ultimate success that must precede the use of these categories. This is the identification of what the *real contribution* of the job is toward the success of the organization—what one really gets paid for doing. It must be done before trying to *establish* MBO objectives with the employee. Without this knowledge one not only cannot write meaningful measurable objectives, but the job incumbent, without this understanding, will not make a proper selection of goals. He/she will therefore fail to make a proper contribution to the organization's success, and they will also fail to optimize career success.

Few people can give you a good answer if you ask them what they get paid for doing. The ordinary response is "I run a turret lathe" or "I manage the quality control function" or "I am the commercial loan officer of the City National Bank," etc. These are all functional statements of little real guidance to the man or woman trying to write significant measurable objectives.

In order to *begin* to discover the real contribution of a job one must ask a series of questions.

[22]Additional information on MBO can be obtained from Morrissey, G., *Management by Objectives,* Addison-Wesley Publishing Co., 1970; and Lasagna, J.B., "Make Your MBO Pragmatic," Harvard Business Review, Nov.–Dec., 1971. However, do not fail to follow the precepts discussed in this chapter if you want to optimize your chances for success.

1. Why do you get paid, and I do not want an answer like nursing, engineering, or accounting?
2. Why should the organization really pay you?
3. How can they afford to pay you?
4. How does the organization make money from what you do?

Recently this kind of thinking was applied by one of the authors to the proper definition of a Vice President and Commercial Loan (C.L.) Manager of a substantial bank. First, he described his job as "making loans." Next as "making commercial loans." The next definition offered was "investing the bank's money." All of these were *functionally* oriented job descriptions. More questions brought forth the fact that about 65 percent of the bank's revenue came from the C. L. activity. That the C.L.M. and his staff adjusted loan interest rate to what he thought they could get and what rate would compensate the bank for the risk involved. This meant that the business(es) of the potential borrower had to be analyzed. The actions of competitive banks also had to be taken into consideration. Additionally, the C.L.M. and his staff had to go out and solicit business. Naturally, losses had to be kept to a minimum, and this sometimes required helping a borrower solve a serious business problem. Only now could one begin to formulate a valid definition of the C.L.M.'s job.

> "The Commercial Loan Manager is responsible for marketing the banks services, especially the Commercial Loan Services, so that the function brings in 65 percent (adjustable to each particular case) of the bank's revenue through investing the bank's funds in loans to professional people and commercial firms without incurring losses in excess of good acceptable practice (losses should not exceed *X* percent of loan revenue)."

Guidelines for the actual writing of measurable objectives are as follows:

1. Start with an action verb.
2. Identify the most logical measurable result for each objective—a result that gives a fair indication of the degree of attainment. Sometimes, two or more results must be used.
3. Identify the exact date each objective is to be achieved.
4. Costs, time, material, and manpower required for each objective are next identified.
5. Relate performance or improvements in terms of past history, company standards, and/or professional standards if at all practical.
6. Finally, only list objectives that are controllable by the individual.

A few examples may help the reader to more fully comprehend what is meant by the foregoing material. We will start with some middle-level managerial positions and end with an engineering worker level position. In each case measurability will be kept to the fore.

Our first example concerns the Quality Control Manager. Logically he might be given as his objective "to achieve the number one quality reputation for our company within our industry." It sounds good, but how would one be able to measure such an objective. A rational and measurable substitute could be the following:

The desired quality reputation for our company will be achieved when the quality control function achieves the following:

1. Field service calls do not exceed X percent in 1 year and Y percent in 2 years.
2. Plant reject rate is reduced to no more than X percent in 1 year and Y percent in 2 years.
3. Warranty costs do not exceed X percent in 1 year and Y percent in 2 years.
4. Rework cost is less than X percent of manufacturing cost in 1 year.
5. The company's product is rated in the first position in *Consumer Reports* within 2 years and rated acceptable within 1 year. (Or, for an industrial product the statement might read "The company's product line has an operational experience or predictable life-reliability rating (or guarantee) at the end of the first year that is above the average for the industry and equal to or better than the best at the end of the second year.")

The first objective and also several of the others involve more than the quality control function. Product design by the engineering department is such a factor, since the design obviously will affect field service calls. However, a good quality control manager and quality control activity will identify design weaknesses, document the involved problems and their effects, and through "pressure" on engineering get the design deficiencies corrected. In addition, public distribution of these objectives by management will also serve notice to engineering and other functional managers that this is the official "word."

In like manner, poor manufacturing could be a major contributor to excess field service calls, even with a good inspection department operation. Again, deficiencies can be properly documented and "pressure" used to bring about corrective action.

Purchasing's selection of poor vendors again could be the cause of or a contributor to excess field service. Again corrective action can be initiated by Quality Control.

The objective is logical and rational if management "signals" by their actions and statements that they do want to achieve the desired objective. Naturally, the quality control manager will have to be knowledgeable, courteous but firm, objective and not subjective, cooperative and not obnoxious, etc. It *is,* properly, a measure of the quality control manager's performance if top management gives him or her reasonable support when he or she acts properly.

Objectives 2, 3, and 4 have similar complicated relationships involved, and indirectly so does the fifth objective.

There would logically be other objectives for the quality control manager. Some possible additional one's could be the following:

Obtain an assistant capable of "taking over" the quality control function by the end of the second year.

With the Industrial Engineering department develop sufficient engineering competence to "counterbalance" the Engineering Department.

Encourage learning and professionalism in quality control staff. This will be evidenced by X percent of the quality control staff being active ASQC members and Y percent having participated in quality control related in-company training courses or in such an off-hours university course by the end of the year.

Complete personal work toward an MBA within the next two years, with emphasis on marketing and finance—as a preparatory step toward general management (incumbent already had an engineering bachelor's and a masters in applied mathematics).

One final example will be given for a Project Industrial Engineer (PIE), but only with regard to the more generally recognized position attributes (as for the C.M. position).

The PIE's job is to provide all required IE services so that the manufacture of the concerned product proceeds on an optimal basis within the cost centers served by the PIE without delays, with costs that do not exceed the department's estimate by more than 6 percent, and with rework held to no more than 10 percent for the first 6 months of operation. This will deemed to be essentially accomplished satisfactorily during the year ahead when the project enters its manufacturing phase if the following are achieved:

1. Processes are written for each required operator position and "in place" before the operator begins the job.
2. Tooling, jigs, fixtures, etc., are all designed and in place before the work commences—and debugging is accomplished without incurring chargeable production loss in no more than 10 percent of the cases. Also such losses do not cause the production schedule to be missed or that the excess manhours involved do not exceed 500.
3. Engineered labor standards in accord with company policy are set via Company Standard Data and MTM before production starts, except where prior data do not permit determination of the process time.
4. Work is accomplished within ±10 percent of the tooling-equipment budget.
5. Labor manhours are within ±10 percent of the labor estimate.
6. All chargeable expenses to the estimated first lot production do not exceed 6 percent of the total costs involved in the IE's pervue.
7. Cost reductions identified and approved prior to the end of the first year's production will amount to at least 15 percent of the manufacturing cost.

Concluding Remarks

Planning and MBO are indeed high-level activities of great complexity which are further complicated by the large number of factors that affect them. While this chapter has covered many important points, there are many more important and significant factors involved.

Index